DATE DUE

Errata

On Fig. 17a "Geology of Burma 1 : 2 000 000" the Kanpelet Schists around 21°N and 94° E are shown as "Metamorphics, undifferentiated" instead of "Low Grade Metamorphics". Mt. Victoria in the same area has been misplaced. It is located at 21°12′N and 93°54′E.

Foreword

Burma begins with the *Indo-Burman Ranges* E of the lowlands of the Ganges and the Brahmaputra. This S-shaped fold belt extends 2,000 km to the S as far as the Andamans. As a physiographic barrier it has been since time immemorial an ethnographic, cultural and political boundary.

Further to the E, in northernmost Burma, other mountain ranges branch off to the S abruptly from the E–W-striking fold bundles of the E Himalayas and run from the headwaters of the Irrawaddy via the Shan Plateau southwards to the Tenasserim Ranges: the *Sino-Burman Ranges.*

Between these two major mountain chains, which issue from a common root in the N, lies the Chindwin-Irrawaddy Basin, the *Inner-Burman Tertiary Basin,* which gradually widens towards the S. This is the fertile heartland of Burma.

In terms of regional geology, Burma is in a key position for studying the phenomena of converging plates: W of and parallel to the Indo-Burman-Andaman–Nicobar Fold Belt runs the S Tethys suture, which separates the oceanic crust of the E Indian Ocean from the continental crust of SE Asia. In its offshore section, the suture is identical with the Indonesian-Andaman Trench, a pronounced foredeep caused by a subduction zone. E of and parallel to the foredeep are (listed from W to E) the "Outer Arc", the "Inter-Arc Trough (Interdeep)", the "Inner Volcanic Arc", and the "Back Arc Basin" with its cratonic foundation in the E, the Sino-Burman Ranges.

Geological research in Burma commenced in the first half of the 19th century and was relatively intensely pursued during the years when Burma formed part of British India (1885–1948), but it ended abruptly during the period of Japanese occupation (1942–1945). Following the establishment of the independent Union of Burma (1948), geological studies were only hesitantly continued. It was not until about 1965 onwards that they were encouraged to some extent by organizational measures introduced by the Burmese government, as well as by the careful opening up of the country in the areas of economic activity and science. It is thus understandable that despite an astonishingly large number of individual papers on its geology in the older literature, and despite its considerable potential as regards natural resources, Burma is still in many respects geologically unknown territory.

Because of the inaccessibility of the region, large areas of the W and E ranges framing the Inner-Burman Tertiary Basin, and also the entire region N of about latitude 25° N, remain some of the geologically least known areas of SE Asia. Obviously, any synthesis based on such sketchy knowledge is bound to be very unreliable. Nevertheless, it makes sense to sort out the knowledge that has so far been gained and to arrange it in accordance with modern geological criteria, because

– it makes it easier for the geologist concerned with special questions to gain some insight into regional geological features,
– it provides an overview for dealing with regional geological problems, and
– it defines gaps in the knowledge and thus indicates the points where further geoscientific research should commence.

In the years from 1969 to 1982, the authors spent various periods of time working as geoscientists in Burma. They got to know and developed an affection for this beautiful country and its friendly, hospitable and understanding inhabitants. In publishing this contribution on the geology of Burma, the authors wish to thank their Burmese colleagues for the co-operative attitude and friendship which the latter displayed towards them.

I owe a debt of thanks to Colonel MAUNG CHO, the Minister of Industry (2) in the Socialist Republic of the Union of Burma, for the great deal of interest he showed in the writing of this book. Without his sympathetic support it would not have been possible to compile this contribution on the Regional Geology of the Earth.

The authors also thank the following staff members of the Bundesanstalt für Geowissenschaften und Rohstoffe (Federal Institute for Geosciences and Natural Resources) and of the Niedersächsisches Landesamt für Bodenforschung (State Geological Survey of Lower Saxony) who have worked on special problems of Burmese geology in Burma and in their home office and have provided numerous data for inclusion in this volume: G. BRASS, Airborne Magnetometry, F.-J. ECKHARDT, Mineralogy, Petrography, W. HANNAK, Metal Ore Deposit Studies, W. HEIMBACH, Geology, K. HILLER, Petroleum Geology, R. HINDEL, Geology, Geochemistry, C. HINZE, Geology, H. HUFNAGEL, Coal Petrography, H. JACOB, Coal Petrography, E. KEMPER, Micropaleontology, E. G. KIND, Mathematics, CH. KIPPENBERGER, Mining and Economic Geology, W. KNABE, Mineralogy, Sedimentology, J. KOCH, Coal Petrography, H. KREUZER, Radiometric Age Determination, H. LENZ, Radiometric Age Determination, A. LEUBE, Metal Ore Deposit Studies, H. MÜLLER, Palynology, P. MÜLLER, Radiometric Age Determination, Petrography, H. PORTH, Petroleum Geology, K.-U. REIMANN, Palynology, D. STOPPEL, Micropaleontology, TH. WEISER, Mineralogy, Petrography, H.-J. ZSCHAU, Geoelectrics.

Last, but not least, my thanks are due to the publishers, Gebrüder BORNTRAEGER, for the excellent cooperation in the publication of this volume and for its generous form and layout.

Hannover, April 1982   FRIEDRICH BENDER

Contents

1. General introduction

1.1 Area and population

Burma covers an area of 678,528 km², which is about three times the area of the Federal Republic of Germany. The country lies between about 10° N and 28° 30′ N and 92° 30′ E and 101° 30′ E. The greatest N–S extent is about 2,200 km and the greatest E–W extent about 950 km. In the 800 km long part of S Burma, the Tenasserim area, the E–W extent is only 50 to 150 km (Fig. 1).

In the W the Union of Burma is bordered by Bangladesh, in the NW by India, in the N and NE by the People's Republic of China, in the E by Laos and in the SE by Thailand. The total length of the country's borders is 4,000 km. The central and S part of the country is bounded in the W by the 2,100 km long Arakan and Tenasserim coastline of the Bay of Bengal and the Andaman Sea.

The Union includes 7 federal states (Karen, Mon, Kaya, Arakan, Chin, Shan and Kachin) and 7 divisions (Rangoon, Pegu, Mandalay, Sagaing, Irrawaddy, Tenasserim, Magwe). In a census taken in 1973, the population was found to be 28.9 million people (43 per km²). Burma is the homeland of a great number of different ethnic and linguistic groups. Three major groups are recognized:

– the Mongolian, SE Asian type, which in Burma includes the Mon and Shan groups
– the Tibeto-Indonesian type including the Burman, Chin, Karen and Kachin groups
– the Polynesian-Micronesian type including the Malayan and Salon groups in the southernmost part of the country.

Reputedly, over 100 different languages – not dialects – are spoken in Burma. They belong to the Sino-Tibetan, Austro-Asian and Austronesian language groups.

The most important cities are (from S to N; approximate population in brackets):

Tavoy	(102,000)	Magwe	(175,000)
Moulmein	(203,000)	Akyab	(143,000)
Rangoon (Capital)	(1,954,000)	Taunggyi	(149,000)
Pa-an	(250,000)	Sagaing	(190,000)
Bassein	(139,000)	Mandalay	(417,000)
Pegu	(255,000)	Haka	(29,000)
Loikaw	(48,000)	Myitkyina	(103,000)

1.2 Geological research, economic development, infrastructure

Probably the oldest description containing some geologically interesting data is that given by R. FITCH (1599) who undertook a very adventuresome journey from Syria to Burma. The

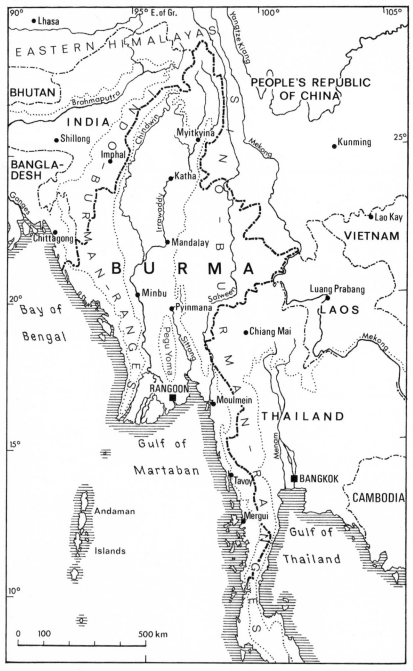

Fig. 1. Situation map.

description of the hand-dug oil wells on the anticline of Yenangyaung (p. 161), which is given by H. COX (1799), and of the famous sapphire and ruby deposits in the region of Mogok are among the earliest geological reports on Burma. For example, in 1833 the English translation of a letter of d'AMATO in which mention is made of a "ruby mass" weighing 80 lb (!), was published in the Journal of the Asiatic Society of Bengal; this ruby was acquired by the King of Ava, the ruling monarch, but not without difficulty because it first fell into the hands of dacoits as it was being brought to him by courier. In 1829, in Asiatic Researches, J. L. LOW described his geological observations of parts of the Malayan Peninsula "and the countries lying betwixt it and 18° North lat.", i.e. including the Tenasserim region. J. W. HELFER (1839) also studied this part of the country. Even before Burma had been fully incorporated into British India, a number of European scientists from the Geological Survey of India, which had been founded already in 1851, had studied the country. Examples of publications from this period are those by T. OLDHAM (1856) on coal and tin ore occurrences of the Tenasserim Provinces; those of W. THEOBALD (1869, 1873) on the Quaternary of the Irrawaddy or on the geology of Pegu; and the publications by F. R. MALLET (1876), who studied coal deposits of the Naga Hills in the districts of Lakhimpur and Sibsagar.

Once Burma became part of British India in 1885, general and, for those days, extensive and well-organized geological research got under way. The very many publications in the

– Memoirs of the Geological Survey of India,
– Records of the Geological Survey of India,
– Journal of the Asiatic Society of Bengal,
– Proceedings of the Asiatic Society of Bengal,
– Transactions of the Mining, Geological and Metallurgical Institute of India,
– Palaeontologia Indica,
– Geological Magazine (London) and
– Nature (London)

report on new facts that were discovered and also describe the difficulties with which the knowledge was obtained. Towards the end of the last century, authors such as R. D. OLDHAM (1883), CRIPER (1885), HUGHES (1889), NOETLING (1890), THEOBALD (1880, 1891) and C. BARRINGTON-BROWN (1895), to name just a few, published a wealth of geological observations which, because of their accuracy and the painstaking attention to detail, have lost nothing of their value right through to the present, even though the data may nowadays be differently interpreted.

General geological research continued unabated until the interruption caused by the Japanese occupation in 1942. The first regional syntheses appeared, such as those by L. V. D. DALTON (1908), G. H. TIPPER (1911), E. H. PASCOE (1912), T. O. A. LA TOUCHE (1913), J. P. DE COTTER (1918), L. D. STAMP (1922a, b, c), J. C. BROWN & A. M. HERON (1923), E. R. GEE (1927), R. S. RAO (1930) and by V. C. BURRI & H. HUBER (1932). A summary review of the status of geological knowledge, based on the literature up to 1933 and supplemented by his own profound knowledge of the country, was published by H. L. CHHIBBER (1934a, b), Head of the Department of Geology and Geography at Rangoon University. In a series of publications (1932, 1933, 1936), REED contributed important knowledge on the Paleozoic and Mesozoic fossils of Burma. CLEGG (1938, 1941) and DE COTTER (1938) dealt with regional geological problems of Central Burma and the stratigraphy of the Cretaceous.

In the years following the Second World War, publications on the geology of Burma became increasingly infrequent, not only in comparison with preceding years, but also compared with other countries in East Asia. Following the founding of the Union of Burma in 1948, efforts were concentrated on exploring for mineral raw materials in the vicinity of known deposits that were being mined. Systematic regional and detailed prospecting and exploration work, using modern methods in association with systematic geologic mapping of certain selected areas, did not commence until 1970. The results of the geological research from this period for the most part take the form of unpublished reports in the archives of the various state corporations which are concerned with mineral raw materials, and as such they are not accessible to the geoscientific public. Of the few syntheses published during this period, special mention should be made of the works by E. H. PASCOE (1950, 1959, 1964), who published a Manual of the Geology of India and Burma, by E. L. G. CLEGG (1953), who wrote about the Mergui, Moulmein and Mawchi series, by T. H. HOLLAND, H. S. KRISHNAN & K. JACOB (1956) in the Lexique Stratigraphique International, and by R. O. BRUNNSCHWEILER (1966, 1970, 1974) who published contributions on the geology of the Indo-Burman Ranges, and on post-Silurian geology in the Northern Shan State and in Karen.

For many years, the **economic development** of Burma has been decisively influenced by the development of agriculture. Judged against the value of goods produced in the individual branches of the economy, agriculture and trade have each accounted for 25% to 30% of the gross domestic product in the last ten years as well. Compared with these, mining ranked only 11th with a share of 0.9% of the gross domestic product (financial year 1969/1970) (Table 1).

According to the goals that were announced in 1971 for the next five four-year plans, agricultural production is supposed to increase annually by 3.8%, industrial production by 4.7% and mining output by 8%. Thus, mining would gradually regain its old pre-war share of the gross domestic product.

From 1961 to 1969 mining did in fact achieve an annual growth rate of about 8.1%, which is thus considerably above the annual growth of the gross domestic product at 3.8%.

Table 1. Value of goods produced in various branches of the economy, after deducting the intersectoral consumption at constant 1964/65 prices, in percentage of the total value*.

	1964/65	1967/68	1968/69	1969/70
Trade	26.8	25.1	25.0	25.0
Agriculture	25.9	24.8	24.6	23.9
Business and commerce	10.3	10.7	10.8	11.0
Government services	8.4	8.2	8.4	9.3
Other services	7.8	8.2	8.1	8.7
Meat and fish production	6.2	7.9	7.5	7.8
Transportation	6.5	5.7	6.4	6.1
Forestry	3.0	3.1	3.0	3.1
Construction industry	2.0	2.6	2.7	2.5
Banks	1.6	1.7	1.5	1.0
Mining	0.8	0.9	1.0	0.9
Electric power industry	0.4	0.6	0.6	0.6
Communication	0.4	0.5	0.4	0.4

* After Bundesanstalt für Bodenforschung 1971, 4, Tab. 2.1.3/2.

In the last ten years, mining's share in foreign trade accounted for 5% to 10% of the total value of exports. The export products are mainly silver, raw lead, hard lead, lead ores, copper matte, nickel speiss, as well as concentrates of tin, zinc, tungsten, tin-tungsten and antimony. Figs. 2 to 6 provide a general view of the development of the production of important resources such as oil, natural gas, coal, tin, tungsten, lead, zinc, silver, copper, cobalt and nickel.

Production of antimony ore (p. 201) was mentioned for the first time in 1937 and the output was given as 160 t (Sb metal content). Following the Second World War, antimony ore mining commenced once more in 1958 with an output of 80 t, reaching about 490 t in 1960 and 1,500 t in 1978.

Iron (p. 193) and manganese ore production (p. 195) was insignificant at about 26,000 t Fe ore (1939) and 6,500 t Mn ore (1952), and in the meantime it has stopped altogether.

For centuries, precious stones (p. 207) and jade deposits (p. 208) have been exploited in "Upper Burma". In 1939, ruby production was in excess of 211,000 carats; the more recent data are unreliable and fluctuate between 23,550 (1963) and 37,700 (1968) carats. The data on sapphire production are also contradictory; in 1959 the output was about 438,000 carats and in recent years probably around 60,000 carats. In the period prior to 1940, the annual output of jade was in the order of 25 t (p. 211). Output peaked at 142 t in 1966 and then dropped back to about 6 to 16 t per annum in the seventies. It is also worth mentioning the sporadic production of amber (p. 211) which reached about 1 t/a in 1929.

From the geological point of view, it is quite possible that new deposits of all these raw materials will be discovered or that the reserves of already known deposits will be expanded. The same can also be said of energy resources, iron, manganese, chromium, nickel, cobalt, Portland cement materials, glass sand, fluorspar, feldspar, baryte and gypsum, rock salt, steatite and sulphur. In addition, there are geological indications of good prospectivity for platinum, gold, niobium, tantalum, monazite, uranium, molybdenum, bismuth, asbestos, graphite, mica and bauxite.

In terms of **infrastructure,** the S and central regions of Burma are relatively well developed. On the other hand, all the Indo-Burman Ranges, the catchment area of the Chindwin River, the areas of the country N of Myitkyina and the E part of Shan State have undergone little infrastructural development.

The most important means of transport is the railway system (Union of Burma Railways). The main routes connect Rangoon with Myitkyina in the N, Lashio in the NE, Prome and Bassein in the W and Moulmein in the SE. The network has a total trackage of about 4,000 km.

The second most important means of transport is inland shipping. The country has an extensive network of efficient inland waterways comprising several rivers which are navigable over great distances (Irrawaddy 1,400 km as far as Bhamo, Chindwin 600 km, Salween 400 km), the branched arms of the delta (2,700 km), and canals between Pegu and Sittang (61 km), as well as between Rangoon and the delta (36 km).

The major roads run from Rangoon in a S–N direction to Mandalay (690 km), via Prome to Myingyan (717 km); from Mandalay via Lashio to the People's Republic of China ("Burma Road") and the "Ledo Road" ("Stilwell Road") running from N Burma to Assam; from Rangoon via Pegu to Martaban and S of Salween from Moulmein via Tavoy to Mergui. The all-weather road network is at present about 14,000 km long. The air traffic network is based on 42 airports.

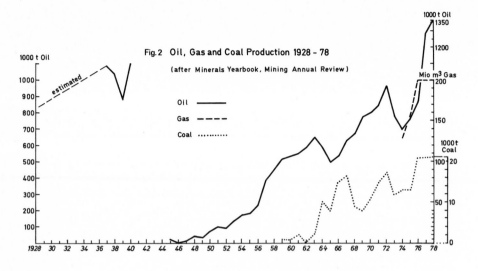

Fig. 2 Oil, Gas and Coal Production 1928 – 78

(after Minerals Yearbook, Mining Annual Review)

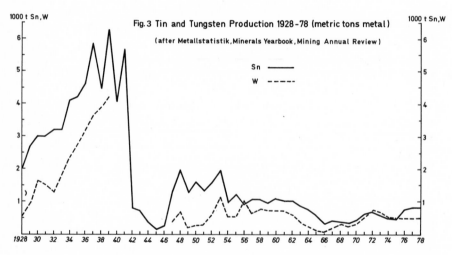

Fig. 3 Tin and Tungsten Production 1928 – 78 (metric tons metal)

(after Metallstatistik, Minerals Yearbook, Mining Annual Review)

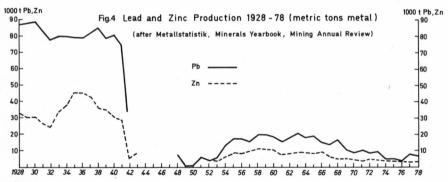

Fig. 4 Lead and Zinc Production 1928 – 78 (metric tons metal)

(after Metallstatistik, Minerals Yearbook, Mining Annual Review)

Figs. 2–4.

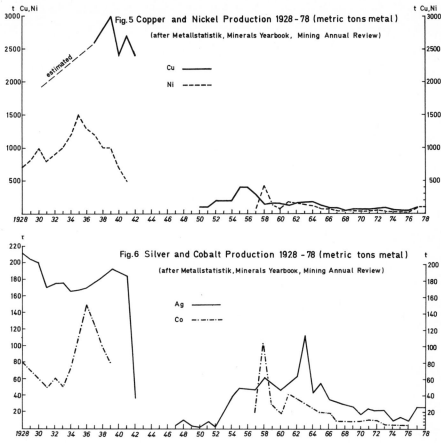

Fig. 5 Copper and Nickel Production 1928-78 (metric tons metal)

(after Metallstatistik, Minerals Yearbook, Mining Annual Review)

Cu ——————

Ni ——————

Fig.6 Silver and Cobalt Production 1928-78 (metric tons metal)

(after Metallstatistik, Minerals Yearbook, Mining Annual Review)

Ag ——————

Co ——————

Figs. 5–6.

1.3 Physiographic units

From W to E, Burma can be divided into four major physiographic units, which run more or less N–S and which are bounded in the northernmost part of the country by a fifth unit striking approximately W–E, namely the E Himalayas (Fig. 7, folder):
– the Arakan Coastal Area
– the Indo-Burman Ranges
– the Chindwin-Irrawaddy Basin, or Inner Burman Tertiary Basin
– the Sino-Burman Ranges
    The narrow **Arakan Coastal Area** on the Bay of Bengal, together with some offshore islands such as Ramree and Cheduba, are geologically part of the Indo-Burman Ranges. Some parts of the coastal strip, e.g. W of Sandoway and particularly in the region of Akyab are a little broader (approximately 70 km at Akyab) as a result of extensive detritus fans and fluviatile sediments brought down from the N by the rivers Mayu, Kaladan and Lemro. Otherwise, the coast is for the most part steep and rocky and is indented by countless bays

with flat, sandy beaches (Fig. 8). The hinterland rises to a hilly landscape covered with dense vegetation and reaching altitudes up to 600 m; this region leads over to the Indo-Burman Ranges in the E.

The **Indo-Burman Ranges** branch off to the S from the E Himalayas in a broad arc whose convex side is towards the Bay of Bengal. They can be followed over a distance of about 1,300 km in Burma (Fig. 9). Towards the S, the ranges decline in height and width before submerging in the sea at Cape Negrais. They continue to run for at least a further 1,700 km below the sea and they include the Andamans and the Nicobar islands. The northern SSW-striking section of the ranges reaches altitudes of 4,000 m (Patkai and Naga Hills). The central section (Fig. 10) bends towards the S with the Chin Hills and reaches altitudes up to more than 3,000 m (Mt. Victoria, 3,201 m). In the S section of the Indo-Burman Ranges, the Arakan Yoma (Padaung, 1,390 m), the height of the mountains declines as they strike SSE down to the Andaman Sea.

In the far N of Burma, the Indo-Burman Ranges link up with the approximately W–E-striking **E Himalayas**. The highest mountains in the country with altitudes above 5,500 m (Hkakabo Razi, 5,885 m) are located in this inaccessible region.

Between the Indo-Burman Ranges in the W and the Kachin Mountain Ranges, which also link up with the E Himalayas, the Kumon Ranges extend from the E Himalayas towards the S into the N section of the Inner-Burman Tertiary Basin. The **Inner-Burman Tertiary Basin** (Fig. 11) is framed in the W by the Indo-Burman Ranges, and in the E by the Sino-Burman Ranges. The basin is here regarded, for mainly geological reasons, as a physiographical macro-unit, although it can be divided into morphological sub-units. From N to S the Inner-Burman Tertiary Basin can be subdivided into (Fig. 7, folder):

– the Putao Basin, framed by the N Kumon Ranges and bounded in the N by the E Himalayas
– the Hukawng Basin, framed by the Patkai and Kumon Ranges in the N and E with the Indawgyi Lake in its narrowing S section
– the Upper Chindwin Basin, bounded in the W by the Indo-Burman Ranges and in the E by the Wuntho Massif
– the Central Basin with the confluence of the Chindwin and the Irrawaddy between the Arakan Yoma and the Shan Massif
– the S Irrawaddy Basin and the Irrawaddy Delta region between the central and S Arakan Yoma in the W and the N–S-striking Pegu Yoma in the E
– the Sittang Basin between the Pegu Yoma and the Shan-Tenasserim Ranges
– the Salween Basin between the N–S-striking ranges of the Tenasserim area.

E of the Putao Basin in northernmost Burma, the Kachin Ranges, as a part of the Sino-Burman Ranges, with altitudes of more than 4,000 m, branch off abruptly towards the S from the E Himalayas. After about 400 km, the **Sino-Burman Ranges** make a sharp elbow-shaped turn to the W and project into the Irrawaddy Basin. Bounded in the W by the Shan Scarp, the Shan Massif (Fig. 12) strikes in a SSW direction for about another 850 km towards the S with the heights of the mountains gradually declining (down to 2,400 m in the Northern Shan State). The mountain chains surround the intramontane basins in the region of Lashio, of Heho-Taunggyi, with Inle Lake, and further E the basin of Namsang. Further S, they frame the Salween Basin and continue towards the S. In the form of the Tenasserim Ranges (Fig. 13), with altitudes up to 2,000 m, they stretch in Burma territory for a further 800 km towards the S and continue to the Malay Peninsula.

Fig. 8. Arakan Coastal Area ca. 5 km S of Ngapali, vertical dipping Eocene flysch sediments. Photo: F. BENDER.

Fig. 9. Arakan Yoma SE of Sandoway, mainly folded and faulted flysch type sediments of Tertiary age; oblique aerial photo: F. BENDER.

Fig. 10. Chin Hills W of Kalewa, ultrabasic rocks of Mwetaung (background, right). Photo: F. BENDER.

Fig. 11. Inner-Burman Tertiary Basin, steeply W-dipping sandstone and shale of Miocene age along N–S-striking thrust fault; approx. 20 km SW of Mt. Popa; oblique aerial photograph, view towards E (Gwegyo Hills). Photo: F. BENDER.

Fig. 12. View from above Bawdwin to the N, Mohochaung in center, background; Northern Shan State. Photo: F. BENDER.

Fig. 13. Tenasserim Ranges; Heinda tin placer from SW; background: granite of Central Ridge. Photo: F. BENDER.

2*

1.4 Climate and vegetation

The **climate** of Burma is controlled by the great monsoon circulation system of S Asia and is influenced in detail by relief-specific peculiarities. The mountain ranges in general run N–S so that they present effective climate barriers for the SW monsoon in the summer and the NE monsoon in the winter. Therefore, the central part of the Inner-Burman Tertiary Basin (Central Dry Zone) lies in rain shadow during the summer monsoon (June to September) and receives less than 500 mm of precipitation (Fig. 14). The considerable differences in relief along the path of the monsoon lead to the formation of the following climatic zones (DRUMMOND 1958):

– tropical monsoon climate (Arakan coast, Irrawaddy Delta, Tenasserim) with annual mean precipitations of more than 2,500 mm and mean annual temperatures of more than 25 °C
– tropical savannah climate (around the Dry Zone) with more pronounced dry seasons between the monsoon rains and thus lower precipitation, but similar mean temperatures to those in the tropical monsoon climate
– tropical steppe climate (Dry Zone), semi-arid climate with less than 1,250 mm of precipitation per annum and mean annual temperatures in excess of 27 °C
– subtropical monsoon and subtropical mountain climates (the higher sections of the Indo-Burman Ranges and of the Shan Massif, N of 23° to 24° N), with mean January temperatures below 18 °C and occasional frost during the winter months in the higher and northerly situated mountain regions
– tundra climate (the most northerly and northeasterly parts of Burma at altitudes above 3,000 m) with mean temperatures during the warmest month of 10 °C or less
– ice cap climate (northernmost part of Burma at altitudes above 4,200 m) with mean temperatures during the warmest month of 0 °C or less.

A number of climatograms (based on the system used by WALTER 1955) from various climatic zones are shown in Fig. 15; in these diagrams the mean monthly temperatures and the mean monthly precipitations (a 3-year average in each case) are compared with each other.

The main type of **vegetation** in the tropical monsoon climate is the tropical rain forest (tropical evergreen forest) which contains a large number of valuable hardwood species such as *Dipterocarpus alatus* (Burmese: kanyin), *Hopea odorata* (thingan), *Parashorea stellata* (thingadu), *Lagerstroemia flos-reginae* (pyinma), *Pentace burmanica* (thitka) and *Swintonia floribunda* (taungthayet).

In the Tenasserim region through to the area N of Rangoon, rubber trees (*Hevea brasiliensis*) are grown, and in the S Irrawaddy Basin mainly rice is cultivated.

The areas near the delta and coast are usually covered by beach and tidal forests (mangrove) or in dune areas by *Casuarina equisetifolia* (pinle-kabwe) and *Albizzia procera* (sit).

The tropical savannah climate is characterized by deciduous forest. The trees lose their leaves shortly after the end of the rainy season; they form open stands with extensive grassy areas, particularly in the high-lying regions. Here one finds timber trees such as *Tectona grandis* (teak), *Xylia dolabriforma* (Burmese: pyinkado), *Pterocarpus macrocarpus* (padauk) and *Terminalia tomentosa* (taukkyan).

The characteristic flora of the Dry Zone are scrub, scrubby growth of small trees (Acaciae, Euphorbiae), cacti and short grass.

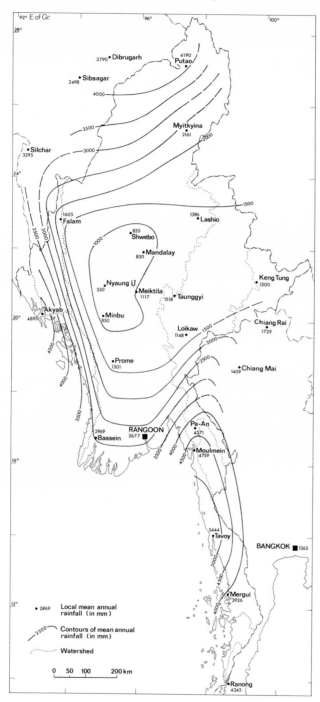

Fig. 14. Contours of mean annual rainfall.

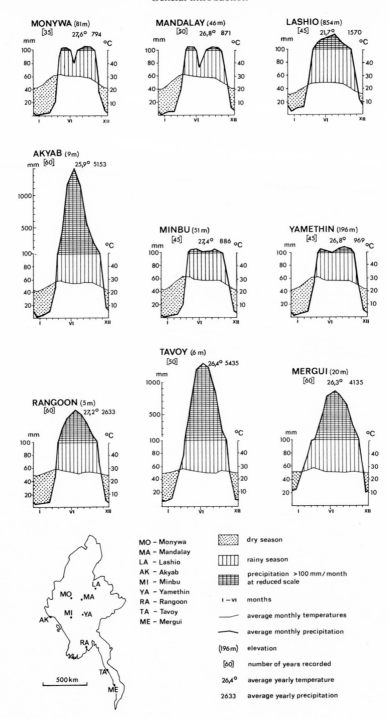

Fig. 15. Climatograms.

In areas where the groundwater is near the surface, toddy palms (*Palmyra*) are widespread. Rice, cotton, sessamum, millet, tobacco and groundnuts are the main crops grown.

The occurrence of oak and chestnuts, coniferous trees (*Pinus merkusii*) and magnolias (*Magnolia pterocarpa*) is typical of subtropical monsoon and subtropical mountain climates. Extensive areas of tall grasses with only sporadic trees are found in the Shan Plateau. At high altitudes coniferous trees, represented by *Pinus insularis,* predominate.

In the tundra climate associated with altitudes above 3,000 m, one finds chiefly stunted conifers, dwarf rhododendrons and birch (*Betula*), juniper and mossy heaths.

DAVIS (1960) differentiated between the following types of forest and other woody vegetation in Burma:

Tropical

Evergreen Hardwood Forests
– Rain Forests
– Coastal Mangrove Swamps
– Inland Swamp Forests

Mixed Evergreen and Deciduous Hardwood Forests
– Lowland Forests
– Upland Forests

Dry Deciduous Hardwood Forests, Scrub Forests, and Scrub
– Indaing (Dipterocarp) Forests and Scrub Forests
– Dry Upland Hardwood Forests and Scrub Forests
– Dry Scrub Forests and Semi-desert Scrub
  Than-Dahat Scrub Forests
  Sha (*Acacia*) Thorn Scrub Forests
  Te (*Diospyros*) Scrub Forests
  Semi-desert *Euphorbia* Scrub

Subtropical

Hardwood Rain Forests
Mountain Forests and Scrub Forests
– Moist Hardwood Forests
– Dry Hardwood Forests and Scrub Forests
– Pine Forests

Temperate

High Mountain Hardwood Forests
– Wet Hardwood Forests
– Moist Hardwood Forests and Scrub Forests
  High Mountain Conifer Forests
  High Mountain Subalpine Scrub

2. Regional geology

2.1 Setting within Southeast Asia

The worldwide investigations into the phenomena of seafloor spreading and of plate movements have also made it possible to gain a better understanding of the geological evolution of the region under discussion here.

The suture of the former Tethys Sea has been identified in Iran and in the Himalayan region (GANSSER 1964, 1966, 1973, 1976, FALCON 1967, WELLS 1969). According to STONELY (1974), this suture bends to the S from the E Himalayas and continues in the Bay of Bengal to the W of the Arakan Yoma of Burma and in the Java (=Indonesian) Trench. The sudden change in direction of the Alpidic structures in the E Himalayas to a general N–S strike takes place in a little-studied region. There is a distinct symmetry between these structures and the structures in the NW part of the Indian subcontinent. The Tethys suture can also be traced W of the Himalayan region in and to the W of the Baluchistan Ranges of Pakistan, between the coasts of Makran and Oman, along the Zagros Crush Zone and via the Bitlis Thrust in SE Turkey through to the Mediterranean. It separates "Eurasia" in the N from the Afro-Arabic and Indian shield regions in the S and it also separates the oceanic crust of the E Indian Ocean from the continental crust of SE Asia (Fig. 16).

As STONELY (1974) pointed out, consistently different geological conditions, which mark the old continental margins that faced each other prior to their collision, exist on both sides of this separating line in the Mesozoic and Cenozoic. Extensive structural elements, which are different on both sides of the suture and which existed prior to the collision of the plates, can also be observed. They probably influenced and were in part responsible for the mode of the orogenic deformation following the collision.

CURRAY & MOORE (1974) and JOHNSON et al. (1976) assumed that the break-up of Gondwanaland, with the separation of India from the then joined land masses of Australia and Antarctica, commenced during the Lower Cretaceous (130–80 m.y. B.P.). New sea floors formed in, among other places, the present Bay of Bengal. Following the separation, flysch series accumulated on all sides of the Indian subcontinent. The Ninetyeast Ridge in the E part of the Bay of Bengal exhibits an echelon topography which indicates the offset pattern of this initial split. During the Upper Cretaceous, the Indian Plate began to move northwards. To the E of the Ninetyeast Ridge, lateral movements took place (motions along a transform fault zone). In the Late Paleocene (approx. 55 m.y. B.P.) this resulted in the collision between the Indian and Southeast-East Asian Plates. As a result, the relative directions of movement changed throughout the entire Indian Ocean, and the intensity of the movement decreased. There is evidence that in the Early Oligocene (approx. 35 m.y. B.P.) the Indian Plate again started to move at a faster rate in a NE direction (MCKENZIE &

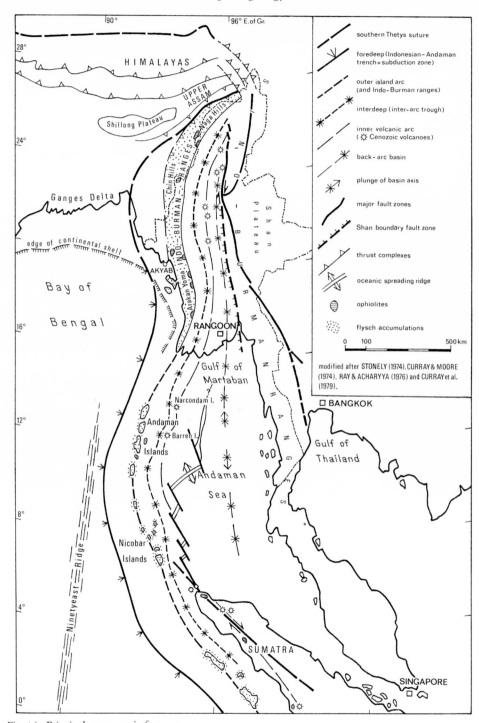

Fig. 16. Principal geotectonic features.

SCLATER 1971, SCLATER & FISHER 1974). The N–S lateral movements to the E of the Ninetyeast Ridge came to an end and probably turned into underthrust motions along the subduction zone which is marked by the present-day Indonesian Trench and its continuation in the N (foredeep in Fig. 16). The first uplift motions in the Himalayan region and in the region of the Indo-Burman Ranges commenced in the Oligocene and Early Miocene. According to the authors mentioned above, the folding of these mountain chains did not take place until the Middle and Late Miocene. Most parts of the region of the present-day Sino-Burman Ranges in E Burma had, however, already emerged from the ocean during the Cretaceous. The axis of the Inner-Burman Geosyncline located to the W of the Sino-Burman Ranges gradually moved further westwards during the Cenozoic.

As is indicated by the zones of increased seismic activity along the Sunda-Andaman Arc through to the E Himalayas and along the Himalayan mountains, the NE movement of the Indian Plate continued through to the present.

The regional geological phenomena in the area under discussion here can be arranged as follows in the general framework that has been sketched out (from W to E; Fig. 16):

According to CLOSS, NARAIN & GARDE (1974, p. 636), the **Ninetyeast Ridge** in the Bay of Bengal was still part of the NW-moving Indian Plate during the Upper Cretaceous. Based on the data obtained by the "Glomar Challenger" during the Deep Sea Drilling Project (Leg 22, 26), the ridge consists, in the N, of Upper Jurassic to Lower Cretaceous shallow-water carbonates which have been covered since the Campanian until the present by pelagic sediments. In the S of the Ninetyeast Ridge, Early Tertiary, lignite- and tuff-bearing series and marine shallow-water sediments were encountered on top of basalt. Pelagic rock sequences were not sedimented here before the Eocene (CURRAY & MOORE 1971, LUYENDYK 1977).

Joining on to the E, and to the N of the Ninetyeast Ridge, thick deposits of the Bengal Fan and its E lobe, the Nicobar Fan, the largest deep-sea fan in the world (CURRAY & MOORE 1971, 1974), were accumulated. Together with the underlying Mesozoic strata they form the Bengal Basin, which, with a maximum thickness of more than 16 km, is one of the deepest basins in the world (MOORE et al. 1974).

The E boundary of the Bengal-Nicobar Fan is formed by the Indonesian-Andaman Trench with its N continuation, the Arakan **Foredeep** (Fig. 16). It marks the subduction zone and, according to STONELY (1974) it denotes approximately part of the route followed by the Tethys suture. Its Miocene and more recent deposits are molasse-like in character.

To the E of the trench comes the **Outer Island Arc** with its N continuation, the Indo-Burman Ranges. The latter consist chiefly of very thick Lower Tertiary terrigenous deposits in flysch facies. Older rocks of the Indo-Burman Ranges are composed of pelagic *Globotruncana* limestone of the Campanian-Maestrichtian, ophiolite and, to a lesser extent, of Early Mesozoic sediments. According to BRUNNSCHWEILER (1966) these older rocks are allochthonous within the major part of the Indo-Burman Ranges. They occur as olistoliths within the flysch series. The region from which they originated is a narrow and only intermittently exposed zone, which also contains low-grade metamorphic rocks, at the E edge of the Indo-Burman Ranges. The flysch sediments accumulated at the margin of the Eurasian continent. Because of the accumulation of ophiolite, it might be possible to place this margin at the E edge of the Indo-Burman Ranges, if one accepts the interpretation of ophiolite belts as put forward by COLEMAN & IRWIN (1974, p. 910). W of the Indo-Burman Ranges, the flysch series are overlain by Cenozoic molasse which was formed after the plate collision (STONELY 1974, p. 889).

The **Inter-Arc Trough** (=Interdeep in Fig. 16) is located E of and parallel to the Outer Island Arc (=partly Indo-Burman Ranges). It was clearly observed also by seismic reflection surveys in the W part of the Gulf of Martaban (Fig. 28). In the S, its filling consists for the most part of marine, argillaceous sand, shale and, locally, of carbonatic sediments of the Tertiary which interfinger towards the N and NE with non-marine deposits of the same age and are mostly overlain by continental, more coarsely clastic deposits of the Upper Miocene to Quaternary. These Tertiary deposits are at the E side of the Indo-Burman Ranges in structural contact with the flysch-type sediments, or they directly overlie Cenomanian to Maestrichtian deep-water sediments and older rocks.

The **Inner Volcanic Arc** separates the Interdeep from the Back-Arc Basin. It is characterized by a chain of Cenozoic volcanoes. It is not always observable onshore in Burma so that a separation between the Interdeep and the Back-Arc Basin adjoining to the E is not clear everywhere. The Inner Volcanic Arc was observed by seismic reflection as well as by airborne magnetometry also in the Gulf of Martaban.

The **Back-Arc Basin** is, like the Interdeep, subdivided into several sub-basins: the "Central Trough" in the E part of the Gulf of Martaban, the Sittang Basin, the Pegu-Yoma Anticlinorium and the Shwebo-Monywa Basin. They are also filled by thick Tertiary marine sediments which interfinger increasingly with continental deposits towards the N.

Towards the E, the Back-Arc Basin is separated by a major fault zone, the Shan (or Sagaing) Boundary Fault, from the tectonically consolidated region of the **Sino-Burman Ranges**. They consist essentially of Late Precambrian to Cretaceous sediments, and of metamorphic and plutonic rocks. Along the Shan Boundary Fault Zone, considerable vertical movements took place. In addition, during Late Tertiary – Recent, lateral movements can be observed at this major structural element. The Paleozoic-Mesozoic sequence clearly identifies the area of the Sino-Burman Ranges as a Eurasian element and as a former part of the Mesozoic Tethys Ocean. Occurrences of the Kalaw Red Beds similar to the Khorat Series are evidence of continental influences on the sedimentation since the Upper Jurassic or Cretaceous.

2.2 Regional structural features

The dominant structural elements of Burma are (from W to E; Fig. 17 a, b=folders):
- the folded and faulted molasse sediments of Late Tertiary age overlying Early Tertiary flysch rocks, locally intruded by basaltic to intermediate dykes and plugs (offshore Arakan and Arakan Coastal Area)
- the alpinotype-orogenetically deformed, essentially Early Tertiary flysch belt (Indo-Burman Ranges)
- the zone of alpinotype-orogenetically deformed pre-Mesozoic (?), Triassic, Cretaceous and Tertiary sequences, of metamorphic rocks and of ophiolite on the E slope of the Indo-Burman Ranges
- the Mesozoic to Cenozoic (?), basic to ultrabasic intrusive rocks along the structural contact between the Indo-Burman Ranges and the Inner-Burman Tertiary Basin (E side of the Indo-Burman Ranges)

– the zone of Mesozoic and Cenozoic, basic, intermediate and acid igneous rocks in the Inner-Burman Tertiary Basin

– the Inner-Burman Tertiary Basin, which is divided into sub-basins with Tertiary sediments and uplift areas with older sediments and crystalline rocks, with block faulting and compressional folding of the basin filling

– the Mesozoic and Cenozoic intrusive and extrusive rocks that are associated with the Shan Boundary Fault at the E side of the Inner-Burman Tertiary Basin or aligned along the W margin of the Sino-Burman Ranges

– the Shan Boundary Fault Zone between the Inner-Burman Tertiary Basin and the Sino-Burman Ranges

– the tectonically consolidated Precambrian, Paleozoic and Mesozoic sediment sequences with their overlying Mesozoic-Cenozoic sediments, with intrusive and extrusive igneous rocks of various age generations and with metamorphic rocks (the Sino-Burman Ranges E of the Shan Boundary Fault Zone)

– the alpinotype-orogenetically deformed rocks of the E Himalayas with their WSW – ENE-striking fold axes and main fault systems. These ranges intersect in E India with NW-striking structural elements (Mishmi Thrust, Lohit Thrust) which border meta- morphic rocks and are associated with ophiolite and granodiorite. The $\pm$ S–N-striking structures of the Indo-Burman Ranges, of the Inner-Burman Tertiary Basin and of the Sino-Burman Ranges turn to an NE to E or an NW to W strike as they approach the E Himalayas.

In the S **offshore area of Arakan** and in the S **Arakan Coastal Area,** structural deformation is much greater than further N. In the offshore area, seismic reflection surveys revealed narrow, 30 km to 80 km long, often faulted and offset anticlinal trends with special culminations of pre-Pliocene rocks. They run generally in S–N and SSW–NNE directions. In the coastal area, the Miocene sequences are frequently steeply tilted, intensively faulted, locally folded and overthrust. The pre-Miocene flysch rocks occurring below the transgressive Miocene beds show the structural style of the Indo-Burman Ranges.

Further N, in the area of Akyab and the Boronga Islands, the anticlines are broader and less faulted both in the offshore and in the coastal area. The trends of the dominant structural elements retain their NNE direction or turn to the NE.

The **W and central parts of the Indo-Burman Ranges,** which consist essentially of Early Tertiary flysch-type sediments, have undergone alpinotype orogenetic deformation. With some intercalated allochthonous Cretaceous rock masses, overthrusts, imbricate structures and decken and with epimetamorphic sediments as well as some ophiolite, these parts of the Indo-Burman Ranges locally display the characteristics of a melange zone. The folds and overthrusts show in general W vergency. The dominant structures strike S–N–NNE–NE and within a distance of about 50 km, in the Patkai Ranges N of the Hukawng Basin, they turn to the E as the E Himalayas are approached.

The structural style of the zone with pre-Tertiary rocks in the **E Indo-Burman Ranges** is not different from that of the central and W parts of this fold belt. Only the composition of the rock sequences differs. As far as is known, Triassic and autochthonous Cretaceous sediments crop out only in the E part of the Indo-Burman Ranges, and they are always in structural contact with the Tertiary flysch series which adjoin to the W. Metamorphic sediments and ophiolite occur more frequently there than in the W part of the Indo-Burman Ranges.

E of the fault contact follow the sediments of the Inner-Burman Tertiary Basin. The fault system, which was almost certainly originated during the pre-Tertiary and rejuvenated during the Tertiary, can be traced from the Gulf of Martaban over more than 1,200 km in a N–NNE–NE direction. From about 25° N onwards, the fault contact between the flysch-type rocks of the Indo-Burman Ranges and the Tertiary sedimentary basin filling is uncertain, and from 27° N onwards (boundary of the Hukawng Basin) normal stratigraphic contacts seem to prevail. The entire fault system is characterized by a **zone of basic and ultrabasic igneous rocks** of Mesozoic-Cenozoic (?) age.

E of and parallel to this fault system with the zone of basic igneous rocks runs a SSW–NNE-striking belt of basic and intermediate to acid volcanics of Late Mesozoic and Cenozoic age within the Inner-Burman Tertiary Basin. These volcanics belong to the **Inner Volcanic Arc,** individual occurrences of which can be traced along a belt extending more than 1,200 km in a SSW–NNE–N direction in the Inner-Burman Tertiary Basin (Fig. 16). The arc has been found by magnetic and seismic reflection surveys in the Gulf of Martaban, also in the course of oil exploratory drilling in the Irrawaddy Delta Basin. It runs via Mt. Popa and the volcanics of the Shinmataung Range, Salingyi, and the Wuntho Massif through to the E border of the Hukawng Basin. The arc is made up of granodioritic, andesitic and basaltic rocks of a Late Mesozoic to Cenozoic volcanism.

The **Inner-Burman Tertiary Basin** is subdivided into a number of S–N or SW–NE-oriented sub-basins up to about 600 km long and about 200 km wide. Probably, these basins did not act as sedimentation troughs before the Miocene. The sub-basins contain up to 10,000 m of Tertiary and Quaternary sediments. Depending on the width of the basins, the sediments were deformed by more or less extensive compression folding which locally lead to overthrusting in the range of more than 1000 m. The anticlinal trends can occasionally be traced over distances of 100 km and more; they exhibit special culminations along their longitudinal axes, they are frequently offset "en echelon", and they are frequently highly faulted. Block faulting is encountered in particular in the boundary zones of the sub-basins. In the uplift areas that separate the sub-basins from each other, Early Tertiary sediments, crystalline and metamorphic rocks crop out. There are indications that some of these uplift areas were swells already before the Miocene and thus source areas of clastic deposits. They run mostly transverse, i.e. more or less E–W, to the general strike of the Inner-Burman Tertiary Basin.

The Inner-Burman Tertiary Basin is bordered in the E by the **Shan Boundary Fault Zone.** It is observable from the E part of the Gulf of Martaban running N–NNW for about 1,000 km to around 23° N, and from there it splits into the SE–NW-striking offset fault lineaments of the "26° uplift" SE of the Naga Ranges and into the NNE-striking faults of the Kumon Ranges. Considerable vertical movements took place along the Shan Boundary Fault Zone with step faults located W of it. The fault system was originated in the Mesozoic, probably earlier, and was rejuvenated during the Cenozoic. Lateral movements in the order of altogether 3 km to 4 km seem not to have occurred before the Late Tertiary and Quaternary.

The **Sino-Burman Ranges** to the E of the Shan Boundary Fault Zone are divided up by major fault zones (Fig. 29):

– the Pan Laung Fault System and its SE extension, which branches off at an acute angle from the Shan Boundary Fault in the area of Kabwet (N of Mandalay) and which can be traced for about 1,300 km towards the SSE and SE, including in W Thailand

– a fault system that branches off to the E from the Shan Boundary Fault N of Kabwet and
  which gradually swings N through Bhamo N of Mogok and can be observed over a total
  length of about 700 km as far as the N Kumon Ranges. It cannot be directly recognized
  everywhere as a major fault zone; its N section is characterized by a number of granitic
  intrusions.

These structures, which break up the Sino-Burman Ranges, also separate regional
geotectonic provinces that are distinguishable on the basis of their lithofacies and they had a
different geological evolution since the Paleozoic (p. 38).

The largely ENE-striking structural elements of the alpinotype deformed fold belt of the **E
Himalayas** intersect with the NW-striking Mishmi and Lohit thrusts in E India. These
overthrust zones frame an area of metamorphic rocks. Ophiolite is associated with the Lohit
Thrust. To the E of the Lohit Thrust, granodiorite crops out striking NW over an area about
250 km long and up to 100 km wide. As they approach the E Himalayas, the structures of
the Indo-Burman Ranges, of the Inner-Burman Tertiary Basin and of the Sino-Burman
Ranges, which in general strike S–N over hundreds of kilometres, change direction or split
up:

– the S–N-running Indo-Burman Ranges turn towards the NE in the Naga Hills and, with
  deflections at fault zones, they also strike W–E in the Patkai Ranges; they form the N
  boundary of the Hukawng Basin with E–W-oriented structural elements
– the uplift areas that separate the sub-basins of the Inner-Burman Tertiary Basin strike
  more or less E–W
– N of latitude 25° N, the Shan Boundary Fault Zone bends towards the NW and splits up
  further N into individual NW to WNW and NE-oriented lineaments
– NW and NNE-striking structures can be observed in the N Kumon Range
– a belt of katametamorphic rocks can be traced, with interruptions, from Mogok via Bhamo
  to E of Putao following a general NE–NNE–N strike. There the belt of these rocks turns
  towards the W and can probably be correlated with the NW-striking metamorphic rocks
  between the Lohit and the Mishmi thrusts, which are located about 100 km WNW of
  Putao at the S edge of the E Himalayas.

2.3 Indo-Burman Ranges, Arakan Coastal Area, Bay of Bengal

The **Indo-Burman Ranges** consist of two different geological units, namely a broad flysch
range, which constitutes the main section, and a narrow, intermittently outcropping zone of
metamorphic rocks, ophiolite, Triassic *Halobia* schist and an Upper Cretaceous sequence at
their E margin.

THEOBALD (1871) assumed the existence of a core of old rocks, which he called "Axials"
in the Indo-Burman Ranges.

However, the oldest rock sequences do not occur in their core but on their E flank. The
rocks in question are graphitic schists, rubellite schists and mica schists (BRUNNSCHWEILER
1966, p. 174 et seq.). These rocks are most widely distributed in the region of Mt. Victoria
and in the E parts of the Naga Hills N of 25° N. In the Naga Hills W of the Chindwin
River, BRUNNSCHWEILER distinguished between two rock complexes:

Naga Metamorphic Complex (pre-Mesozoic)
Naga Hills Flysch (Late Cretaceous and Eocene).

The Naga Metamorphic Complex contains meso-metamorphic rocks, i.e. dark and sometimes graphitic phyllite, biotite-muscovite and biotite-muscovite-sillimanite schist, frequently in association with banded paragneiss, sericitic quartzite and the carbonatic Pansat Series.

BRUNNSCHWEILER (1966, Fig. 14) also included large areas N of the Hukawng Valley in the Metamorphics Complex. Satellite image interpretations of the area in question indicate that this view is correct (Fig. 22); however, appropriate ground checks have not been carried out.

The oldest fossil-dated rocks belong to the **Triassic** (THEOBALD 1871). From this complex he mentioned rocks bearing *Halobia* in the region W of Prome near the village of Natoung (95° 01′ E/18° 55′ N). In the vicinity of Mt. Victoria, GRAMANN (1974) observed Upper Triassic *Halobia* schists interbedded with quartzitic sandstone and slate discordantly overlying metamorphic mica schists and basic rocks. Apart from *Halobia,* these rocks contain *Gondollela polygnathiformis* BUDUROV & STEFANOV in subordinate thin limestone layers (Carnian). Flysch-type sediment characteristics, such as graded bedding, rhythmic interbedding of sandstone and schists, and sole marks can be observed in this rock sequence.

A series of hard greywacke and sandstone running from the area of Mt. Victoria on the W side of the Myittha Valley to Kalemyo in the N is also assigned to the Triassic. This series is several hundred metres thick in places ("Thanbaya Formation"). So far no fossils have been found. In the direction of the strike towards the N, a further occurrence of *Halobia-* or *Daonella*-bearing rocks was discovered W of Kalemyo (MYINT LWIN THEIN 1973).

So far, no **Jurassic** sediments have been found anywhere in the Indo-Burman Ranges.

In the **Cretaceous,** a neritic facies region with carbonate and detrital deposits and a facies region with a trough filling of flysch-type sediments are observed.

In the E parts of the Indo-Burman Ranges the Triassic strata of the Mt. Victoria area are discordantly overlain by fossiliferous sediments (GRAMANN 1974, p. 284), the lower sections of which date from the **Cenomanian.** The sediments consist of interbedded limestone, marl, shale and sandstone. In the upper part of these strata, a **Maestrichtian** *Globotruncana* fauna was obtained from marl (p. 89).

These rocks have to be distinguished from the pelagic *Globotruncana* limestone which BRUNNSCHWEILER (1966, pp. 162 et seq.) described as Senonian limestone from the area between Gangaw and Kalemyo. 32 km NW of Gangaw the Lungrang Klang Hill (782 m) consists of three 60 to 150 m thick anticlinal limestone sequences separated by sandstone and sandy marl. It might be possible to assign this facies to the Cretaceous to Early Tertiary sequence of flysch-like rocks that follow on to the W of the metamorphics of Mt. Victoria, for the limestone of the Lungrang Klang area seems to be associated with hard, argillaceous shale, dark, silicified limestone and quartzitic sandstone, which are found in the Chin Hills between Kalemyo and Falam. Flysch-like rocks with intercalated Maestrichtian limestones occur about 30 km W of Gangaw on the road to Nabung. To the W, in the gorge of the Manipur River N of Falam, layers of exotic pebbles containing *Globotruncana* limestone are exposed in the flysch series (BRUNNSCHWEILER 1966, p. 171).

Ammonite-bearing rocks from the Upper Cretaceous occur at the S end of the Indo-Burman Ranges about 20 km W of Bassein (GRAMANN 1974, p. 286). To the W follow extensive occurrences of Upper Cretaceous *Globotruncana* limestone.

The major part of the Indo-Burman Ranges consists of a rock series that was compared with flysch-type sediments already by NOETLING (1901, p. 6). COTTER (1938, p. 37), however, regarded the non-fossiliferous series as pre-Mesozoic, if not Precambrian. He believed that the few fossiliferous strata, including the Triassic, had been interfolded.

Flysch sediments, i.e. alternating beds of greywacke, sandstone, siltstone, claystone and shale, constitute the major part of the Indo-Burman Ranges. Rhythmic bedding, graded bedding, flow-marks and sole-marks, as well as olistolithic horizons (BRUNNSCHWEILER 1966, pp. 145 et seq.) are evidence of a rapidly sedimented trough filling. Locally, these sediments have an epi-metamorphic character, e.g. as described above W of Kalemyo. Fossils are only occasionally found in the series. THEOBALD (1872) mentioned one Cenomanian *Ammonites inflatus* SOWERBY which was found near the village of Mai-I in the W part of the Arakan Yoma (p. 87). However, nothing is known about the circumstances of the find, so the possibility cannot be excluded that it was an exotic component of the flysch (CHHIBBER 1934a, pp. 205–206).

The limestone of Campanian-Maestrichtian age which is widely distributed in the flysch rocks is regarded, in accordance with BRUNNSCHWEILER (1966), as allochthonous components (Fig. 18). Their primary sedimentation region was probably located at the E edge of the Indo-Burman Ranges. To the W of Bassein, an Upper Cretaceous foraminiferal fauna was found in limestone and marls of the Mai-I Group (THEOBALD 1873).

The Upper Cretaceous flysch-type sedimentation continued in the **Tertiary** without any recognizable interruption and without lithological change. Fossils were found occurring up to the Upper **Eocene** (p. 41), but no younger Tertiary sediments could so far be located in the Burman section of the Indo-Burman Ranges. It is therefore assumed that uplift and folding of the geosynclinal sediments commenced during the Oligocene; then erosion occurred and the weathering material was transported into the adjoining sedimentation areas of the

Fig. 18. Allochthonous Upper Cretaceous limestone in Tertiary Arakan Flysch; Taungup Pass Road S of milepost 71.1, Arakan State. Photo: D. BANNERT.

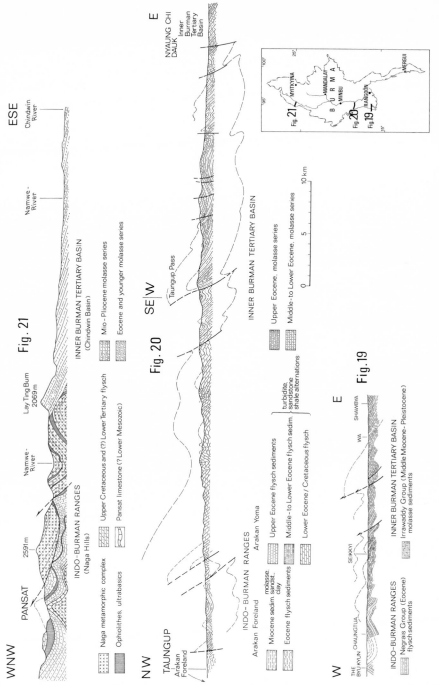

Fig. 19. Schematical cross-section through Indo-Burman Ranges, S part (SW of Bassein), not to scale. – Fig. 20. Schematical cross-section through Indo-Burman Ranges, central part (Arakan Yoma, Taungup Pass Road), not to scale. – Fig. 21. Schematical cross-section through Indo-Burman Ranges, N part (Naga Hills, W of Tamanthi), not to scale (after BRUNNSCHWEILER 1966).

Inner-Burman Tertiary Basin to the E and of the Arakan Coastal Area to the W. Extensive occurrences of Oligocene flysch are found in the Arakan Coastal Area and these are probably the equivalent of the Barail Series of Assam, which are located in a comparable geological position. In the Arakan Coastal Area NW of Sandoway and Taungup, the folded Arakan Yoma flysch (Cretaceous to Early Tertiary) is discordantly overlain chiefly by sandstone and claystone of the Miocene.

The knowledge of the structure of the Indo-Burman Ranges is also incomplete. Some typical cross-sections through the ranges are shown in Figs. 19, 20 and 21.

According to BRUNNSCHWEILER (1966), extensive and far-reaching overthrusts of metamorphic schists over non-metamorphic flysch sediments occurred in the Naga Hills (Fig. 21). On the E side of these imbricate structures come the sediments of the Inner-Burman Tertiary Basin, and here again it is probable that a continuous structural contact exists (p. 110).

According to aerial photo and satellite imagery interpretation, the flysch-type sediments and the rock sequences of the Inner-Burman Tertiary Basin bordering the Hukawng Valley (27° N) appear to have been influenced by the same structural processes (no ground checks). Interpretations of Landsat-1 imagery show that sediments of the Hukawng Valley overlap rock sequences of the Indo-Burman Ranges (Patkai Ranges) at the N edge of the basin S of the Gedu River and possibly also overlie flysch sediments in that area. No angular unconformity could be observed. The boundary between the two units should probably be sought on the S slope of the Gedu Valley (Fig. 22).

It is of interest to note the change in the fold axes that takes place within a distance of only 50 km from an SW – NE direction in the Naga Hills to a W – E orientation in the Patkai Ranges N of the Hukawng Valley (Fig. 17 b = folder).

To the E, towards the Inner-Burman Tertiary Basin, the rocks of the Indo-Burman Ranges border on a fault zone that runs for about 1,000 km from the Irrawaddy Delta in the S to 25° N. Along this fault zone many intrusions of Mesozoic to Cenozoic (?) basic and ultrabasic magmatic rocks (p. 128) occur.

The ultrabasic rocks to the E of Mt. Victoria also mark this fault zone. The Triassic and Cretaceous sediments adjoining further to the E therefore form most probably part of the basement of the Inner-Burman Tertiary Basin, although morphologically they belong to the Indo-Burman Ranges.

Thrusts and folds of the Indo-Burman Ranges show generally W-vergency (Figs. 19–21).

A hilly landscape with a few islands offshore – the **Arakan Coastal Area** – extends W of the Indo-Burman Ranges. Geologically, this area has much in common with the Arakan Yoma because it is part of the same fold belt. Upper Cretaceous (?) and Eocene, and also – unlike in the Arakan Yoma – Oligocene and Miocene sediments are exposed. The sedimentary sequences are cut in places by volcanic dykes (Fig. 23; pp. 29, 128). On the islands of Ramree and Cheduba, as well as on smaller neighbouring islands and further N on the mainland, natural gas seepages have formed mud volcanoes of sometimes considerable size (p. 123; Fig. 23).

The stratigraphic sequence differs from that of the Arakan Yoma because **Cretaceous** sediments are present in flysch facies (Fig. 24). BRUNNSCHWEILER (1966, pp. 140 et seqq.) mentioned Senonian shale and sandstone sequences from the coastal area of Sandoway, transgressively overlain by Eocene sediments. The alternating shale/sandstone series contain many olistostromes, together with reworked Upper Cretaceous *Globotruncana* limestone.

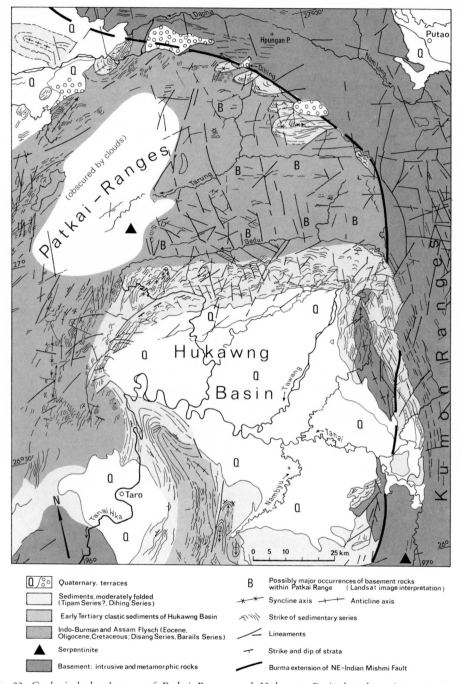

Fig. 22. Geological sketch map of Patkai Ranges and Hukawng Basin based on interpretation of satellite imagery.

Fig. 23. Zaingchaung mud volcanoes, Ramree Island; oblique aerial photograph. Photo: D. HELMCKE.

It is doubtful, however, whether any Upper Cretaceous flysch sediments are present in the Arakan Coastal Area. When a "Senonian" olistostrome was examined near Mazin Point (BRUNNSCHWEILER 1966, Figs. 3 and 4) redeposited Eocene foraminiferal limestone was observed as allochthonous material in addition to Upper Cretaceous limestone. Since the foraminiferal limestone was found approximately at the crest of a saddle of the allegedly Senonian rocks, there is some doubt about BRUNNSCHWEILER's interpretation.

Red radiolarian chert with subordinate sandy limestone, sandstone and silicified greenstone, which were observed in an isolated occurrence in Kyauknimaw near the SW tip of the island of Ramree, have been dated as **Upper Cretaceous** (HELMCKE & RITZKOWSKI, in press).

In the W part of the Arakan Coastal Area, on the islands of Ramree and Cheduba, new data on stratigraphy and structure have been obtained in recent years. According to HELMCKE & RITZKOWSKI (in press), alternating sandstone, conglomerate, siltstone and shale accumulated in the **Lower Eocene.** Apart from this clastic facies, remnants of a calcareous sedimentation are found as reworked material in these strata. They contain Paleocene as well as Eocene fossils (nummulites).

The sedimentary environment underwent a change in the **Middle Eocene.** Turbiditic sandstone, with occasional intercalations of chert, is observed in the autochthonous sequences. Slip horizons also occur. This type of sedimentation continues in the **Upper Eocene.** On Cheduba it provides the oldest sediments so far observed there.

On Ramree, **Oligocene** sediments occupy a large area. Middle and Upper Oligocene sequences were discovered by GRAMANN (1973) in the Konbwe Bay of Ramree and later

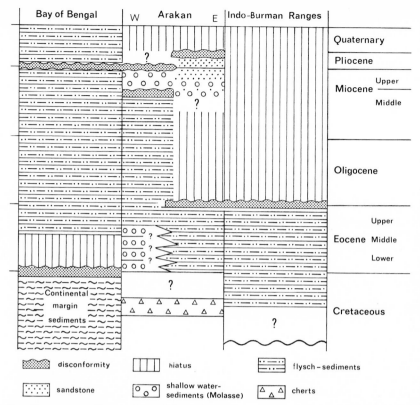

Fig. 24. Post-Cretaceous sedimentation and facies development: Bay of Bengal – Arakan Coastal Area – Indo-Burman Ranges.

recorded from many places on Ramree and Cheduba Islands by HELMCKE & RITZKOWSKI (in press).

The flysch sedimentation ended in the **Middle Miocene**. The "Nga-Ok Conglomerate" is locally intercalated in flysch layers that contain a **Lower Miocene** microfauna. The Nga-Ok Conglomerate does not extend laterally very far; it is in turn overlain by a flysch-like sequence of shale and sandstone of the Middle Miocene. Following their deposition, there was a short but strong phase of folding and erosion after which the sedimentation changed to molasse-like sequences.

Upper Miocene sediments are limited to the W part of the island of Ramree. Their lower portion consists of shale with intercalated lenses of sandstone and siltstone. In NW Ramree, the upper portion consists of massive sandstone. In Cheduba, the Upper Miocene shale sequence with sandstone and siltstone lenses forms the N and E parts of the island.

Towards the E, onshore in the Arakan Coastal Area, sandstones also dating from the Miocene transgress on folded flysch sediments of Eocene age. BRUNNSCHWEILER (1966, pp. 148 et seqq.), about 7 km S of Taungup (94° 15′ E/18° 15′ N), described a transgressive sandstone sequence with a large amount of carbonaceous matter which he believed belongs to the Miocene. The carbonaceous sandstone is overlain by light-coloured, tuffitic siltstone and shale in which exotic blocks from the Upper Cretaceous and also flysch rocks occur. The sequence is concluded by coarse-grained feldspar-containing sandstone.

To the N, in the area of the Boronga Islands and in the region between Akyab and the border with Bangladesh and India, Miocene/Pliocene shale, claystone, silty clay, siltstone with occasionally interbedded lignitic fine-grained sand, as well as limestone and dolomite occur. The great thickness of more than 4,000 m points to a rapid sedimentation.

The **structure** of the N Arakan Coastal Area differs clearly from that of the Indo-Burman Ranges. Instead of steep dips as observed in the S, the Miocene series exhibit relatively minor tectonic deformation in the area of Akyab and on the islands of the Boronga Archipelago. Faults along the general strike of long folds indicate overthrusting approximately parallel to the fold axes. In the area of Sandoway, Early Tertiary and (?) Cretaceous flysch rocks, which reveal the structural style of the Arakan Yoma, occur under overlapping Miocene sandstone series.

In the Ramree and Cheduba Archipelago, which follows to the S of the Boronga Islands, a local structural change occurs in the long fold trend coming from the N. In this area it is replaced by almost circular and crescent-shaped synclines (FAY LAIN & WIN MAW 1971). More recent investigations by HELMCKE & RITZKOWSKI (in press) revealed that the circular structures are not restricted solely to the main island of Ramree but are also found on several islands to the N and E (Figs. 25, folder, 26).

BRUNNSCHWEILER (1966, pp. 158 et seqq.) believed that the circular synclines were originated by structural movements in connection with a sedimentary diapirism of Eocene sediments. The 160° orientation of the fold axes was influenced by a more recent tectonic stress with a 70° orientation in the Ramree-Cheduba Archipelago, and the circular structures were formed as a result. HELMCKE & RITZKOWSKI (in press) thought that the presence of overpressured shale was a presupposition for the origin of these structures but that the formation of the circular structure was initiated by tectonic deformation. This structural deformation took place prior to the transgression of the Upper Miocene molasse. The time of this deformation is probably equivalent with the period of the uppermost (Upper Miocene) discordance in the Bay of Bengal. FAY LAIN & WIN MAW (1971) explained the circular synclines by subsidence resulting from mass deficiency caused by sedimentary volcanism.

W of the Arakan Coastal Area lies the **NE part of the Bay of Bengal.**

JOHNSON et al. (1976) assumed that the oceanic crust that surrounds this section of the Bay is younger, at approximately 80–64 m.y. B.P., than the sections located further to the W of the Bay of Bengal (130–100 m.y. B.P.).

The oceanic crust is overlain by an extremely thick sedimentary series which formed since the Lower Cretaceous (MOORE et al. 1974). This sequence can be subdivided into three portions that are separated by discordances (Fig. 24).

The lowermost portion consists of consolidated sandstone or shale, pelagic limestone, intercalated volcanic rocks and, to a lesser extent, also of evaporites. This series is interpreted as deposits of a midplate continental margin rise (MOORE et al. 1974, p. 411).

The oldest known unconformity represents a stratigraphic gap from the Paleocene to the Middle Eocene.

The middle unit consists of turbiditic sediments, as can be concluded from burried channel and natural levee deposits. This type of sedimentation lasted through into the Upper Miocene.

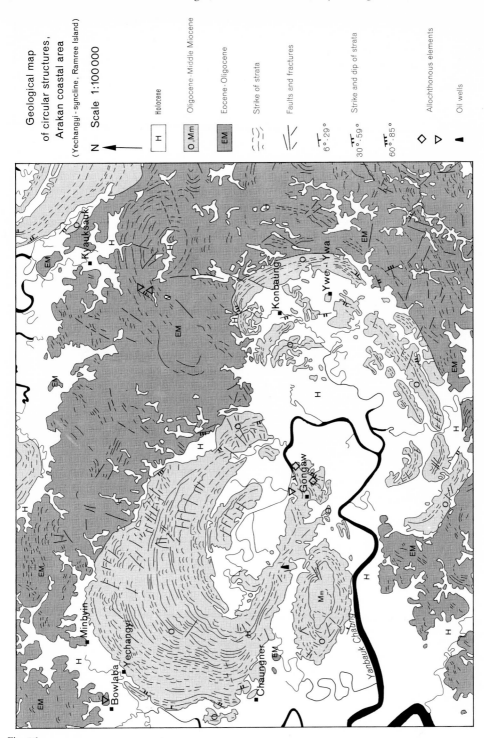

Geological map
of circular structures,
Arakan coastal area
(Yechangyi - syncline, Ramree Island)

N

Scale 1:100 000

H   Holocene

O.Mm   Oligocene-Middle Miocene

EM   Eocene-Oligocene

         Strike of strata

         Faults and fractures

6°-29°   Strike and dip of strata

30°-59°

60°-85°

◇ ▷   Allochthonous elements

◀   Oil wells

Fig. 26.

During the uppermost Miocene, the most recent discordance was formed as a result of another interruption in the process of sedimentation. It is overlain by recent sediments of the Bengal Fan. These sediments consist also of turbidites.

While the sediments in the Bay of Bengal were in recent times mainly distributed over the submarine canyon "Swath of No Ground" S of the mouth of the Ganges-Brahmaputra river system, other turbidity channels were also active somewhat earlier (CURRAY & MOORE 1974). Bathymetric studies of the Bay of Bengal showed that a pronounced turbidity channel had formed in the E part of the Bay of Bengal. Its sediment load was mainly transported to the small E Nicobar lobe of the Bengal Deep-Sea Fan.

Generally, structural deformation in the NE part of the Bay of Bengal can be observed only in the two older sediment sequences (MOORE et al. 1974).

The flysch sedimentation in the Indo-Burman Ranges, in the Arakan Coastal Area and in the Bay of Bengal, indicates that the sediment trough was shifted during the Tertiary from E to W: In the Indo-Burman Ranges, the flysch sedimentation commenced in the Upper Cretaceous (THEOBALD 1871, BRUNNSCHWEILER 1966) and lasted through to the Upper Eocene (Figs. 24, 31). In the E sections of the Arakan Coastal Area near Taungup and Sandoway, the stratigraphic sequence also includes Eocene beds. In the W sections sedimentation continued during the Oligocene and Lower Miocene. After a brief phase of folding, a molasse sedimentation followed, which in turn transgressed over the by now folded Eocene of the Arakan Coastal Area. According to MOORE et al. (1974), the flysch sedimentation in the Bay of Bengal commenced only after the unconformity-gap which lasted from the Paleocene to the Middle Eocene. In the Bay of Bengal, the flysch-type sedimentation continues to the present.

2.4 Inner-Burman Tertiary Basin

With an onshore length of about 1,100 km, the Inner-Burman Tertiary Basin extends from the Gulf of Martaban to the NNW, N and NNE as far as the N surroundings of the Hukawng Basin. In the S Chindwin-Shwebo Basin, it attains its largest E–W extent of around 200 km. To the S and SW, it can be continuously traced for about 3,000 km in the Andaman Sea, and as far as the N, Central and S Sumatra Tertiary Basins (CCOP Newsletter 1976). In Burma, it is bounded in the W by the Indo-Burman Ranges, in the E by the Sino-Burman Ranges (Figs. 16, 17 a, b, folders, p. 22, p. 36), and in the N by the E Himalayas (p. 107).

The sedimentary filling of the Inner-Burman Tertiary Basin consists of Miocene, of Oligocene, of Eocene and to a lesser extent of Paleocene sequences (p. 91) overlain by Middle Miocene-Quaternary clastic rocks and underlain by Cretaceous, probably also by Jurassic, Triassic and older rocks. The Tertiary sediments that are developed exclusively in marine facies in S Burma indicate a continental influence, becoming increasingly intense towards the N and NE, on the depositional environment. The mostly argillaceous and fine-sandy Tertiary sediments were deposited in a shelf region off the continent then located in the E (Sino-Burman Ranges). The shelf merged to the W into the Indo-Burman Geosyncline. The sediments attain overall thicknesses of more than 10,000 m (p. 162). The non-sedimentary rocks in the area of the Tertiary shelf region consist of ultrabasic, basic, intermediate and acid extrusives and intrusives of various Cenozoic, Mesozoic and Paleozoic ages and of metamorphic rocks (p. 128, p. 131).

In accordance with plate tectonic criteria, the Inner-Burman Tertiary Basin can be divided into an Inter-Arc Trough (or Interdeep), an Inner Volcanic Arc and a Back Arc Basin (Fig. 16). The Inter-Arc Trough and the Back Arc Basin can be subdivided into a number of N–S or NNE–SSW oriented sub-basins which are separated from each other by uplift areas consisting of Early Tertiary and older rocks running at approximately right angles to the general N–S strike. Isopachs of Tertiary sequences in the Salin Sub-Basin (Figs. 68 a–f) show that this sub-basin was not formed before the Middle Miocene.

The pre-Upper Miocene sediments of the Inner-Burman Tertiary Basin have undergone regionally varying tectonic deformation at various periods. The dominant structural deformations were caused by tangential compression and block-faulting. They resulted in structures ranging from broad anticlines (e.g. in the W Irrawaddy Delta Basin; p. 110) to narrow fold trends over 100 km long with culminations less than 1 km wide and exhibiting clear vergence and considerable overthrusts (e.g. Chauk-Yenangyat in the Central Basin; p. 169).

The subdivision of the Inner-Burman Tertiary Basin into sub-basins and uplift areas probably did not begin before the Oligocene. Some of the uplift areas, however, may have acted on the facies and thickness of the sedimentation as swells already from the Eocene onwards. The following structures can be distinguished from N to S in the Interdeep (Fig. 27, folder):

- Putao Basin (?)
- "28° N uplift area"
- Hukawng Basin
- "26° N uplift area"
- Chindwin Basin-North
- "24° N uplift area"
- Chindwin Basin-South
- "22° N uplift area"
- Central Basin
- "20° N uplift area"
- Prome Embayment
- W Irrawaddy Delta Basin
- W Gulf of Martaban Basin

in the Back Arc Basin (from N to S):
- Putao Basin (?)
- Bhamo Basin
- Shwebo Basin
- Pegu Yoma-Sittang Basin
- E Gulf of Martaban Basin

Many of the sub-basins are or were sites of hydrocarbon production or they can be regarded as hydrocarbon prospective areas; they are therefore described in more detail in Chapter 5.11 "Hydrocarbons".

The structural elements that separate the sub-basins run more or less at right angles to the general S–NNW–N–NNE–NE strike of the Inner-Burman Tertiary Basin, in the Interdeep as well as in the Back Arc area of sedimentation:

At 28° N an uplift area with a crystalline rock complex separates the Putao Basin from the Hukawng Basin. The latter ends in the SW at the "26° N uplift area", which is also characterized by an extensive complex of crystalline and sedimentary rocks. To the S and SE the Tertiary basin sediments of the Hukawng Basin are linked with those of the Back Arc area of sedimentation.

The Chindwin Basin, in which a structural elevation is indicated at 24° N by the course of the isopachs of Upper Eocene to Miocene sediments and by the configuration of the magnetic basement, commences SW of the 26° N uplift area. Towards the E the Chindwin Basin is separated from the Back Arc Basin by the "Igneous Wuntho Massif" (p. 136), while towards the SE it is connected with the Shwebo Basin of the Back Arc area of sedimentation.

The "22° N uplift area" is characterized by outcropping Eocene rock sequences. From here on, the axis of the Chindwin Basin plunges to the N and the axis of the Central Basin with exposed Late Tertiary sequences plunges to the S (Fig. 27, folder). The Central Basin is divided into two parts:

– the approximately 220 km long N–S oriented and up to 70 km wide Salin Syncline in the W, whose Early Tertiary rock sequences have been lifted up in the E on the steep and narrow anticlines of Suwin, Kyaukwet, Sabe, Yenangyat-Chauk and Yenangyaung, which are offset "en echelon" (Fig. 27, folder)

– the portion between these narrow structures and the Inner Volcanic Arc which is less deeply troughed than the Salin Syncline. This portion of the Central Basin also exhibits NNW–SSE-striking and offset anticlinal trends. The Tertiary sediments of this part of the Central Basin ("Taungdwingyi Syncline") are connected with those of the Back Arc area of sedimentation E of the Inner Volcanic Arc.

At about 20° N, the axis of the Central Basin gradually rises towards the SSE without, however, any Early Tertiary sediments coming to the surface as they do in the "22° N uplift area". The Prome Embayment, a further Tertiary sub-basin with an NNW–SSE–S–SSW-striking axis, commences S of 19° 45′ N. Towards the S it gradually widens to about 100 km from E to W and, after approximately 200 km, merges into the W Irrawaddy Delta Basin. In the W, the Prome Embayment borders with a fault contact on the Indo-Burman Ranges, and in the E on the anticlinal Miocene sequences of the Pegu Yoma (Back Arc Basin). In the area of the boundary between the Prome Embayment and the Pegu Yoma, the Inner Volcanic Arc can be observed in several occurrences of Cenozoic volcanics. The Tertiary rock sequences of the Prome Embayment and of the W Irrawaddy Delta Basin continue towards the E into the Back Arc area of sedimentation in the Pegu Yoma and in the Sittang Basin (Fig. 17 a, folder). In the narrower N Prome Embayment the folds show very narrow culminations, an E-oriented vergence and considerable block faulting (e.g. the Prome Anticline), while towards the S they are broader and much less block-faulted.

The axis of the Interdeep gradually bends from the S Prome Embayment into the SSW-striking W Irrawaddy Delta Basin. In the W, the W Irrawaddy Delta Basin borders with a fault zone on the flysch rocks of the S Arakan Yoma. The fault system runs SSW (∼ 200°) and thus no longer parallel to the E edge of the Indo-Burman Ranges. The Early Tertiary sequences of the "Western Outcrops" of the Inner-Burman Tertiary Basin further to the N have subsided at this fault zone; the only other place where occurrences of Eocene and older strata crop out again is SW of Bassein. Towards the SSW the W Irrawaddy Delta Basin continues offshore in the W Gulf of Martaban (Fig. 27, folder). To the E it borders on the S-pitching folds of Miocene rocks of the Pegu Yoma. Basic volcanic rocks of the Inner

Volcanic Arc were found during oil exploration drilling on the E edge of the W Irrawaddy Delta Basin. Here, too, the Tertiary sequences continue into the area of the Back Arc Basin (Sittang Basin).

Airborne-magnetometric and offshore-seismic surveys have to a certain extent clarified the structural conditions in the Gulf of Martaban. From a total of 16 exploratory oil wells (1979), data have also been gained on the stratigraphy of this southern portion of the Inner-Burman Tertiary Basin. It was clearly apparent that the Outer Island Arc, the Interdeep and the Inner Volcanic Arc continued to the SSW and SW, and the Back Arc Basin – while widening in the W–E direction – continues towards the SSE in the Gulf of Martaban (Fig. 28). The W Gulf of Martaban Basin, which narrows to less than 30 km in an E–W direction towards the SSW, is located in the Interdeep area of sedimentation. It is relatively shallow (estimated thickness of the Tertiary sediments approximately 6,000 m) and relatively undeformed. Transverse structural elements (WNW–ESE and E–W-striking normal faults and E–W striking anticlines and synclines) are found at 15° 15′ and at 15° N.

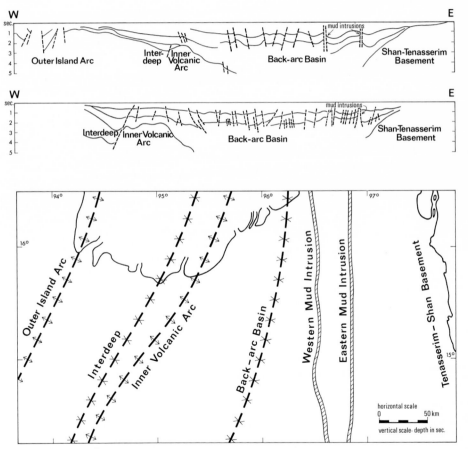

Fig. 28. Principal structural features, Gulf of Martaban; cross-sections along approx. 15° 15′ N (above) and along approx. 14° 45′ N (seismic reflection phantom horizons).

On the basis of seismic surveys, the Tertiary sediments of the W Gulf of Martaban can be correlated with those E of the Inner Volcanic Arc in the Back Arc Basin. The E Gulf of Martaban Basin is approximately 200 km wide and up to 8 km deep at 15° N. The Late Tertiary rock sequence wedges out towards the E above the Tenasserim basement rock complex with the respective younger stratigraphic members overlapping the underlying members (Fig. 28).

According to CURRAY et al. (1979), the S–N-striking axis of the Tertiary North Sumatra Basin rises in the area between 10° N lat. and 11° N lat., in the Andaman Sea W of the southern Tenasserim coast. N of this structural high or "uplift area", it plunges again and continues in the E Gulf of Martaban Basin. W of the uplift area which terminates the North Sumatra Basin in the N, a spreading center occurs (Fig. 16). The open central part of the corresponding rift valley is according to CURRAY et al. (1979, p. 195) 5 km wide, with a total opening rate of the spreading center of 3.72 cm/y. The oldest paired upturned edges of the rift valley are approx. 65 km apart. The spreading commenced ca. 10.8 m.y. B.P. in the Middle Miocene.

The trough axis of the E Gulf of Martaban Basin runs in a N–NNW direction. The Tertiary sequence reveals numerous anticlines and synclines, several kilometres wide, mostly striking N–S. They are transected by a network of E–W, ENE–WSW and N–S-striking faults with vertical displacements of up to 300 m. The structural transverse elements ($\pm$ E–W) occur particularly clearly between 15° N and 14° N, as in the Interdeep. In the E Gulf of Martaban Basin a number of diapiric structures were located by seismic reflection surveys; they were interpreted as mud intrusions. They attain a maximum width of 5 km and N–S lengths of approximately 10 km to more than 250 km ("western and eastern mud intrusion" in the E part of the E Gulf of Martaban Basin, Fig. 28). They continue towards the N in major fault zones onshore.

Drilling revealed a total of approximately 4,200 m of Tertiary sedimentary rocks (Miocene, Oligocene (?), Eocene) in the W Gulf of Martaban Basin (p. 160), and about 5,000 m in the E Gulf of Martaban ("Central Trough"). Here, the Lower Cretaceous/Jurassic was also reached (p. 166). Regardless of the stratigraphic level, there is a sudden and pronounced increase in the degree of diagenesis, hydrocarbon maturity and coalification in the depth range between approximately 2,500 m and 3,600 m (Fig. 71).

2.5 Sino-Burman Ranges

The Sino-Burman Ranges (Kachin-Shan-Tenasserim Highlands, "Eastern Highlands") represent the largest tectonic unit in Burma and occupy the E part of the country. Its main components are consolidated, partially low-grade metamorphic, Paleozoic and Mesozoic sediments of the Burmese-Malayan Geosyncline and its substratum consisting of Precambrian crystalline rock of the Mogok Series (p. 47) and of the pre-Upper Cambrian Chaung Magyi Group (p. 51). Cretaceous sediments occur chiefly in the W part of this tectonic domain, while Late Tertiary and Quaternary sequences surround its W marginal blocks and form the fillings of intramontane basins. Igneous rocks are represented by various generations of acidic and basic intrusive and acidic, intermediate and basic extrusive rocks.

Fig. 29. Tectonic domains of eastern Burma.

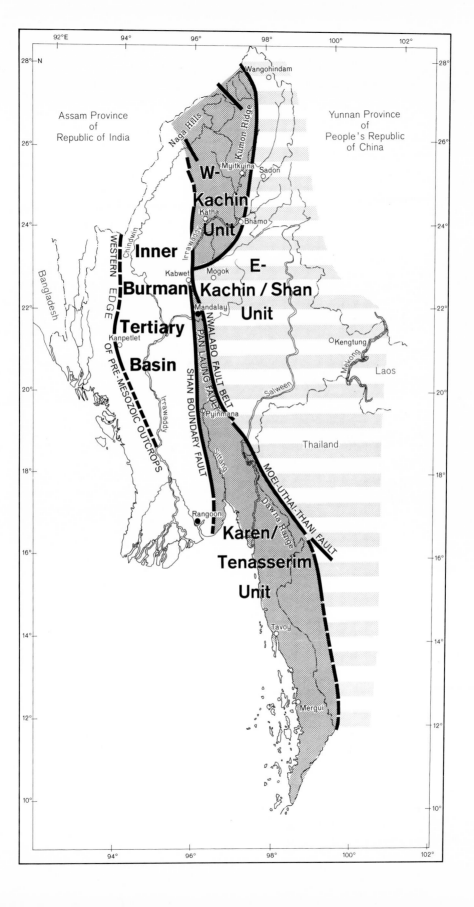

The Sino-Burman Ranges form part of the land mass of the Indo-Chinese Peninsula (Yunnan, Thailand and Malaysia), which extends to the S in Sundaland (HUTCHISON 1973). CHHIBBER (1934a) and MAUNG THEIN (1973) placed the W boundary of this section of continental crust at the generally N–S-striking Shan Boundary Fault, which accompanies the Shan Escarpment in the W and is presumed to continue to the S in the Gulf of Martaban beneath the delta deposits of the Sittang (Fig. 29).

Various interpretations have been made of the character of the Shan Boundary Fault. WIN SWE (1970, 1972) regarded it as a transform fault; according to COOPIOC (1974, p. 33), the fault system shows evidence of old and considerable vertical movements as well as of Quaternary right lateral movements. It can be interpreted as a system of normal faults with strong vertical displacements and only slight lateral movements, which occurred in the more recent history of the fault. It has shaped the morphological E margin of the adjoining Tertiary Basin. The Shan Boundary Fault strikes SSE–NNW and extends towards the N into the area of Katha (approximately 25° N). Further N it peters out in approaching the Naga Hills as a NW–SE-striking fault lineament. In this region, the E–W-striking Paleozoic rocks of the W Kachin Highlands are linked with the metamorphic rocks of the Naga Hills. Bending from the E–W to the N–S-strike direction, these rocks ("Kanpetlet Schists", p. 77) together with the *Halobia*-bearing Triassic sediments (p. 81) at the E margin of the Arakan Yoma, possibly form the substratum of the N Inner-Burman Tertiary Basin.

So far, it has not been possible to undertake a structural division of the Sino-Burman Ranges because of lack of detailed geological information. An attempt is made here to provide a general view based on new data on the stratigraphy and structure (Fig. 29). It is assumed that in the W a "foreland" is located ahead of the E part of the Kachin-Shan Highlands at a morphologically lower level. In the area of Neyaungga this foreland is bounded in the W by the Shan Boundary Fault and in the E by the "Pan Laung Fault" or the "Nwalabo Fault Belt". GARSON et al. (1976, Fig. 4) traced the Pan Laung Fault, which separates the Jurassic-Cretaceous Pan Laung Formation from the Ordovician to Jurassic sedimentary series of the Shan Plateau in the area of Neyaungga, towards the S as far as latitude 20° N. The fault appears clearly still further S on satellite images, then it changes direction to the SE as a bundle of lineaments E of Pyinmana and, where the Salween forms the border between Burma and Thailand, it runs into the territory of Thailand. Here, it corresponds to the Moei-Uthai Thani Fault Zone (CAMPBELL & NUTALAYA 1975). It seems to split off from the latter fault zone and to continue in a southerly direction parallel to the Dawna Range (Fig. 29).

Towards the N, the Pan Laung Fault intersects with the Shan Boundary Fault at an acute angle and appears to join with it in the area of Kabwet, which is characterized by Cenozoic basalt volcanism (p. 140). GARSON et al. (1976) were able to show that the movements along the Pan Laung Fault commenced in the Jurassic but took place for the most part during the Tertiary. However, the region in the immediate vicinity of the Pan Laung Fault was already tectonically active during the Paleozoic, because two areas with different lithofacies and paleogeographic development adjoin each other along this zone. The areas in question are the **Karen/Tenasserim Unit** located to the W and SW of the fault and containing mainly flyschoid clastic sediments of the Silurian to the Carboniferous (possibly also of the Early Paleozoic Lebyin Group, Mawchi Series, Taungnyo Group, Mergui Series; Fig. 30), and the **Shan-Kachin Unit** located E and NE of the fault with mainly carbonate and silty-sandy shelf facies as well as euxinic graptolite shale facies (Naungkangyi Formation, Panghsa-Pye Formation, Zebingyi Beds, Maymyo Dolomite Formation; Fig. 30, folder).

The regions with different lithofacies development extend further to the S on the S peninsula towards Thailand and Malaysia. Clastic Paleozoic sediments, characterized by turbidites, are found there in the Thong Pha Phum area (KOCH 1973, 1978), in the Phuket area ("Phuket Group", GARSON et al. 1975) and in W Malaysia (Machin-chang, Setul and Singa Formation; GOBBETT & HUTCHINSON 1973). On the S peninsula, too, an area with mainly argillaceous-silty and carbonate sedimentation adjoins.

The division into a marine shelf region with a marginal flysch zone W of it, which existed during the Paleozoic, disappeared in the Permian and the Triassic (Permian transgression facies–Nwabangyi Dolomite Formation, Moulmein Limestone), and it was not very prominent at all during the Jurassic either. Strong fault movements and thus a "resurrection" of the division that existed in the Paleozoic did not take place before the Cretaceous and Tertiary along the Pan Laung Fault at the boundary between the Paleozoic sedimentation regions. These movements resulted in an uplift of the Eastern Kachin/Shan Unit in the E in which, from then on, only continental sediments were deposited in intramontane basins. Marine Cretaceous and Tertiary sediments were laid down in the subsided zone – here referred to as the Karen/Tenasserim Unit – W of the Pan Laung Fault.

As in SE Burma, the Sino-Burman Ranges can also be divided up into two sections N of the Kabwet basalt area (N of Mandalay), namely the "foreland" to the E of the Shan Boundary Fault, and the crystalline rock complex of the E Kachin Highlands that follows to the E. The area E of the N section of the Shan Boundary Fault and its N extension is for the most part unexplored. On the basis of the scant information available (STUART 1919, 1923 a), the geological map (Fig. 17 a, folder) shows Precambrian gneiss in the area of Haungpa (26° N uplift area, p. 113) and an extensive area with Early Paleozoic rocks between Katha and the Kumon Ridge; furthermore marine Cretaceous sediments and extensive areas with the Late Tertiary to Quaternary Irrawaddy Formation and post-Pleistocene unconsolidated sediments.

Besides granite intrusions, this area is characterized by gabbro, peridotite and serpentinite bodies which cover up to approximately 100 km^2. In regional geological terms, the zone referred to here as the **West Kachin Unit** corresponds to the Karen/Tenasserim Unit of SE Burma (Fig. 29). It is possible that the clastic sediments of the West Kachin Unit, which are placed in the Cambrian, also contain younger Paleozoic rocks and thus the lithofacies-paleogeographical criterion for delineating this unit from the East Kachin/Shan Unit is also satisfied.

The E boundary of the West Kachin Unit is located in the broad Quaternary belt which lies in front of the Precambrian crystalline rocks of Mogok-Bhamo-Sadom-Wangohindam. Like the Pan Laung Fault it is characterized, particularly in the area between Bhamo and Myitkyina, by numerous granite intrusions. The area with crystalline rocks of the eastern Kachin State was named by GOOSSENS (1978, p. 452), together with what is here referred to as the West Kachin Unit, as a "North Eastern Belt", separated off by the "Lashio Fault System". This division does not take account of the fact that the Precambrian crystalline belt in Yunnan in the E is accompanied by the same Paleozoic to Mesozoic sedimentary sequence of the Burmese Malayan Geosyncline as in the region between Lashio and Mogok (J. C. BROWN 1923a, map on Pl. 9). Thus, the Lashio Fault System does not have a general separating function. Instead, all the paleogeographic and structural features in the area of the

Shan State can be traced right through to the area of the E part of Kachin State as well as to the E Himalayas and W Yunnan. As a result, the E part of the E Sino-Burman Ranges is grouped together as the East Kachin/Shan Unit.

At the E edge of the Kachin-Shan Highlands, SEARLE & BA THAN HAQ (1964) distinguished the almost 1,400 km long and tens of kilometres wide unit of the so-called Mogok Belt. They assumed that the "Mogok Series" newly defined by them consists of a Precambrian to Triassic rock sequence which was metamorphosed in various grades in the course of the Himalayan orogeny and intruded by Tertiary granite. GARSON et al. (1976) contradicted the view that all the rock units grouped together under the term "Mogok Series" belong to one single metamorphic assemblage. In addition, it seems questionable that the crystalline rocks of Mogok, as the component of an assumed alpidic-folded Mogok Belt, can be considered separately from its structural surroundings. There is a link along the strike towards the NE and extending into the area of Yunnan/China. In W Yunnan (J. C. BROWN 1923 a,b,d) and in the West Kachin Unit, parametamorphic rocks or Chaung Magyi sediments of lower metamorphism are overlain by non-folded Triassic red beds or by marine Cretaceous sediments and thus they cannot have been folded or metamorphosed during the Tertiary Himalayan orogeny.

It is more likely that the folding events that affected the Burmese part of the Burmese-Malayan Geosyncline took place entirely during the Late Paleozoic to the Mesozoic. They affected not only the Paleozoic flysch zone of the Karen-Tenasserim Unit (to which the West Kachin Unit also possibly belongs) but also the marine shelf region of the East Kachin/Shan Unit adjoining to the E. The intensity of structural deformation varied both in time and in space, without forming any fold belts that are chronologically fixed on specific events which led to high-grade metamorphism. The orogenic folding and block faulting commenced locally in the Carboniferous; this is directly apparent at fold discordances and at erosion discordances extending down to the Ordovician, and also it can be observed in conglomerates at the base of the Permian. The movements continued throughout the Mesozoic, with interruptions by tectonically relatively inactive periods (Permo-Triassic reef and fusulinid limestone facies). They culminate in the Jurassic, after which they are followed by the sedimentation of thick sequences of the Kalaw Red Beds. These orogenic events manifest themselves in the Karen/Tenasserim Unit as narrow isoclinal folding and in a low-grade regional metamorphism; in the East Kachin/Shan Unit through broad fold structures of the thick marine shelf carbonates. Granite intrusions with their thermo-metamorphic aureoles are common to both units. Probably, it is true in the case of Burma, as well as for Thailand, that the more recent Tertiary granite generations become more frequent towards the W (GOBBETT & HUTCHINSON 1973). Radiometric age determinations carried out on the granites along the Pan Laung Fault and the Hermingyi granite have yielded Rb/Sr ages of 30–50 m.y. (p. 118), while the granites of Keng Tong are probably Triassic in age, as was proven by datings carried out in their S sections located in Thailand (VON BRAUN et al. 1976).

During the Late Mesozoic and Tertiary, the E part of the Inner-Burman Tertiary Basin subsided along the Shan Boundary Fault, and the West Kachin and Karen/Tenasserim Unit subsided along the Pan Laung Fault, which significantly is located in the boundary area of the Paleozoic shelf and its adjoining flysch zone. The process of subsidence was accompanied by granite intrusions along the Pan Laung Fault, among other places, and at the same time marine Cretaceous and Tertiary sediments were laid down in the West Kachin Unit and in the N parts of the Karen/Tenasserim Unit, while terrestrial unconsolidated sediments were locally deposited in intramontane basins in the "uplifted fault block" of the East Kachin/Shan Unit. Acting as a foreland, the West Kachin Unit and the Karen/Tenasserim Unit thus play an intermediary role between the deeply subsided Inner-Burman Tertiary Basin (Back Arc Basin) and the East Kachin/Shan Unit. The latter includes the Paleozoic shelf sea region of the Burmese-Malayan Geosyncline and was uplifted during the Late Mesozoic and Tertiary. Since the Cretaceous, no marine depositional environment has developed anymore in the area of the East Kachin/Shan Unit.

2.6 Eastern Himalayas

So far only STUART (1919, 1923a) has carried out geological reconnaissance in the Burmese section of the E Himalayas. Metal ore reconnaissance prospecting in the area NE of Putao up to the Chinese border was undertaken by B. ZALOKAR (JURKOVIC & ZALOKAR 1961).

The region E of Putao is built up of highly metamorphosed metasediments (gneiss, amphibolite, quartzite and marble). As geological reconnaissance surveys and interpretations of Landsat imagery have shown, this belt of crystalline rocks can be traced via Bhamo to the area of Mogok. There is thus a connection between parts of the E Himalayas and the East Kachin/Shan Unit of the Sino-Burman Ranges. SEARLE & BA THAN HAQ (1964) named the crystalline rocks of the Mogok area the "Mogok Belt" (p. 49).

STUART (1923a) mentioned from this zone, at about 26° 30′ N, a sandstone series with conglomerate at its base, discordantly overlying the metamorphic series. The sandstone series dips gently towards the WNW. This basal conglomerate contains syenite-gneiss-detritus; syenite-gneiss is intrusively intercalated in the metamorphic series. The age of the clastic series was given by STUART (1923a), on the basis of lithological comparisons, as equivalent to the Tipam sandstones (Middle Miocene). SALE & EVANS (1940) were of the opinion that these conglomeratic sandstones were more likely to be the equivalents of the Dihing Series (Pliocene).

MACLAREN (1904) mentioned that the axes of the Burmese section of the mountain chains W of Putao could perhaps be traced through to the Lohit District of the Indian State of Arunachal Pradesh. The regional geology of the Lohit District is briefly discussed here, because the geological conditions there are better understood and it is possible to draw conclusions about the geology of the region around Putao.

According to KARUNAKARAN (1974) the structural features of the Lohit District differ from the Siang District adjoining to the W chiefly because of the following factors:
- the dominant strike direction is NW–SE in the Lohit District and NE–SW in the Siang District
- Siwaliks and Gondwana rocks are lacking in the Lohit District, or they play only a very secondary role, while further W they assume their characteristic form
- there is an approximately 100 km long, relatively broad zone of ultrabasic rocks between the Tidding Valley and the Twelang Valley in the Lohit District. So far no ultrabasic rocks have been observed in the Siang District.

In the Lohit District (KARUNAKARAN et al. 1964) the metamorphic rocks occur along the NW–SE-striking Mishmi Fault in the contact with the Quaternary sediments of the Upper Assam Plain, and a narrow belt of Siwaliks occurs only in the area of the Dihing River.

To the NE of the Mishmi Fault, quartzite, quartz-rich schist and coarse-grained marble crop out; further to the NE come biotite, graphite and garnetiferous schists, then more highly metamorphosed schists, which contain locally biotite, staurolite, garnet and locally sillimanite or graphite. The next zone is built up of quartzite, pink-grey marble, quartz-rich schist and chlorite schist. In the Tidding Valley, and from here on further to the SE, the intensively folded marble is flanked by highly stressed serpentinite. These are followed by chloritic and graphitic schists and metasediments, which are locally rich in calcareous matter.

This rock complex is bounded towards the NE by the Lohit Fault which strikes NW–SE. Opinions differ on the internal structure of the metamorphic rocks between the Mishmi and Lohit faults. According to KARUNAKARAN et al. (1964), some authors believe that the ultrabasic rocks come from a eugeosynclinal facies which is overthrust towards the SW on miogeosynclinal rocks (quartzite, marble).

The serpentinite and marble of the Tidding Valley could not be followed to the NW all the way through to the Dibang Valley. Nor could they be found any longer towards the SE in the Lohit-(Tellu) Valley.

To the NE of the Lohit Fault comes a diorite-granodiorite complex consisting of banded dioritic gneiss, granodiorite, biotite gneiss and syenite. Metanoritic and doleritic dykes, lamprophyres, pegmatites and quartz veins transect these rocks. In addition, this complex also contains lenses and bands of chloritic and amphibolitic schists and marble. The tectonic stress is severe and manifests itself, among other ways, in the strong and extensive epidotization of the rocks. No equivalents of this complex are known further to the W. The entire rock complex, including its structural deformation, however, shows many similarities with the rock complex of the N extension of the Mogok Belt in N Burma.

3. Stratigraphy, tectonics and magmatism

3.1 Stratigraphy

Following the discussion of the regional geological and regional structural features, an overview is given of the stratigraphic sequence. Here, again, it must be pointed out that the knowledge of the stratigraphy is very incomplete. Terms such as "series", "group", "formation", "unit", "bed", "stage" and the names of the corresponding sequences have in most cases been adopted from the literature. Detailed stratigraphic analyses of the rock sequences has been carried out only for limited areas. A stratigraphic scheme applicable to the whole of Burma and using modern nomenclature is still in the early stages. An attempt has been made in Figs. 30 (folder), 31 and 32 (folder), to correlate the major stratigraphic sequences.

3.11 Precambrian and Paleozoic

3.111 Precambrian

The metasediments and crystalline rocks at the W edge of the Sino-Burman Ranges are assigned to the **Precambrian**. They are arranged in a more or less coherent belt having a general N–S trend from Putao in the N to Martaban (Fig. 17a, folder).

Because of the rich occurrences of rubies and sapphires, numerous geological studies have been published on the metamorphic and igneous rock complex of Mogok, e.g. BROWN & JUDD (1895), LA TOUCHE (1913), BROWN & BANERJI in FERMOR (1932, 1933, 1935, 1936). IYER (1953) prepared a synthesis. On the basis of petrographic comparisons with the crystalline basement of India and Ceylon, these authors placed the crystalline rock of Mogok in the Precambrian.

The metasediments of the "Mogok Series" consist for the most part of pelitic and semi-pelitic gneiss, marble, calc-silicates and quartzite intruded by a number of acid, intermediate and basic igneous rocks. The great variety of the crystalline rocks of Mogok is apparent from the following lists (IYER 1953, pp. 14, 15):

Near Mogok and Mong Long

basic and ultrabasic intrusives: dolerite, peridotite, hornblende-pyroxene and hornblende rocks, hornblende, and aegirine-scapolite rocks

Urtite Series: hornblende-nepheline rocks and aegirine-nepheline rocks

tourmaline-granite

4*

		Indo-Burman Ranges BRUNNSCHWEILER 1966		**Inner-Burman Tertiary Basin** GRAMANN 1974		**West Kachin Unit** CLEGG 1936, 1941	
System	Stage	Name	Lithology	Name	Lithology	Name	Lithology
CRETACEOUS	Maastrichtian	↑ **Indo-Burman Flysch**	Turbidites, laminites, sandstone-shale-alternations with exotics (Limestone, *Globotruncana*-Limestone, chert, volcanics)	↑ *Siderolites* Lst. (**Kabaw Group**)	Bioclastic and algal-reef Lst. (100 m)		
	Campanian			*Globotruncana* Lst.; (volcanoclastic series) (**Kabaw Group**)	Graded volcanoclastics (200 m)		
	Turonian			Ywahaungyi Lst.; Paung-Chaung Lst. (**Kabaw Group**)	Lst. with Turrilites; Laminated bituminous Lst. (100 m)		
	Cenomanian	**Kyauknimaw Cherts**	Chert, volcanic Tuff			*Orbitolina* **Lst.**	Limestone, shale sandstone
	Albian						300 m
	Early Cretaceous	unknown or missing		unknown or missing		unknown or missing	
JURASSIC	Bathonian			Hiatus ?			
	Aalenian						Clastic and turbididic sedimen of Mesozoic and Paleozoic age
TRIASSIC	Rhaetian					unnamed	
	Norian						
	Carnian			*Halobia* **Group** = **Thanbaya** Fm.	Sandstone-shale-alternations with subordinate nodular Limestone		
	Ladinian						
	Anisian						
	Scythian						
PERMIAN CARBONIFERIAN	Dzulfian			**Kanpetlet Schist** Rubellite Schist of Panmun Chaung (Possible equivalents of Lebyin Group, Mawchi Series, Mergui Group)	Mica-schist, talc-schist, green-schist, quartzite conglomeratic schist		

Fig. 31. Correlation of Mesozoic sedimentary rocks.

East Kachin / Shan Unit				Karen-Tenasserim Unit		
Northern Shan State BRUNNSCHWEILER 1970, LA TOUCHE 1913		Southern Shan State AMOS 1975, GRAMANN, FAY LAIN & STOPPEL 1972, BRÖNNIMANN et al.		v. BRAUN & JORDAN 1976, BRUNNSCHWEILER 1970		
Name	Lithology	Name	Lithology	Name	Lithology	
						CRETACEOUS
Hsipaw Red Beds	Red shale, siltstone, sandstone (1200 m)	Kalaw Red Beds	Red conglomerate, sandstone, mudstone (1000 m)	Red Beds of Mergui = Khorat Group	Red sandstone and conglomerate	
Tati Limestone	Limestone, alternating with marl (650 m)	Thigyit Beds	Shale with *Tmetoceras*			JURASSIC
		Loi-An Series Ma-U-Bin Fm.	Shale, sandstone, coal seams (600 m)			
Napeng Fm.	Shale, marl, limestone (100 m)	Mudstone Fm.	Mudstone, laminated, calcareous, siltstone (100 m)			
Pango Evaporite	Gypsum, residual clay, mudstone (200 m)				Limestone, black and grey (up to 1000 m)	TRIASSIC
		Natteik Fm.	Calcarenite (500 m)	Kamawkala Limestone		
Nwabangyi Dolomite Fm.	Banded dolomite and limestone, micritic to sparitic (2500 – 5000 m)	Nwabangyi Dolomite Fm. Pinme-Kondeik Lst.	Dolomite, brecciated, laminated Micritic Limestone, Chert layers and nodules (2500–5000 m)	Martaban Beds	Sandstone, shale, calcareous shale	
Otoceras Beds of Na Hkyan (Nankam)	Shale with limestone layers	Thigaungdaung Dolomite, Limestone		Moulmein Limestone	Biohermal and biostromal Limestone, crystalline Limestone, (1000 m)	CARBONIFERIAN PERMIAN
Thitsipin Lst. Equivalents	Limestone, algal micrites with Fusulinids, Bryozoa, Brachiopods	Thitsinpin Limestone	Limestone, algal micrites			

Namyau Group / Bawgyo Group / Shan Dolomite Group (left unit)

Mae Moei Group / Ratburi Group (Karen-Tenasserim unit)

Kabaing Granite: pegmatite, aplite, leptynite, minor intrusions

syenite: nepheline syenite

augite and hornblende granite

quartzite (quartz sillimanite rock of Nammi)

calc-granulite: scapolite gneiss and pyroxene granulite

crystalline limestone and calciphyre (rubies and spinel), bands of calc-gneiss

khondalite: feldspathic garnet sillimanite gneiss with graphite and hybrid rocks

biotite gneiss, garnet gneiss and biotite garnet gneiss: injected intrusives as pegmatite, feldspathic veins and quartz veins

West of Kin Chaung (Thabeikkyin Township)

Tawng-Peng System: biotite schist and biotite hornblende schist

basic and ultrabasic intrusions

Kabaing Granite

tourmaline granite

augite and hornblende granite

syenite and syenite gneiss

quartzite

calc-granulite (scapolite gneiss to pyroxene gneiss)

crystalline limestone and calciphyre

hornblende-schist and epidiorite

garnet sillimanite gneiss (khondalite)

unclassified crystalline rocks

While LA TOUCHE (1913) grouped the scapolite and garnet biotite gneisses together with the marble and calc-silicate rocks as the "Mogok Series", IYER restricted this term to the paragneisses and "unclassified crystallines". SEARLE & BA THAN HAQ (1964, p. 138) divided the rock complex of Mogok as follows:

Intrusive rocks	Kabaing Granite
	pegmatite and aplite
	mafic and ultramafic rocks
	alaskitic suite
Mogok Series	calcareous and arenaceous rocks
	gneissose rocks
	migmatite.

Close to the S of Mogok, the gneisses build up a continuous ENE-striking zone, consisting of garnet-biotite, biotite and garnet-graphite-sillimanite gneiss ("khondalite", after the tribe of the Khonds in S India; WALKER 1902).

The fine- to medium-grained and banded gneisses exhibit leucocratic streaks and nests of cryptoperthite. They are frequently migmatized. In that case, the rock consists of a medium-to coarse-grained quartz-feldspar mass which penetrates into the banded gneiss structure and breaks it down. In contact with marble, pyroxene and amphibole were formed during the migmatization while the marbles remained unaffected. The calc-silicate gneiss occasionally contains feldspar porphyroblasts and less frequently quartz-feldspar nests.

Above the gneisses follows a series of **marble and calc-silicate rocks,** the most important host rock of corundum and spinel (p. 208). The marble occurs as rock masses several tens of metres thick extending over several kilometres, or as lenses which are often strung out in festoon fashion along the general strike. IYER (1953) described the following four belts of marble that belong to different stratigraphic levels.

– Letha Taung-Kyetnapa in the Bernardmyo area
– Chaunggyi Valley to Nampheik Chaung near Nampheik and Pazunzeik
– Pinpyit to Onbin-Yedwet Taung and Oksaung Taung in the upper Mogok Valley
– N of Yebu via Pyopon to the Kyaukpyathat-Mgainggyi-Taung Range.

The **marble** (Figs. 33 a, b) is coarse-grained (idiomorphic, up to 10 cm large calcite rhombohedra). In the transition to calc-silicate rock, it becomes more finely grained. The colour varies between bright white and grey to yellow, occasionally pink. Graphite occurs as an accessory mineral. In the vicinity of the contact to the countless intrusions the following accessory minerals are also found: spinel, corundum, forsterite, tremolite, diopside, topaz and titanite.

BROWN & BANERJI in FERMOR (1932) described a marble from the region NE of Mogok that had been contact-metamorphically changed by acid palingenetic melts as an apatite-containing nepheline-diopside-calcite-feldspar rock.

The **calc-silicate rocks** (calciphyre and calc-silicate granulite) are interbedded with marble, garnet-biotite gneiss and quartzite. They are medium- to fine-grained and markedly banded as a result of the alternating mineral composition comprising the main components diopside, oligoclase, titanite, calcite, scapolite, garnet and graphite. The intercalated quartzite is medium-grained to coarse-grained. It occurs as decimetre to 10 m thick beds. Along the Mogok-Thabeikkyin road, between Gwebin and Kyaukkyan, it forms pronounced cliffs (IYER 1953, p. 33).

BROWN & JUDD (1895) believed that the marble was magmatic, while LA TOUCHE (1913, pp. 43, 44) left no doubt about its sedimentary origin. He pointed out the petrographic similarity between the carbonate sediments of Mogok and the sequence of marble, quartz-pyroxene, gneiss and calciphyre of the Precambrian Dharwar System from the Chhindwara Region of India, and therefore also assigned a Precambrian age to the metasediments of Mogok. IYER (1953, p. 79) described the Mogok Series as a sequence of gneisses that were tightly folded with the more recent marble and calc-silicates and he placed them also in the Precambrian together with the intrusives that were affected by the folding. A Precambrian position of the Mogok Series was also assumed by FERMOR (1932, p. 81) who drew attention to the similarities with the rocks of Vizagapatam, Tennevelly (India) and Ceylon.

SEARLE & BA THAN HAQ (1964, p. 138) defined the crystalline at the W edge of the Kachin/Shan Highlands, together with less metamorphosed Paleozoic and Mesozoic sediments, as the "**Mogok Series**". In their opinion, the folding and metamorphism of the metasediments comprising the approximately 1,400 km long and approximately 30 km wide

Fig. 33 a. Coarse crystalline limestone, graphite- and gemstone-bearing; Precambrian. Approx. 5 km N of Mogok. Photo: F. BENDER.

Fig. 33 b. Surface karst phenomena in same limestone. Approx. 7 km S of Mogok. Photo: F. BENDER.

"**Mogok Belt**" were connected with the Himalayan orogeny. They based this view on their own observations and on the opinion expressed by CLEGG (1936b, 1953), who believed that the metasediments occurring on the latitude of Kyaukse were of the same age as the Jurassic coal measures (Loi-an Series; p. 84), or that because of its immediately adjacent position, the marble was of the same age as the Mesozoic carbonates (CLEGG in HERON 1937, 1938). SEARLE & BA THAN HAQ (1964, p. 138) assumed a continuous decline in the grade of metamorphism of the Mogok crystalline rocks from W to E, via the Mong Long mica schist to the greywacke and schist of the Chaung Magyi Formation. GARSON et al. (1976), on the basis of new stratigraphic classifications and structural observations, contradicted this view and showed that the various rocks of the Mogok Belt must be assigned to different metamorphic assemblages.

It is probable that the Mogok Series, together with the Chaung Magyi Formation, form the basement of the Paleozoic-Mesozoic sedimentary sequence of the Shan Plateau. The contact extends several hundred kilometres from Mogok to Namhkam and further NE to the region of Yunnan/China (BROWN 1923, Pl. 9). There the basement is overlain with a pronounced unconformity by Paleozoic-Mesozoic sediments of the Burmese-Malayan Geosyncline (BROWN 1923 d, p. 49). At its S boundary, the crystalline basement rock complex of T'sang Shan is covered by unfolded Triassic red bed series (PASCOE 1950, p. 340). This therefore disproves the theory of alpidic folding of the metasediments.

In the **region of Putao,** STUART (1923 a) described metamorphic series consisting of graphite schist, calc-silicate gneiss and schist veined with red jasper. These rocks resemble the Mogok Series. On the other hand, the less metamorphosed phyllite and quartzite bear a greater resemblance to the Chaung Magyi Series and Pangyun quartzites (PASCOE 1950, p. 340). At the Nany Tamai this series has a wedge-shaped configuration and occurs in a granite massif; it is overlain by Miocene to Pliocene conglomerates and sandstones.

In the **region between Myitkyina and Hkamti,** three occurrences of biotite syenite gneiss have been reported in a slightly metamorphic series of phyllite and quartzite, and locally also calc-silicate gneiss (PASCOE 1950, p. 341). NE of Myitkyina, granite-gneiss with several interfolded marble intercalations occur which are associated with schistose, phyllitic and hornblendic varieties of gneiss (GRIESBACH 1924). They extend to the Lagwi Pass, at the border with China.

Marble intruded by granite is also known from the **NW Myitkyina District.** The grey-white rocks contain a wide spectrum of accessory minerals which formed during the contact metamorphism: corundum, forsterite, serpentine pseudomorph after forsterite, phlogopite, chondrodite, red garnet (hessonite), pyrrhotite, apatite, tourmaline, spinel, ruby and sapphire.

In addition, in this region crystalline schists (hornblende schist, glaucophane schist, chlorite, kyanite and graphite schists) are exposed which are probably associated with gneiss and marble. They surround the large Late Cretaceous to Eocene serpentinite intrusive body of Tawmaw. The serpentinite contains jade deposits (p. 210). The different varieties of magmatic rocks such as hornblende-granite, quartz-augite-monzonite, quartz-biotite-monzonite and diorite, probably correspond to the Alaskite Suite of the Mogok area and are probably also closely connected with the migmatization.

The approximately 130 km wide SE–NW-striking belt of highly metamorphic rocks **between Bhamo and the border with Yunnan** (China) is regarded as the N continuation of the Mogok Series. J. C. BROWN (1917d) had traced these rocks through into the Yunnan Province and recognized the distinct discordance which separates the paragneiss here from the overlying Paleozoic sediments (p. 115). In the E part of the Bhamo District as far as the Salween River, and to the N into the border territory of Kachin State, the basement rock complex consists of banded garnet-biotite gneiss and beds and lenses of amphibolite. There are intercalations of crystalline limestone consisting of saccharoidal calcite with accessory minerals such as diopside, forsterite and graphite. Further to the N between the Irrawaddy and the Salween, light-coloured mica schist and hornblende gneiss predominate. In Yunnan the Tali Fu Marble, which is characterized by its brown and green mottling, belongs to these metasediments. Together with gneiss they form mountain massifs up to 4,000 m high and an approximately 15 km wide zone, which is overlain in the S by non-folded Mesozoic red beds.

Near Wabyudaung in the **region of Shwebo-Sagaing-Mandalay** the general E–W strike of the metasediments of Mogok turns approximately N–S. Close to the N of this locality a small graphite deposit occurs at the contact between scapolith gneiss and marble (p. 203).

To the S, crystalline rocks occur also on the W bank of the Irrawaddy as horst-like uplifts between sediments of the Inner-Burman Tertiary Basin. Near Thayetkon, BROWN & SONDHI (1933a) described E-dipping mica schist, crystalline limestone and epidiorite, which are intruded by granites. IYER (1934) mentioned blue-grey, dolomitic crystalline limestone, gneiss and hornblende gneiss from Male.

The **Sagaing Hills** located further to the S are also, in part, built up of metasediments, such as gneiss and quartzite with marble beds. Biotite-rich mica schist occurs as intercalations.

Between **Mandalay and Thabeikkyin** there is a zone of E-dipping marble and gneiss which are intruded by granites. Some occurrences, such as the Bo Ywa Mountain, form isolated horst-like uplifts that are surrounded by Cenozoic sediments. The same holds true also for the Mandalay Hill that is built up of marble and gneiss which are intruded by granite on the E flank of the hill.

Highly metamorphic rocks also occur S of Mandalay. They accompany the Shan Plateau in a belt which is locally more than 18 km wide and occasionally includes the entire escarpment, which is defined by the Shan Boundary Fault Zone at the W edge of the plateau.

Near Kyaukse, approximately 40 km S of Mandalay, biotite gneiss, diopside-gneiss, calc-silicate rocks and marble occur. The latter is mined in large quarries. These marbles were described by CLEGG (1936b) as the metamorphic equivalents of the Mesozoic carbonates of the Shan Plateau. THA HLA & BA THAN HAQ (1960), however, observed a transitional metamorphosis into overlying phyllite of the Chaung Magyi Series.

Further S, in the **Meiktila and Yamethin District,** a broad zone of gneiss and migmatite forms the basement of the Paleozoic sequence. Still further S, gneiss with intercalated calc-silicates has been observed at **Taungoo and in Thaton District** where THEOBALD (1873d) grouped it together with clastic sediments overlying low-grade metamorphic schist as the "Martaban Group". A further belt of paragneiss splits off in an SE direction near Papun and extends, with interruptions, in the Salween Valley as far as the Dawna Range (Fig. 17a, folder).

The **Pawn Chaung Series** (BA THAN HAQ & SEARLE 1961), a series of chlorite schist, phyllite and marble, which is exposed NE of Baw Lake between the Salween and Pawn Chaung, also possibly belongs to the Precambrian.

3.112 Late Precambrian and Cambrian

In NE Burma, the Precambrian basement rock complex is overlain by several thousand metres of clastic rock sequences composed (from bottom to top) of the flysch-type **Chaung Magyi Group** (LA TOUCHE 1913), the **Pangyun Formation** with intercalated rhyolitic volcanics of the **Bawdwin Volcanic Formation** (J. C. BROWN 1918a), and the **Molohein Group,** which correlates with the Pangyun Formation (MYINT LWIN THEIN 1973). With one exception, the age of these units has so far not been established: MYINT LWIN THEIN (1973) discovered an Upper Cambrian trilobite fauna from sandstone and quartzite in the upper section of the Molohein Group in the Pindaya Range. This fauna is comparable with trilobite faunas, also in sandy-quartzitic sediments, from Yunnan (SUN 1939) and from Thailand (KOBAYASHI 1957). These Upper Cambrian sediments are so far the oldest dated strata in the **Burmese-Malayan Geosyncline** (KOBAYASHI 1973a, b: Yunnan-Malayan Geosyncline, BURTON 1967a, b, SUN 1945: Sino-Burmese Geosyncline) which extended from Yunnan, Burma-Thailand via the Malayan Peninsula to Borneo and acquired thick eugeosynclinal and miogeosynclinal sediment series during the Paleozoic.

Chaung Magyi Group

The flysch-like, approximately 3,000 m thick sediments of the Chaung Magyi Group represent the major portion of the fossil-free sequences (Fig. 34). They border in a belt several kilometres wide on the Precambrian Mogok Series. MITCHELL et al. (1977, pp. 14, 15) distinguished between three main rock types:

- **greywacke,** with a laminated texture of coarse-grained to fine-grained layers; the grain components, with diameters up to 2 mm, are composed of quartz, Na- and K-feldspar, mica, tourmaline, zirconium, leucoxene, magnetite and hypersthene
- **feldspathic sub-greywacke,** with a coarse sand fraction of quartz, microcline, Na-feldspar, also fragments of quartzitic rocks and of volcanic glass in a fine sericitic quartz-feldspathic matrix
- **pelitic rocks,** silty mudstone or shale with quartz as the main mineral, in addition chlorite, sericitic mica and illite.

These clastic sediments exhibit all the characteristics of turbidity current deposits that were laid down in a rapidly subsiding geosynclinal basin: graded bedding, flute casts, groove casts and wet-sediment deformation structures. The Tawngma Siltstone Member in the upper part of the Chaung Magyi Group, which is composed of cross-bedded, burrowed siltstone, indicates sedimentation in the shallow-water environment of the gradually filling basin. The microcline components of the greywacke are evidence of granitoid source rocks, possibly from the metamorphic Mogok Belt or from a Precambrian land mass in the W.

GARSON et al. (1976, pp. 19, 20) described the following lithological facies types from the Chaung Magyi Group in the area of Yengan in Southern Shan State, where the group is also about 3,000 m thick:

- **cross-laminated sandstone and mudstone,** alternating with cm-thick layers of fine-grained sandstone with decimetre-thick cross-laminated, mostly brown-grey sandstone and greenish-grey, cm-thick mudstone layers. The sedimentary structures point to a shallow subaqueous environment, possibly a delta topset bed ("facies a")
- **thick-bedded massive and cross-bedded sandstone;** it usually consists of 5 m thick beds which are overlain by "facies a" or they overlie the latter with a distinct erosion contact. Their

sedimentological association with "facies a" permits the conclusion that they are delta top or delta slope channel deposits
– **dolomite;** it is intercalated with the "facies a" and occurs as up to 200 m thick units with beds up to several metres thick exhibiting cross-bedding textures. The dolomite was probably deposited in a shallow-water environment which had an intermittent supply of carbonate
– **green limestone;** it occurs as up to 2 m thick beds and is intercalated between the dolomites and the "facies a"
– **turbidites;** these sediments attain a thickness of approximately 2,700 m and are composed of m-thick sandstone beds with sharp lower contacts and parallel top and bottom surfaces, alternating with thin shale layers. The individual sandstone beds with sedimentary structures, such as flute casts and erosional contacts, are turbidites which were laid down near their source area.

In N Burma, rocks of the Chaung Magyi Group were described from the Pyehpat Range in the Myitkyina District, on the Htawgaw Road to the Chinese border, near Chuchho on the Fen Shui Pass Road – where they are unconformably overlain by limestones – and in the highlands located between Myitkyina and Putao (PASCOE 1959, p. 465).

Between the Shan Plateau and the Salween, Chaung Magyi sediments form the over 2,000 m high Loi Ling E of the Nam Peng Valley and also N–S-striking arched uplifts which rise like islands between the "Plateau Limestone" further S (LA TOUCHE 1913, p. 51, Pl. 24). BROWN & SONDHI (1933a, Pl. 5) believed that the sediments that form the 2,200 m high mountain massif of Loi Pan in the state of Moeng Tung and extend E of the Salween to the frontier with Thailand, also belong to the Chaung Magyi Group. It seems probable, however, that parts of the clastic sediments should be assigned to the Carboniferous (Fig. 30, folder).

Fig. 34. View from NW of Maymyo to NW to the Irrawaddy Valley; hills in middle ground are built up by greywacke and shale of Chaung Magyi Group; Late Precambrian to Cambrian. Photo: J. BRINCKMANN.

Pangyun Formation and Molohein Group

The chiefly purple, sandy and quartzitic sediments that overlie with an angular unconformity the Chaung Magyi rocks were referred to by J. C. BROWN (1918a, p. 145) as the "**Pangyun Beds**" after the river Pangyun which flows into the Namtu. At the type locality in the Nam Pangyun Valley near Bawdwin, BRINCKMANN & HINZE (unpubl. report, 1976) distinguished the following litho-facies types in the Pangyun Formation which is here about 2,000 m thick:

- brown-red and violet-red sandstone (Fig. 35), fine-grained to medium-grained occasionally occurring with red, decimetre-thick bedded to unbedded quartzite; frequently with ripple marks; locally cm- to decimetre-thick red and light greenish-grey siltstone and shale intercalations
- red siltstone with intercalated shale and friable, muscovite-rich sandstone

Fig. 35. Pangyun Formation and Bawdwin Volcanic Formation; upper Pangyun Valley, N of Bawdwin lead-zinc mine; layer of volcanoclastic sediments is intercalated in sandstone of Pangyun Formation (top and base); Cambro-Ordovician. Photo: J. BRINCKMANN.

- white-grey sandstone and quartzite, fine-grained to medium-grained, decimetre-thick beds to massive, frequently cross-bedded
- greenish-grey banded shale, siltstone with mm- to cm-thick banding and decimetre-thick fine sandstone beds
- greenish-brown silt and claystone, flakey, with bedding planes having a silky lustre.

MITCHELL et al. (1977) described very similar rock types from the Pangyun Formation in areas S of Bawdwin:

- **Kyaukme-Longtawkno area**
  Mudstone conglomerate, thin-bedded quartzite, thin dolomite
  Cross-bedded purple and white quartzite
  Massive and cross-bedded quartzite; acid tuff
  Purple conglomerate
  Thickness: 1,300 m
- **Yadanatheingyi area**
  Purple and white quartzite with pitted surface, cross-bedded dolomite and limestone, pink quartzite and shale
  Green siltstone
  Lenticular purple polymictic polymodal orthoconglomerate, quartzite and grit
  Thickness: approximately 1,200 m

The often cross-bedded sandstone and quartzite of the Pangyun Formation were probably formed in a coastal area. MITCHELL et al. (1977, p. 16) assumed that the intercalated dolomite was deposited in the marine shelf area and that the sandy facies of the quartzite with large-scale cross-bedding and mudflake casts were deposited in a fluviatile environment.

Pebbles and boulders of the Pangyun Formation occur as purple, cross-bedded sandstone and siltstone in river deposits in the Southern Shan State (GARSON et al. 1976, p. 20).

From the **Pindaya Range**, MYINT LWIN THEIN (1973, pp. 152, 153) described reddish, fine-grained, micaceous sandstones, which overlie the Chaung Magyi sediments with an angular unconformity and pass into Middle Ordovician strata. MYINT LWIN THEIN (1973, p. 152) observed **Cambrian** trilobites in this sandstone sequence, which is taken together as the **Molohein Group** and is correlated with the Pangyun Formation (see also PASCOE 1959, p. 593). The following trilobites were identified:

Saukiella junia (WALCOTT) var. A. WINSTON & NICOLS 1967
Eosaukia buravasi KOBAYASHI 1957
Saukiella sp. A
Drumaspis texana RESSER 1942

MYINT LWIN THEIN (1973) compared this fauna with the Upper Cambrian fauna of the *Saukiella* subzone of the Trempealeauan Stage (Upper Cambrian) in the Wilberns Formation in Central Texas (WINSTON & NICOLS 1967) and with the likewise Upper Cambrian Tarutao fauna from reddish mica standstones on the W coast of Tarutao Island, Thailand (KOBAYASHI 1957).

The Molohein Group, which is named after the Molohein Peak in Lawksawk Township, is about 1,100 m thick (MYINT LWIN THEIN 1973). Conglomerates with quartzite detritus from the underlying Chaung Magyi Group frequently occur at its base. Rather like in Yadanatheingyi and Kyaukme, blue-grey, thick-bedded, fine crystalline dolomite is intercalated in the upper part of the sandstone facies. The Molohein Group forms the peaks in the Pindaya Range, and in the Bawsaing Range it forms, for example, the Hethin Hill N of Heho.

Saukiid trilobites were also found in the Ngwetaung Sandstone in the Northern Shan State (Explanatory Notes on the Geological Map of Burma 1977, p. 5). LA TOUCHE (1913, p. 66) had named this brown-red micaceous sandstone after the Ngwetaung Hill E of Mandalay (Fig. 38) and he placed it in the Ordovician. From the sandstone, which he believed also occurs on the S slope of Loi-hen E of Lashio, he described "non-characteristic fossils, such as *Orthis (Dalmanella) testudinaria* DALMAN".

In the Northern Shan State rhyolitic volcanoclastic sediments of the **Bawdwin Volcanic Formation,** which are up to 1,000 m thick, are intercalated in the sandstone and quartzite of the Pangyun Formation (p. 125).

Age relationships and comparisons with neighbouring countries

The dating of the Molohein Group and the Ngwetaung Sandstone on the basis of Upper Cambrian trilobites provides a minimum age for the underlying Chaung Magyi Group which is separated from the Molohein Group by an angular unconformity. The Pangyun Formation, which can be correlated with the Molohein Group, and the Bawdwin Volcanic Formation, which is intercalated in the Pangyun Formation, date therefore from the Upper Cambrian to Lower Ordovician, because they pass concordantly into the overlying Middle Ordovician sediments of the Naungkangyi Formation.

The lithofacies of the sedimentary succession of the Chaung Magyi Group and of the Pangyun Formation correspond to the Machingchang Formation of NW Malaysia which, according to GOBBETT & HUTCHINSON (1973), consists in its lower part of greywacke, siltstone, flaggy mudstone and shale about 1,100 m thick and in its upper part of a purple sandstone sequence about 900 m thick. This sandstone sequence, which corresponds to the Pangyun Formation/Molohein Group, is the source of the Upper Cambrian to Lower Ordovician trilobite faunas of the islands of Langkawi and Tarutao (KOBAYASHI 1957, GOBBETT & HUTCHINSON 1973).

The Upper Cambrian to Lower Ordovician sandstone was named the Tarutao Series (formerly the Phuket Group) in S Thailand. Equivalent formations are the several hundred metres thick quartzite overlain by Ordovician rocks in N Thailand (BAUM et al. 1970) and a quartzite-phyllite series in Kanchanaburi Province (KOCH 1973, HAGEN & KEMPER 1976). In addition, there are lithological similarities between the Chaung Magyi Group and the Shillong Series of the Satpurn Group in E India (LA TOUCHE 1913, p. 53, MAUNG THEIN 1973, p. 94).

3.113 Ordovician

During the Ordovician, E Burma was located in the W shelf region of the Burmese-Malayan Geosyncline in which several thousand metres of variegated finely clastic as well as carbonate sediments of the **Naungkangyi Stage** (LA TOUCHE 1913) were deposited. The fauna of the Naungkangyi facies (KOBAYASHI 1959, 1973a, b) is characterized by echinoderms (Cystoidea, Crinoidea), trilobites and brachiopods and to a lesser extent graptolites and orthoceratites (REED 1906, 1915). Conodonts have also been observed in Ordovician limestones of the Mawson Group in the Southern Shan State (STOPPEL 1974, unpubl. report).

In the Shan States, the Ordovician sediments overlie concordantly with a lithological break the sandy-quartzitic series of the Pangyun Formation/Molohein Group. In general, the lithofacies of the Ordovician deposits of the Naungkangyi Stage in the Northern Shan State, with variegated marly-fine-sandy-silty and argillaceous deposits, are less calcareous than the sedimentary series of the Southern Shan State. There the thickest stratigraphic unit, the Wunbye Formation (MYINT LWIN THEIN 1973) or the Doktoye Limestone Formation (GARSON et al. 1976), consists of 2,000 m thick limestones (Fig. 30).

Naungkangyi Group

The first subdivision of the Paleozoic strata of Burma was published by LA TOUCHE (1913) for the region of the former Northern Shan States. Previously, Paleozoic rocks had been identified merely on reconnaissance traverses (NOETLING 1890, v. RICHTHOFEN 1882). REED (1906, 1908, 1915, 1929, 1936a) carried out biostratigraphic classification of the numerous fossil assemblages and also conducted comparative regional studies.

LA TOUCHE (1913) subdivided the Ordovician into two parts. Two further groups are special facies developments and are locally limited in extent (PASCOE 1959, p. 608). The groups established by LA TOUCHE (1913, pp. 65, 66) for the Ordovician are (Fig. 30)

Nyaungbaw Limestone (local)
Hwe Mawng Purple Shale, homotaxial with the
Upper Naungkangyi Stage
Lower Naungkangyi Stage
Ngwetaung Sandstone (local; newly classified in the Upper Cambrian).

MITCHELL et al. (1977) confirmed the lithostratigraphic division of the Ordovician into two parts. In the N sector of the Kyaukme-Longtawkno area they divided the Naungkangyi Group into the older **Taungkyun Formation** and the younger **Lilu Formation** (Fig. 30).

The rocks of the **Lower Naungkangyi Stage** (named after a village about 4 km N of Maymyo) are composed of fossiliferous yellow or buff-coloured fine sandy limestone (Fig. 36), which frequently weathers into sandy marl and is intercalated with lense-like layers of spathic limestone. Outcrops where these rocks concordantly overlie the Pangyun Formation occur in the Nam-Pangyun Valley E of Bawdwin (BROWN 1918a, pp. 149, 150, Pl. 6) and in the Kyaukme-Longtawkno area (MITCHELL et al. 1977, pp. 16, 17).

Various publications exist on the Middle Ordovician fossils of the Lower Naungkangyi Stage: LA TOUCHE (1913, pp. 68, 69), REED (1906, 1915, 1936a), PASCOE (1959, pp. 609, 610). Cystoidea with the genera *Aristocystis, Heliocrinus, Echinocrinus, Cheirocrinus, Caryocrinus* and *Protocrinus* were observed. The genus *Diplotrypa* of the family of the Bryozoa occurs frequently. Brachiopods are represented by genera such as *Rafinesquina, Sowerbyella, Orthis* and trilobites by the genera *Calymene, Remopleuritis, Asaphus, Megalaspis* and *Cheirurus*.

The rocks of the **Upper Naungkangyi Stage** are composed of a variegated shale and claystone series which is red, orange, yellow and lavender in colour and concordantly overlies the Lower Naungkangyi Stage.

LA TOUCHE (1913, pp. 84, 85) and PASCOE (1959, pp. 613, 614) distinguished between two areas of different lithofacies in the Upper Naungkangyi Stage:

Fig. 36. Fossiliferous, calcareous fine sandstone, Yadanatheingyi; Ordovician. Photo: F. BENDER.

– W of Lashio, fringing the E and S borders of the old Chaung Magyi land surface: variegated shale and claystone
– E of Lashio: bright purple claystone (Hwe Mawng Beds)

The **Hwe Mawng Beds** occur locally also to the W of Lashio as a metre-thick persistent band of dark purple shale that appears in the uppermost portion of the variegated shale-claystone series of the Upper Naungkangyi Stage. In the Kyaukme-Longtawkno area MITCHELL et al. (1977, Map 2) mapped the Upper Naungkangyi Stage as the "Lilu Formation" which is composed of phacoidal limestone, siltstone and mudstone.

The **Nyaungbaw Limestone** (LA TOUCHE 1913, p. 119, PASCOE 1959, p. 616), a red-brown to blue-grey limestone-claystone series (calcareous "knotenschiefer" facies), which has a strikingly phacoidal texture, is a special facies of the Upper Naungkangyi Stage containing numerous orthoceratites as well as brachiopods and the echinoderm *Camarocrinus asiaticus* REED (REED 1913, 1936a). This series, which was placed in the Ashgillian (PASCOE 1959, p. 617), was correlated by BROWN & SONDHI (1933b, p. 220) – on the basis of lithological similarities – with the *Orthoceras* beds of the Southern Shan State, which MYINT LWIN THEIN (1973), however, has shown are Silurian in age (Linwe Formation, p. 61). BERRY & BOUCOT (1972, p. 20) also considered the possibility of a Lower Silurian age for the Nyaungbaw Limestone.

The total thickness of the Ordovician is about 1,600 m. LA TOUCHE (1913) and PASCOE (1959) noticed less thick sections in the vicinity of uplifts of the substratum and explained this observation as an irregular relief of the "old Chaung Magyi surface", characterized by islands, swells and basins.

As in the Lower Naungkangyi Stage, Cystoidea with the genera *Heliocrinus, Echinocrinus* and *Caryocrinus* frequently occur in the Upper Naungkangyi Stage. Of the brachiopods, the main genera represented are *Orthis, Clitambonites, Porambonites* and *Rafinesquina,* and of the trilobites, among others, the genera *Ampyx, Dionide, Asaphus, Birmanites, Ptychopyge, Encrinurus, Megalaspis, Illaenus, Calymene, Phacops* and *Agnostus* (REED 1906, 1915, 1936a, PASCOE 1959) are represented. Of the trilobites, the species *Birmanites birmanicus* (SUN 1939), which REED (1906) determined as *Ogygites birmanicus* and *Encrinurella insanguinensis* in the Hwe Mawng Beds, have a regional significance because they also occur in finely clastic sediments of the Ordovician in Thailand, where they are used as Upper Ordovician index fossils for the ostracod stratigraphy (TRAPP 1975).

LA TOUCHE, REED and PASCOE (op. cit.) have on several occasions stressed that the faunal assemblage of the Naungkangyi Stage, which extends as far as W Yunnan, covers the entire Ordovician, with the exception of the Tremadocian, and bears an unmistakable affinity to the Baltic fauna province of N Europe. It is represented, among other places, in the Echinosphaerite limestone of Scandinavia and Russia. MITCHELL et al. (1977, p. 17) remarked: "most of the group is late Caradocian in age, indeed many fossils can be matched with material from the Caradoc type area in Shropshire, England." TRAPP (1975), on the basis of the Ordovician ostracod faunas of Thailand, also confirmed the close relationship with the N European province. Nevertheless, the genera *Eurochilina* and *Drepanella* are elements that are typical of the North American-Arctic region.

REED (1906, 1936a) separated the faunal assemblages of the Naungkangyi Stage from the Boreal province as well as from the Ordovician faunas of the Himalayas (from the Spiti Valley in the Kangra District), and the latter were regarded as "highly individualized faunas but with a strong American" or "Pacific element" (PASCOE 1959, p. 633). The sharp distinction between the Naungkangyi fauna and those from the Himalayas is no longer made. For example, WHITTINGTON & HUGHES (1972) combined the trilobite faunas of E Asia, the S Himalayas and China and Australia as Hungaiid calymenid (Lower Ordovician) and as Pliomerina calymenid (Upper Ordovician) and placed them together with the *Selenopeltis* or Trinucleid fauna around the old Gondwana Continent. The Russian, Baltic and North American trilobite faunas, which are distributed around the Canadian and Baltic Shelf, were separated from these (WHITTINGTON 1973, Fig. 3).

BROWN (1932) and BROWN & SONDHI (1933a, b) mapped the Paleozoic stratigraphic sequence in the former Southern Shan State. They named the Ordovician rock sequences, of which the Pindaya and Mawson (Bawsaing) Ranges are built up, the **Pindaya Beds** and the **Mawson Series** respectively. They assumed that these sediments replace each other at least partially (see also PASCOE 1959, pp. 617 and 623). The **Orthoceras Beds** are placed in the upper part of the Ordovician. These beds are a sequence of variegated argillaceous limestone, in which a phacoidal texture has developed and whose fossil content consists of cephalopods from, among others, the genera *Orthoceras, Leurorthoceras, Actinoceras, Aspidoceras, Ormoceras, Cameroceras* and *Cyrtoceras* (BROWN & SONDHI 1933b, p. 219, REED 1936a, PASCOE 1959, p. 616). This cephalopod fauna makes the Ordovician in Burma correlatable with the "Thungsong limestone facies" of S Thailand and Malaysia (KOBAYASHI 1959, 1973a). According to MYINT LWIN THEIN (1973, p. 149), this Ordovician *Orthoceras* fauna does not originate in the mapped *Orthoceras* beds but from limestone intercalations of the Pindaya Group which exhibit a very similar phacoidal texture. Apart from proving the existence of several horizons of phacoidal textured limestone containing orthoceratids of different age, MYINT LWIN THEIN (1973) also found Silurian graptolites of the genus *Monograptus* in the *Orthoceras* beds of BROWN & SONDHI (1933b), for which he now introduced the name **Linwe Formation** (THEIN 1973, p. 149). He placed this formation in the Lower Silurian (p. 61).

MYINT LWIN THEIN (1973, p. 153) combined the Ordovician strata in the W part of the Southern Shan State under the **Pindaya Group** which overlies the Upper Cambrian Molohein Group. It is composed of approximately 2,500 m thick, thick-bedded, burrowed, pelletal or silty limestone with irregular silt specks or laminae, and grey or yellow siltstone and it is subdivided as follows (from top to bottom):

Tanshauk Member
Nan-on Formation
Wunbye Formation
Lokeypin Formation.

Lokeypin Formation (village of Lokeypin, approximately 4.5 km N of Myaing in Ye-ngan Township): approximately 500 m thick medium- to thick-bedded, grey to buff, soft to indurated, micaceous siltstone. The sequence contains orthid brachiopods.

Wunbye Formation (Wunbye Hill, approximately 1.5 km SW of Linwe, Ye-ngan Township): approximately 1,600 m thick, thick-bedded limestone, siltstone and dolomite. This sequence contains orthid brachiopods, *Actinoceras, Ormoceras, Armenoceras, Endoceras, Receptaculites, Lophospira, Helicotoma* and stromatoporids.

Samples collected from limestone in the lower section of the Wunbye Formation in the area of Bawsaing yielded the following conodonts at 3 localities. STOPPEL (1974, unpubl. report) identified these faunas:

1. Roadcut approximately 1 km N of the Theigon mineshaft, ENE Bawsaing

Acodus deltatus LINDSTRÖM
Acontiodus alveolaris STAUFFER
Coelocerodontus sp. *indet.*
Cordylodus prion LINDSTRÖM
Distacodus expansus (GRAVES & ELLISON)
Distacodus cf. *peracutus* LINDSTRÖM
Drepanodus sp. *indet.*
Loxodus sp.
Oistodus inaequalis PANDER
Oistodus cf. *linguatus* LINDSTRÖM
Oistodus triangularis LINDSTRÖM
Prioniodus? sp. (fragment)
Scolopodus quadraplicatus BRANSON & MEHL
Scolopodus warendensis DRUCE

Age: Tremadocian to Lower Arenigian

2. Pathway cut on "Hill 1", approximately 2 km SW of Bawsaing

In light grey limestone
Oistodus inaequalis PANDER
Oistodus parallelus PANDER
Oistodus venustus STAUFFER
Scolopodus asymmetricus DRUCE & JONES

Age: Lower Ordovician

3. Roadcut between Bawsaing and Pindaya, approximately 1.5 km NNE of Ywan Aung

Drepanodus homocurvatus LINDSTRÖM or *Drepanodus planus* LINDSTRÖM
Oistodus inaequalis PANDER

Age: Lower Ordovician

These datings indicate that the thick Ordovician carbonate sequence commenced earlier than was hitherto assumed (Tremadocian to Arenigian).

Nan-on Formation (village of Nan-on, Ye-ngan Township): approximately 110 m thick yellow to buff or light orange, thin- to medium-bedded siltstone, mudstone and marlstone. This formation is highly fossiliferous and contains orthid brachiopods, Bryozoa, sponges, and trilobite genera such as *Illaenus* and *Sphaerocorypha*. MYINT LWIN THEIN (1973, p. 157) correlated the Nan-on Formation with the Thittetkon Sponge Beds, which hitherto had been compared as a special facies development with the Lower Naungkangyi Stage (REED 1936a, PASCOE 1959).

Tanshauk Member (of Nan-on Formation) (village of Tanshauk, 1.5 km S of Nan-on, Ye-ngan Township): uppermost section of the Nan-on Formation; approximately 60 m thick purple or pink, soft, laminated, thin- to medium-bedded siltstone, calcareous shale and calcareous mudstone.

According to MYINT LWIN THEIN (1973, p. 161) volcanic rocks are intercalated in the Ordovician formations. For example, the Wunbye Formation contains rhyolitic tuff and volcanic ash layers. Near Wabya and Menedaung, rhyolite and rhyolitic tuff occur along fault zones (p. 127).

In the Ye-ngan area, GARSON et al. (1976, pp. 21 et seq.) combined the Ordovician strata as the **Mawson Group**, i.e. an approximately 3,000 m thick, mainly calcareous sequence, which they subdivided into the following three formations (Fig. 30, folder; from top to bottom):

Kinle Siltstone Formation
Doktoye Limestone Formation
Ngwetaung Formation.

Ngwetaung Formation (hill E of Mandalay) (LA TOUCHE 1913): yellow or multicoloured clay, marl and fine-grained calcareous sandstone, changing in thickness from several 100 m to approximately 700 m, probably due to the irregular surface of the Chaung Magyi rocks on which it was unconformably deposited. The sequence contains trilobites:

asaphid free-cheek
hystricurid free-cheek

and brachiopods:

dalmanellid
syntrophiid aff. *Thaumatrophia* sp.

This fauna indicates an Arenigian age.

Doktoye Limestone Formation (village of Doktoye): sequence of blueish-grey limestone with intercalated sandstone, maximum thickness of approximately 2,000 m. The limestone passes without lithological break into the underlying Ngwetaung Formation. It is well stratified and contains numerous cylindrical sinuous burrows, which run parallel and perpendicular to the bedding. Algal structures occur locally. Because of the major lateral changes in thickness, GARSON et al. (1976, p. 22) assumed that the facies of the Ngwetaung and Kinle Formations were locally replaced by the limestone sequence. The fauna contains brachiopods of the genera *Aegiromena* and *Leptellina* and trilobites of the genera *Basilicus* and *Neseuretus* cf. *birmanicus* (REED) and is definitely Ordovician, and probably Caradocian in age.

Kinle Siltstone Formation (Kinle River): 100–500 m thick deeply weathered yellow, brown or buff-coloured shale and marl, laminated siltstone and cross-laminated calcareous silty sandstone from which the following fossils have been described:

Crinoids:
Heliocrinites cf. *qualus* (BATHER)
Protocrinites cf. *sparsiporus* (BATHER)

Brachiopods:
  Leptellina sp.
  Nicolella sp.
  orthid
  enteletacean

Bryozoa:
  Diplotrypa sp.

The fauna indicates an Ordovician, presumably Caradocian age.

The **Thapanbin Sponge Bed** (village of Thapanbin): a maximum 2 m thick lenticular inclusion in the upper portion of the Kinle Siltstone Formation (GARSON et al. 1976, p. 23). The bed is built up of sponges of the genera *Aulocopium* and *Calathium*.

So far no Ordovician fossils have been found in S Burma. On the basis of lithofacies comparisons with the Paleozoic of W and SW Thailand, it can be assumed that Ordovician strata are perhaps contained in carbonate-sandy sequences of the Mawchi Series in Kayah State (Biya Limestone, HOBSON 1941) and of the Mergui Series in the Tenasserim Division (p. 70).

The **Thung Song Limestones** in Thailand, which are several hundred metres thick (BROWN et al. 1951, IGOE & KOIKE 1967), and the **Setul Limestones** of Malaysia (JONES 1968) correspond to the limestone sequence of the Naungkangyi Formation. These shallow-water shelf carbonates, which according to KOBAYASHI (1973 a, b) belong to the Thung Song limestone facies, contain mainly shelly faunas and cephalopods; Cystoidea and graptolites have not so far been observed. There is a great degree of similarity between the Naungkangyi fauna and the Ordovician Cystoidea, brachiopod and trilobite faunal assemblage which is contained in the approximately 1,000–15,000 m thick sandstone, sandy claystone and limestone sequence of W Thailand (WOLFART, STOPPEL in BAUM et al. 1970, KOCH 1973, HAGEN & KEMPER 1976). This facies, which is similar to the Naungkangyi Stages, extends to W Yunnan (BROWN & REED 1913, REED 1917, BROWN 1950). In E Yunnan, the euxinic graptolite-schist facies, the Shihtien facies (SUN 1945, p. 3, KOBAYASHI 1973 a, b), marks the axial trough of the Burmese-Malayan Geosyncline; to it belongs the Tambuk Mudstone (KOBAYASHI & HAMADA 1970) in the lower section of the Ordovician-Silurian Mahang Formation of Malaysia (BURTON 1967 b).

3.114 Silurian

During the Silurian, the sedimentation of fine clastics and carbonates continued in the shelf region of the Burmese-Malayan Geosyncline. In lithostratigraphic terms, the Silurian strata, which are identified by a large number of different local names, can be subdivided into three portions: at the base there is a red-mottled shale and mudstone sequence associated with phacoidal limestone which probably extends into the Ordovician (Linwe Formation). Above, a grey graptolite shale facies (Pangsha-Pye-Formation) of the Llandoverian follows. A sand and marl facies (Namhsim Formation) which is several hundred metres thick forms the upper portion of the Silurian (Fig. 37 a).

The **Linwe Formation** of MYINT LWIN THEIN (1973) (village of Linwe, Ye-ngan Township) is an approximately 550 m thick sequence of purple, pink and grey coloured, phacoidally textured limestone, argillaceous limestone, calcareous mudstone and shale (Fig. 37 b).

a

Fig. 37a. Ridge of crystalline limestone (foreground) as viewed from the Maymyo-Mandalay Road to NW, Silurian; middle ground: limestone of Permian age; background: Mandalay Hill, metamorphic limestone, Precambrian. Photo: F. BENDER.

Fig. 37b. Linwe Formation near Linwe, Southern Shan State. Red and greyish-green phacoidal limestone; b   Silurian. Photo: J. BRINCKMANN.

This sequence corresponds to the **Orthoceras beds,** which REED (1936a) placed in the Ordovician (p. 58). MYINT LWIN THEIN (1973) was able, however, to find the Silurian graptolite genera *Monograptus* and *Climacograptus* in the *Orthoceras* beds that had been mapped by BROWN & SONDHI (1933b, Pl. 11). He correlated the Linwe Formation, which concordantly overlies the Ordovician Pindaya Group and which passes into the overlying graptolite-bearing shale of the Wabya Formation, with the **Nyaungbaw Limestone** of the Northern Shan State. The latter underlies the **Pangsha-Pye Formation** of the Llandoverian in that area (Fig. 30, folder).

In the area of Ye-ngan, GARSON et al. (1976) described considerable lateral thickness changes of the Linwe Formation between 150 m and 650 m within only a few km or less. They subdivided it into a **Red Shale Member** at the base, consisting of an approximately 300 m thick red-mottled sequence of phacoidal limestone and a **Sandy Marl Member** consisting of fine-grained calcareous sandstone with interbedded siltstone.

The **Pangsha-Pye** (village approximately 12 km NW of Hsipaw) **Graptolite Band** (LA TOUCHE 1913, p. 125) follows above the purple shale of the Upper Naungkangyi Stage. According to MITCHELL et al. (1977, p. 17) it is max. 60 m thick and consists of grey to black shale and micaceous siltstone containing numerous graptolites. It is subdivided into a lower or trilobite unit and an upper or graptolite unit (REED 1915, PASCOE 1959, p. 643). The trilobite unit contains the genera *Acidaspis, Phacops, Bollia, Kloedenella, Primitiella, Turrilepas.* The graptolite unit, in which 3 graptolite zones were observed (PASCOE 1959, p. 644), contains the genera *Diplograptus, Climacograptus* and *Monograptus,* which resulted in the Pangsha-Pye Formation being classified in the **Llandoverian.** The joint occurrence of *Monograptus millepedia* and *M. triangularis* permits the graptolite band to be assigned to the *gregarius* zone of the Llandoverian. BERRY & BOUCOT (1972, p. 22, Fig. 1), in their revision of all the graptolites and brachiopods found by LA TOUCHE (1913), also assigned an early Llandoverian age to the Pangsha-Pye Graptolite Band.

The Pangsha-Pye Formation is widely exposed in the Shan States (BROWN 1918a, Pl. 7, LA TOUCHE 1913, MITCHELL et al. 1977, Maps 1 and 2). In the Ye-ngan region it consists of 20–30 m thick pale grey fossiliferous shale and thinly laminated siltstone with graptolites of the Upper *gregarius* and *convolutus* zone (GARSON et al. 1976, p. 15).

In the Southern Shan State, graptolites of the Llandoverian have been described by BROWN & SONDHI (1933a, b) from the following locations:

Wabya Graptolite Bed (Pindaya Range): Shale with *Monograptus cyphus, M. incommodus, M. andersoni, Orthograptus vesiculosus, Climacograptus medius, C. rectangularis, Glytograptus tamariscus* var. *incertus,* discordantly overlain by "Permo-Carboniferous Limestone". BERRY & BOUCOT (1972, pp. 29–31) assumed a Llandoverian age of this fauna (*M. cyphus* zone). For another graptolite fauna, which was listed by PASCOE (1959) from the Wabya Beds, they assumed a Lower and Middle Llandoverian age.

Kyawkyap Graptolite Bed (fauna from the E slope of the Kyawkyap-Pagoda, approximately 15 km N of Heho): Bleached shale (PASCOE 1959, p. 653) containing *Orthograptus mutabilis, Climacograptus medius, C. innostatus,* which are evidence of Llandoverian age (*Monograptus gregarius* zone) (BERRY & BOUCOT 1972).

Graptolite Beds of Mebyataung (approximately 15 km S of Heho): Limestone and mudstone with 3 graptolite beds on top of each other containing *Monograptus gemmatus, M.*

concinnus, M. jaeculum and *Rhaphidograptus tornquisti* in the lowest horizon, with *M. sedgwicki, M. tenuis,* and *Climacograptus scalaris* in the middle horizon, and *M. tenuis, M. millepeda, C. scalaris* and *R. toernquisti* in the uppermost horizon. These faunas point to a Llandoverian age of the *M. gregarius, M. convolutus, M. sedgwicki* zones (BERRY & BOUCOT 1972).

Panghkawkno Graptolite Bed (Loilem Area): Shale and limestone containing *Glyptograptus serratus, Monograptus sedgwicki, M. regularis, M. distans, M. lobiferus, Climacograptus scalaris,* for which a **Middle Llandoverian** age was assumed (zones 20–21, BERRY & BOUCOT 1972).

The above faunas are observed in the **Wabya Formation** (MYINT LWIN THEIN 1973), which consists of an approximately 300 m thick sequence of light grey, soft to subindurated, micaceous or non-micaceous shale and silty shale, with subordinate black slaty shale, slate, bentonitic and volcanic ash beds, and at some places, anthracitic coal (p. 170).

The graptolite shales of the Llandoverian are overlain by the mainly sandy deposits of the **Namhsim Formation** (Namshin River, LA TOUCHE 1913, p. 130). This formation is composed of a sequence of cross-bedded and massive, white and brown quartzite about 500 m thick (LA TOUCHE 1913, Pl. 24). In addition to brachiopods, gastropods and orthoceratites, LA TOUCHE (1913, p. 131) and PASCOE (1959, p. 646) mentioned a trilobite fauna with the genera *Illaenus, Encrinurus, Calymene, Cheirurus* and *Phacops* which REED (1906) placed in the **Wenlockian.** A brachiopod fauna of the Wenlockian was also identified (REED 1906, PASCOE 1959, p. 647, HAMADA 1964) in coarse-grained, pink sandstone, the **Manaw Beds** (NNE of Kyaukme), which LA TOUCHE (1913, p. 143) placed with the Kongsha Marl. BERRY & BOUCOT (1972, pp. 16, 17) classified the faunas collected by LA TOUCHE from the Namhsim Formation in the **Upper Llandoverian and later.**

In the Southern Shan State, the Namhsim Formation contains approximately 1,000 m of fine-grained sandstone with intercalated phacoidal limestone in the lower portions (GARSON et al. 1976).

The **Konghsa Marl Member** is composed of sandy marl with lenses of compact limestone. LA TOUCHE (1913, p. 139) assumed that this sequence, which is widely exposed in the Northern Shan State, is the facies equivalent of the Namhsim Sandstone. While the latter was deposited in near coast areas, the Konghsa Marl Member was deposited towards the deeper part of the basin (PASCOE 1959, p. 648). Faunas with brachiopods of the genera *Bilobites, Dalmanella, Platystrophia, Pentamerus, Atrypa* and *Spirifer* (LA TOUCHE 1913, PASCOE 1959) are known from the Namtu Gorge, N of Hsipaw, and from the Namhsim Valley near its junction with the Namtu Gorge, as well as from the railway line W of Kyaukme. MITCHELL et al. (1977, p. 18) described a fauna with *Nikiforovaena?* sp., *Duvillina* and trilobite fragments as well as with Bryozoa, bivalves and brachiopods of "Devonian or very latest Silurian aspect". On the basis of the incrinal trilobite fragments, they assigned an **Upper Ludlowian** age rather than a Lower Devonian age to the Konghsa Marl Member.

The **Taungmingyi Member** (village of Taungmingyi, approximately 20 km E of Myinkyado, Ye-ngan Township) is the youngest stratigraphic element in the Wabya Formation and consists of 100 m thick fine- to medium-grained, very poorly cemented, massive quartzite, sandstone and shale. The sandstone forms lense-shaped bodies, which are intercalated in the upper section of the shale of the Wabya Formation (MYINT LWIN THEIN 1973).

Silurian graptolites are also known from the Loikaw area. They were found in deep-red mudstone of the **Loikaw Beds** (HOBSON 1941, p. 127) and in limestone and calcareous mudstone of Balu Chaung (HOLLAND et al. 1956, p. 73). Further finds of Silurian graptolites from the Mawchi Series were mentioned by BA THAN HAQ & SEARLE (1961, p. 20) from "Mile 14" along the Kermapyu-Mawchi Road.

No Silurian faunas have so far been identified in S Burma. It is possible that a sequence of sandstone, shale, purplish mudstone and phacoidally banded limestone described by CLEGG (1953, p. 174) from the Eastern Thaton District corresponds to the rocks of the Linwe Formation and would therefore belong in the Lower Silurian.

The fine sandy-argillaceous-carbonate facies of the Upper Silurian extends into NW Thailand (BAUM et al. 1970). The Silurian is developed mainly in W Thailand in the form of black graptolite shale (BASTIN et al. 1970, 1977, KOCH 1973, HAGEN & KEMPER 1979). In Yunnan (China) the Silurian is represented by shale and calcareous "knotenschiefer" of the "Lower Jenhochiao Series" (SUN 1948).

In Malaysia, the separation into several geosynclinal facies regions, which started already in the Upper Ordovician, becomes clearer during the Silurian. GOBBETT & HUTCHINSON (1973) distinguished the following approximately N–S-striking zones (from W to E):

- mainly shelf facies: Lower Detrital Member (graptolite shale facies) of the Lower Silurian transgression, intercalated in the shelf sea carbonates of the 1,400 m thick Ordovician-Silurian Setul Formation
- mainly basin facies: graptolite shale of the Mahang Formation (miogeosynclinal deposits)
- mixed facies: rhyolite and micaceous schist of the Hawthorden Schists, 920 m thick; overlain by 1,830 m thick Kuala Lumpur Limestones and Chemor Limestones
- mixed facies with geanticlinal rocks: rhyolitic tuff and rhyolite (commencing already in the Upper Ordovician), overlain by Grik Siltstone and by 1,200 m thick carbonate schists and phyllite
- basin facies with eugeosynclinal rocks; 1,500 m thick flysch-like sediments with ophiolite.

BERRY & BOUCOT (1972) separated the carbonate suite (platform carbonates), to which the Silurian shelf-sea deposits of Burma and of parts of W Thailand and W Malaysia belong (as well as the Silurian of Yunnan, the Himalayas and Afghanistan), from a mudstone suite (platform mudstones), which occupies the area of the euxinic graptolite schist facies of Malaysia, Thailand, China and Vietnam. From the existence of the platform carbonates on the E side of the Bay of Bengal. BERRY & BOUCOT (1972, p. 6) assumed that the shelf extended further to the W than its present W margin.

3.115 Devonian

In contrast to the Thai and Malayan section of the Burmese-Malayan Geosyncline, no field evidence has been found of regional lithostratigraphic transitions from the Silurian into the Devonian in Burma. However, the presence of the Lower Devonian graptolite shale facies of the Zebingyi Beds, the Middle Devonian Padaukpin Beds and the Upper Devonian Wetwin Shales, indicates local continuous sedimentation from the Silurian into the Devonian. Frequently, however, gaps and erosion discordances were developed above the regression facies of the Upper Silurian Namhsim Formation. Locally, Permo-Triassic carbonates overlie directly Ordovician formations.

LA TOUCHE discovered the first graptolites in SE Asia (LA TOUCHE 1913, pp. 163, et seqq.) in the approximately 60 m thick sequence of grey limestone and shale of the Zebingyi Railway Station between Mandalay and Maymyo. The **Zebingyi Beds**, which PASCOE (1959, p. 649) described as a sequence of grey thin-bedded limestone followed by black limestone with black shale and white flaggy thin-bedded unfossiliferous limestone, is a formation of limited extent. It is found along the W edge of the Late Paleozoic carbonates which form the escarpment of the Shan Boundary Fault zone E of Mandalay (Figs. 38, 39).

The faunas of the Zebingyi Beds (LA TOUCHE 1913, pp. 173 et seqq.) are composed of graptolites of the genus *Monograptus,* of brachiopods with the genera *Stropheodonta, Meristella, Pentamerus,* of Lamellibranchiata with *Modiolopsis* and *Vlasta,* of Pteropoda with Tentaculites and Styliolinae, of cephalopods with *Orthoceras* and of trilobites with the species *Dalmanites swinhoai* (REED). In general, these faunas were placed in the Upper Silurian together with their Hercynian elements. Some, however, for example the fossils found along the Lashio-Mang-pyen Road, were determined by REED (1906) as "Lower Devonian in aspect" (PASCOE 1959, pp. 650, 651, 679). A **Lower Devonian (Siegenian-Emsian)** age for the Zebingyi Beds is now regarded as certain since the graptolite species *Monograptus* cf. *M. riccartonensis* from Pyintha (LA TOUCHE 1913, p. 168) was found to belong to the *M. hercynicus* group and *M. dubius* was re-identified as *M. atopus* (ANDERSON et al. 1969, p. 114, BERRY & BOUCOT 1972).

Below the Zebingyi Beds, in the area of Maymyo, a small section of Silurian sediments is probably developed, although the Upper Silurian Namhsim Formation is missing.

Towards the top, the Zebingyi Beds merge into the "ordinary crushed type of the Plateau Limestone, the passage being apparently quite conformable" (LA TOUCHE 1913, p. 166).

Evidence of the existence of Lower Devonian shale was provided by a re-investigation of the graptolite faunas of BROWN & SONDHI (1933 b, p. 218) by BERRY & BOUCOT (1972). The graptolite fauna in the shale SE of Taung-ni occurs together with *Tentaculites elegans.* A further fauna observed in limestone with intercalated shale E of Pon (BROWN & SONDHI

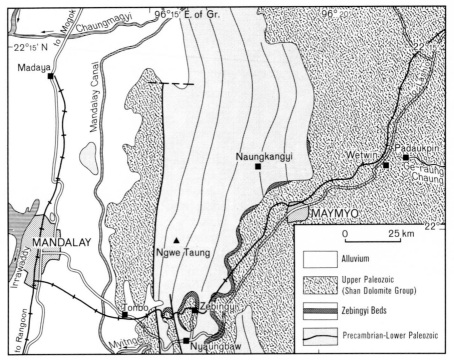

Fig. 38. Geological sketch map showing Zebingyi Beds E of Mandalay (after LA TOUCHE 1913 und AMOS 1975).

Fig. 39. Zebingyi Beds in a stream cut E of the village of Pyintha (E of Mandalay), E-dipping alternating argillaceous limestone and shale; Devonian (Siegenian-Emsian). Photo: D. BANNERT.

1933b, pp. 226, 227) contains *Orthoceras, Tentaculites elegans* and *Phacops ponensio.* BERRY & BOUCOT (1972, p. 15) assumed a **Gedinnian** age for this fauna, and they correlated the rock sequence with the Zebingyi Beds.

Also in Thailand, graptolite beds with tentaculites, formerly placed into the Silurian, are to be regarded as Lower Devonian in age (JAEGER et al. 1968, 1969, BASTIN et al. 1970, 1977).

The **Maymyo Dolomite Formation** (AMOS 1975, p. 68) is a part of the several thousand metres thick Upper Paleozoic to Lower Mesozoic carbonates that is mainly identified by Devonian faunas. The faunas in question are the "Padaukpin and Wetwin faunas" which have been known since REED (1906, 1908) and LA TOUCHE (1913).

The "Padaukpin fauna" (LA TOUCHE 1913, pp. 196 et seqq.) is so far the only evidence of **Middle Devonian** strata in Burma. It was found in well stratified dolomitic limestone close W of the village of Padaukpin, about 17 km ENE of Maymyo, and it was described by REED (1908), who classified it in the **Eifelian** (see also PASCOE 1959, pp. 683 et seqq.). ANDERSON et al. (1969) reported on new collections of fossils at the known locality and at a second locality nearby, and they newly described the fauna observed by REED (1908). They confirmed the classification as Eifelian (*Reticulariopsis eifeliensis, Devonaria minuta, Plectospira ferita, Cimicinoides struvei, Indospirifer, Alatiformia, Calceola*). They reinforced the classification by citing conodont faunas. In addition to brachiopods, conodonts and corals (tetracorals and tabulates), sponges, stromatoporoids, lamellibranchs, gastropods, tentaculites, cephalopods, annelids, ostracods, crinoids, blastoids and echinoids were found.

ANDERSON et al. (1969) compared the "padaukpinensis" fauna with the *Acrospirifer tonkinensis* fauna from N Indo-China (MANSUY 1916), which is also widely distributed in S China and occurs, among other places, in the Pochiao Shale of E Yunnan. The padaukpinensis and tonkinensis faunas contain genera and species which are characteristic of the Rhenish province. They lack elements of the Bohemian and Uralian subprovinces as defined by BOUCOT, JOHNSON & TALENT (1967).

According to ANDERSON et al. (1969), the fossiliferous strata with the Padaukpin fauna form a biostrome (not a reef, as LA TOUCHE 1913 assumed) which was not affected by the diagenetic processes of dolomitization.

A further Middle Devonian fauna was discovered in the former Lawksawk State, approximately 1.6 km E of the village of Htantabin. It was collected from well stratified limestone, shale and sandstone (Fenestellae, brachiopods, gastropods and trilobites, which also occur in the Padaukpin Limestone (SAHNI 1936, PASCOE 1959, p. 693)).

The fauna from the **Wetwin Member** was found in a layer of shale which presumably overlies the Padaukpin Limestone. (Wetwin, approximately 1.5 km WSW of Padaukpin, Fig. 38; LA TOUCHE 1913, p. 241, PASCOE 1959, p. 689). It contains Bryozoa, brachiopods of the genera *Lingula, Chonetes, Camarotoechia* and *Athyris,* also Lamellibranchiata, including *Janeia birmanica* REED, *Paleoneilo* cf. *plana* and *Chonetes subcancellata* and gastropods of the species *Bellerophon shanensis* REED. According to REED (1908), this fauna, which contains N American elements rather than European, belongs in the **Givetian to Frasnian.**

A lower Late Devonian *Philipsastraea* assemblage was also found in limestone in this area, in a tributary of the Ge-raung Chaung, about 5 km SSE of Padaukpin. In addition to many species of corals, this limestone contains brachiopods of the species *Stropheodonta (Douvillina) imitatrix, Orthothes (Schwellwienella) umbraculum* and a new variety (*subauriculata*) of *Atrypa reticularis* (PASCOE 1959, p. 689).

MITCHELL et al. (1977, pp. 18, 19) reported on further possibly Devonian fossils near Hkai Hsin, "to within 25 km of the Yadanatheingyi area". The coral fauna in pinkish-grey dolomitic limestone is composed of *Thamnospora* sp. (Silurian-Permian) and *Coenites* sp. (Silurian-Devonian).

According to the nomenclatural definition given by AMOS (1975), the Devonian dolomites, which were dated with the fossils in the above-mentioned "inliers", extend to the transgressively overlying Permotriassic Nwabangyi Formation, whose stratigraphic contact is very difficult to observe lithologically within the monotonous dolomite sequence.

3.116 Carboniferous

So far no Carboniferous fossils have been identified in NE Burma. THAW TINT (1972) indicated the existence of Carboniferous strata there. He named six limestone formations, of which he placed two in the Ordovician. The superjacent "Tangtalone Limestone" is Lower Devonian in age. The "Ohn-hne-ye Limestone" and the "Sitha Limestone" have Visean and Middle Carboniferous ages respectively. He placed the "Dattawtaung Limestone" in the Middle to Upper Permian (in an unpublished brief report without fossil data or lithological descriptions).

However, some fossils were found in the Carboniferous strata of the Tenasserim Division of S Burma, which are built up largely of monotonous, mainly clastic sediments several thousands of metres thick. These sequences are known as the Lebyin Group, Taungnyo Series,

Fig. 40. Taungnyo Group (Carboniferous) and Moulmein Limestone (Permian), Zwekabin Range S of Pa-an, Karen State (after BRUNNSCHWEILER 1970, p. 64). 1: Reef facies of Moulmein Limestone, 2: Bedded biostromal facies of Moulmein Limestone, 3: Fossiliferous calcareous sandstone of Taungnyo Group, 4: Shale of Taungnyo Group, 5: Alluvial plain of Salween River with coastal, Martaban Range in far distance on right.

Mergui Series and Mawchi Series. Because they are all overlain by the Permian Moulmein Limestone, they have been grouped together here although parts of the sequences may possibly extend into the Lower Paleozoic (Figs. 30, folder, 40).

A sequence of argillite, quartzite and sandstone several thousands of metres thick was grouped together by LEICESTER (in PASCOE 1930, p. 94) as the **Taungnyo Series** (after the Taungnyo Range S of Moulmein). In the Zwekabin Range S of Pa-an, an Upper Carboniferous fauna was recovered by BRUNNSCHWEILER (1970, pp. 63–66, Figs. 2–4) from calcareous sandstone of the Taungnyo Series. It is composed of gastropods, brachiopods, bryozoa, corals and ostracods. According to ÖPIK (quoted in BRUNNSCHWEILER 1970, p. 6), the Carboniferous age of this fauna is identified by the productid genus *Dictyoclostus* and the chonetid genus *Mesolobus*. The fauna can be positively classified in the Lower Pennsylvanian if the identification of the Nebraskan form of *Mesolobus* is correct. The fauna was found in a 660–670 m thick sequence of shaly marl, brecciated pinkish limestone, mud- and siltstone, sandy shale, "Pa-an Sandstone", shale and fossiliferous calcareous sandstone. This sequence according to BRUNNSCHWEILER (1970) forms no more than half the thickness of the Taungnyo Group.

A Carboniferous fauna from the clastic sediments of the Mergui Series in the Tenasserim Valley was discovered by OLDHAM (1856). The fauna is now lost. It consisted of corals, gastropods, brachiopods and Crustacea and was found in a sequence of reddish, weakly calcareous sandstone and grey shale. OLDHAM (1856) grouped this sandstone-shale sequence together with the overlying massive limestone of Permian age as the "Moulmein Beds" (Fig. 30).

The flysch-like sequence E of Yamethin is referred to as the **Lebyin Group** in the Explanatory Brochure to the 1 : 1 million geological map of the Union of Burma (1977, p. 8). It occurs between the area of the Plateau Limestone of the Shan Plateau and the igneous and metamorphic rocks of the lowlands in the W, and it consists of slate, quartzite, greywacke with intercalated conglomerate and limestone. In a few localities Carboniferous fossils were found in this sequence. No details or fossil names, however, are mentioned in the Explanatory Brochure of the Geological Map of Burma (1977).

The **Mergui Series,** whose thickness is assumed to be more than 3,000 m, are composed of the following rock types (BROWN & HERON 1923, RAU 1933):

– blue-grey to black, reddish brown weathered, splintery argillite with rarely detectable stratification and schistose structure, and with carbonate intercalations several metres thick containing finely disseminated graphite in addition to andalusite, sillimanite and pyrite; intercalations of quartzite and sandstone several tens of metres thick. In the contact zone to granite the argillite has been metamorphosed into phyllite and locally into quartzite
– dark grey greywacke and/or fine-grained agglomerates with angular fragments of quartz, slate quartzite, feldspar in an argillitic matrix
– conglomerate intercalations in the greywacke and agglomerate containing rounded granite components. Coarse conglomerates are known from Lampi Island with individual granite, quartz and quartzite components up to 25 cm in diameter
– limestone intercalations mainly in argillites:
  thin-bedded, generally impure, homogeneous type of red and black colour
  mostly thick-bedded, coarsely spathic type, of white colour with irregular knots and streaks of ferrigenous and clayey material (CHHIBBER 1934 a, p. 182).

Volcanic rocks within the Mergui Series are known mainly from the Mergui Archipelago (p. 127).

The sequence exposed W of Loikaw/Kayah State, which follows the general NNW–SSE strike of the mountain ranges in S Burma, was given the name **Mawchi Series** by HOBSON (1941, p. 127).

It consists of:

– fine-grained mudstone, siltstone, quartzite, sandy limestone ("normal Mawchi Rock Series")
– limestone intercalations several 100 m thick
– red sandstone and conglomerate as well as limestone conglomerate occurring at various horizons, but particularly in one or two cases closely along the borders of the limestone masses ("Red Beds", HOBSON 1941, p. 129)
– alternating series of limestone and calcareous marl, sandstone and mudstone ("Namphe Beds")
– a series of compact limestone generally of very light colour, with partings of green or grey schistose matter, and fine-grained bright green to greyish schists, which are described as being older than the Mawchi Series ("Biya Limestone Series", HOBSON 1941, p. 140).

The age of the Taungnyo, Mergui and Mawchi Series, comparisons with neighbouring countries

The discordant overlying of the Permian carbonates (Moulmein Limestone) opened up the possibility of classifying the thick clastic sediments in the Paleozoic in one of two different ways, namely as:

- Late Precambrian to Early Paleozoic, or
- Late Paleozoic

The first of these classifications is based on comparisons of the lithofacies with the thick, similarly monotonous-clastic Chaung Magyi Series of NE Burma and their similarly developed metamorphism ("phyllitic shales", RAU 1933, p. 14, J. C. BROWN & HERON 1923, p. 180, HOBSON 1941, pp. 133, 134, SAITO 1964a). The second and more probable classification of these sediments in the Late Paleozoic, which was discussed by, among others, CLEGG in HOBSON (1941, pp. XIV, XV), SCRIVENOR (1934, pp. 56, 57) and CLEGG (1953), is substantiated by the discovery of Upper Carboniferous fossils in sediments of the Taungnyo Series near Pa-an and in sediments of the Lebyin Group. Since the fossils were found in the upper portion of the Taungnyo Series, BRUNNSCHWEILER (1970) concluded that lower portions of this series may extend down into the Lower Carboniferous or Devonian.

In BRUNNSCHWEILER's (1970) view, the Taungnyo Series of the Moulmein District is younger than the rock sequence of the Mergui Series, and he assumed that it is discordantly underlain by the latter. From the region of the Tenasserim Valley, OLDHAM (1856) described a (?) Carboniferous fauna from a shale-sandstone sequence overlying the Mergui Series, which he grouped together with the overlying thick carbonates of Permian age as the Moulmein Beds. However, the findings reported by IYER (1938, p. 66) and PASCOE (1959, p. 725) indicate that the Taungnyo and Mergui Series should be regarded as homotaxial and thus at least parts of the Mergui Series also belong to the Carboniferous.

This is substantiated by fossils from the **Tanaosi Group** which according to JAVANAPHET (1969) consists of parts of the former "Phuket Series" of Thailand. Between Ranong and Bang Saphan this group is linked with the Mergui Series and it yielded Devonian and Carboniferous faunas (BAUM & KOCH 1968, YOUNG & JANTARAIPA 1970, GARSON et al. 1975).

Clastic, Upper Devonian and Carboniferous sedimentary sequences, which SCRIVENOR (in CHHIBBER 1934a) had compared on the basis of the lithofacies with the Mergui Series, are widespread also in the areas of Perlis and the Langkawi Islands/Malaysia. These areas are located in the S extension of the general strike of the Mergui Series. The sequences in question are the "Basal Red Beds" and the 1,500 m thick sequence of black mudstone commonly with pebbles, silty shale, siltstone and sandstone of the "Singa Formation" which, like the Mergui Series, are overlain by thick limestone of the Permian "Chuping Formation" (GOBBETT & HUTCHINSON 1973, p. 63).

Furthermore, there is a striking lithologic similarity between the Mergui Series and the flysch-like turbidite facies of the Carboniferous pebbly shale, sandstone and conglomerate from the Thong Pha Phum area (Thailand) described by KOCH (1973, p. 181). Also, the clastic sediments designated as "Normal Mawchi Rock Series" and the "Red Beds" probably correspond partly to the Devonian and Lower Permian sequences of Thailand.

Apparently, however, these clastic sedimentary sequences of S Burma also contain older Paleozoic components. In the area of the geological map of Thong Pha Phum (KOCH 1978) a sedimentary sequence extending from the Silurian into the Permian, strikes into Burmese territory that is occupied by the Mergui Series, Taungnyo Series and Moulmein Limestones. The sequence can be traced on aerial photographs through to the area of Moulmein (Fig. 30, folder).

The Biya Limestone Series (HOBSON 1941) at the base of the Mawchi Series, and the limestone intercalated in the Mergui Series, may correspond to the Ordovician limestone sequences in the Shan States and W Thailand (p. 61); the purple mudstone with phacoidal banded limestone, which CLEGG (1953, p. 174) described from E Thaton, may be assignable to the Upper Ordovician or Lower Silurian. Fossil evidence of a Silurian component in the Mawchi Series is provided by graptolites found in the Loikaw area (Loikaw Beds, HOBSON 1941, p. 127, BA THAN HAQ & SEARLE 1961, p. 20, THEIN & HAQ 1967, p. 281, BRUNNSCHWEILER 1970, p. 67).

The presence of the thick accumulation of clastic sediments reveals a Paleozoic flysch zone which extended from Malaysia (detrital members of Setul Formation, Singa Formation, older arenaceous series of W Penang, psammitic slate of Mallaca and Negri Sembilan and the feldspathic sandstone of the Kubang Pasa Formation; GOBBETT & HUTCHINSON 1973) via Thailand, SE Burma to at least the area E of Yamethin. The flysch-like sediments are evidence of tectonic movements that caused distinctive paleogeographic changes of the sedimentary environment of the Burmese-Malayan Geosyncline during the Late Paleozoic. Locally, erosion discordances extended into the Ordovician. An undisturbed sedimentary environment with a complete carbonate series extending from the Devonian to the Permian existed, however, in the shelf region of the Burmese-Malayan Geosyncline, namely in Kinta Valley/Malaysia (SUNTHARALINGAM 1968), and also in Thailand, Annam, Laos and Yunnan (MANSUY 1919b, YABE & HAYASAKA 1920, PASCOE 1959, BROWN 1917, 1923a, MUIR-WOOD 1948, HAMADA 1961, KOBAYASHI 1964a).

3.117 Permian

In E Burma the Lower Paleozoic strata are overlain by limestone and dolomite several thousand metres thick, which cover large areas of the Shan Highlands and for which LA TOUCHE (1913) introduced the name "Plateau Limestone". He subdivided this thick carbonate sequence into two units: a Devonian section of chiefly "shattered dolomite", and a Permo-Carboniferous or Anthracolithic section of crystalline limestone.

Fossil evidence (esterian shale from Kyaukme, LA TOUCHE 1913, Fig. 6; the "Na-hkam fauna", SAHNI 1933) and lithostratigraphic correlations in the Southern Shan State (BROWN & SONDHI 1933b, p. 211) made it possible to divide the "Plateau Limestone" into a lower (Devonian) and an upper (Triassic) dolomitic portion which are separated by a Permo-Carboniferous calcareous portion (PASCOE 1959).

This scheme was modified by new fossil finds (GRAMANN et al. 1972, AMOS 1975, GARSON et al. 1976, BRÖNNIMANN et al. 1975, MITCHELL et al. 1977). The name "Plateau Limestone" was abandoned because it was defined as a carbonate rock sequence that was originally placed in the Permo-Carboniferous. The several thousand metres thick carbonates are now taken together as the **Shan Dolomite Group** (AMOS 1975), which is divided into a Devonian and possibly Carboniferous **Maymyo Dolomite Formation** (p. 67) and a Permo-Triassic **Nawabangyi Dolomite Formation** (Fig. 41).

DIENER (1911) had identified Permian faunas in the Shan Dolomite Group of NE Burma. They include the "Namun fauna" from the Namtu Valley with Bryozoa of the genera *Polypora* and *Fenestella* and brachiopods of the genera *Spiriferina*, *Streptorhynchus*, *Productus* and *Chonetes*. Having regard to the cold-water habitat of the brachiopods, WATERHOUSE (1973) assumed that this fauna with *Polydiexodina elongata* (SHUMARD), which may be referable to *Skinnerella*, is **Filippovian** in age. PASCOE (1959, p. 699) mentioned a number of further sites of Permian fossils (Tonbo, about 20 km SE of Mandalay; the Namtu Valley S of Hsipaw near the Tong-Ang ferry; Namlan and the Loi-lan Range, about 18 km E of Lashio). Waagenophylliid corals and brachiopods such as *Spirigerella, Leptodus* and *Notothyris,* as well as fusulinids of the genus *Polydieoxodina* (DIENER 1911, REED 1931, SMITH 1941) were observed in dark algal micrites of the Tonbo locality. The brachiopod fauna corresponds to the Warga and Chidru Formations (Middle and Upper *Productus* limestone) of the Salt Range (GOBBETT 1973).

The carbonate rocks which, in the Yadanatheingyi area, overlie the Naungkangyi Formation (pp. 55, 56) with a low angle unconformity and, in the Kyaukme-Longtawkno area, overlie the Silurian Namhsim Formation (p. 64), are approximately 3,000 m thick and composed of brecciated dolomite and limestone, mostly massive but locally laminated, micritic and bioclastic. BRÖNNIMANN et al. (1975) identified Triassic Foraminifera of the genera *Glomospira* and *Glomospirella* only in the upper sections of the dolomitic sequence. The authors therefore assign these dolomites to the Permo-Triassic Nwabangyi Dolomite Formation, although MITCHELL et al. (1977) do not exclude the possibility that a Devonian section, which thus belongs to the Maymyo Dolomite Formation, may be hidden in the lower sections of these dolomites.

In the Southern Shan State also, Permian fossils have been known since DIENER (1911) from the "fossiliferous Upper Plateau Limestones" (BROWN & SONDHI 1933b, p. 210). These include the "Kehsi Mansam Fauna", a productid fauna from a dark, thin-bedded fusulinid limestone containing *Fusulina elongata,* (Nam Hen river, about 1.6 km SW of Kehsi Mansam). Many fossils have also been found in limestone along the old road leading from Taunggyi to Mong Pong. These faunas were collected by MIDDLEMISS between miles 20–28.3 and identified by DIENER (1911) (see also PASCOE 1959, p. 701). REED (1933, pp. 83 et seqq.) reported on a fauna found by J. C. BROWN in the locality of Htam Sang. It comes from very fossiliferous, dark, thin-bedded limestone with intercalated marl and contains brachiopods, including the genera *Streptorhynchus, Orthotetes* and species such as *"Schuchertella" semiplana,* Chonetids, *Krotovia hanus, Notothyris exilis, Lyttonia,* Bryozoa of the genera *Fenestella* and *Polypora* and corals of the genera *Sinophyllum* and *Michelinia* (WATERHOUSE 1973, p. 202).

According to AMOS (1975, p. 57) and GARSON et al. (1976, p. 26), in the area of Ye-ngan, the approximately 800 m thick Permian **Thitsipin Limestone Formation** overlies transgressively with an angular unconformity an irregularly eroded surface of different Lower Paleozoic sequences.

This formation is subdivided into three types of facies:

- massive limestone facies: massive micrite with large recrystallized shells of brachiopods, individual corals and colonies of corals, often in the in-situ position (reef limestone)
- massive cherty limestone facies: characterized by lenses and beds of dark-grey cherty limestone in massive micrite
- well-bedded calcarenite facies: the beds are several centimetres to several tens of centimetres thick and occasionally cross-bedded; the detritic material consists of fine organic reef fragments in a fine-grained matrix.

Locally the limestone has been dolomitized and it is intensely brecciated. The Thitsipin Limestone Formation was dated on the basis of Foraminifera (GARSON et al. 1976, pp. 27, 28). **Lower Permian** fossils, namely *Pseudoschwagerina, Pseudofusulinella* and *Parafusulina* cf. *Kattaensis* (SCHWAGER) of the *Pseudoschwagerina* zone (see also AMOS 1975,

Table 1), have been found in particular in the dolomitic section of the Thitsipin Formation. Higher up in the section the faunas contain *Yangehienia, Shubertella, Verbeekina* and *Parafusulina* together with *Plathyploia kattaensis* thus indicating an **Upper Permian** age (upper *Parafusulina* to *Neoschwagerina-Verbeekina* zone). In general, the Burmese fusulinid genera belong to the Tethyan genera (GOBBETT 1973).

GARSON et al. (1976) assumed that the Thitsipin Limestone was deposited in an open shallow environment. Very frequently, the Thitsipin Limestone frames uplift areas ("islands") of older rocks (BROWN & SONDHI 1933 b, p. 210). Locally, reefs or bioherms were formed, as indicated by corals in growth position. The bedded limestone probably represents a near-reef facies. The occasionally observed cross-bedding indicates locally strong currents while the massive cherty limestone facies probably formed in a still-water area.

In the region of Ye-ngan the Thitsipin Limestone Formation is overlain by the **Nwabangyi Dolomite Formation** (GARSON et al. 1976, p. 29). This formation is characterized by an intensively dolomitized, shattered and brecciated carbonate sequence, which is at least 2,500 m and possibly 5,000 m thick and contains the following facies types:

- thin-bedded foraminiferal limestone facies with occasional gastropods and compound corals
- laminated and turbiditic limestone facies: an alternating sequence of light grey, fine-grained calcarenites and dark grey micrites. The calcarenite beds have a sharply defined lower lithostratigraphic contact and frequently merge into micrite towards the top. Intercalated channel fillings occur frequently in the form of the
- sedimentary breccia facies: these breccias are not thicker than 6 m and they contain angular and rounded inclusions of laminated calcareous siltstone and calcilutite
- light and dark grey, fine-grained limestone facies; it constitutes the approximately 170 m thick middle portion of the formation.

Because of intense dolomitization, few fossils are found. GARSON et al. (1976, pp. 30, 31) described a Late Permian foraminiferal fauna from two locations SSE of Mandalay. They originate from the lower part of the Nwabangyi Dolomite Formation: *Shanita amosi, Paraglobivalvulina mira, Hemigordius reicheli, Bisalina pulchra, Pachyphloia cukurkoyi, Nankinella* sp. are among the Foraminifera, found together with the calcareous algae *Mizzia veleloitana* and *Permocalculus*. These fossils indicate an **Early Dzhulfian** (= **Tatarian**) age.

More complete lists are given by ZANINETTI et al. (1979) and WHITTAKER et al. (1979).

The upper part of the Nwabangyi Dolomite Formation contains Triassic fossils (p. 79), thus indicating that the Permian sedimentation continued without lithological break into the Triassic.

The widespread dolomitization of the Upper Paleozoic carbonates of the Shan Plateau resulted in the destruction of fossil remains. Fossils are found only in places where the primary carbonate structure has remained largely intact. A diagrammatic representation of stratigraphic relationships (Fig. 41) shows the "relict" character of individual fossiliferous portions of the sequence within the dolomitized Upper Paleozoic and Mesozoic carbonates.

The limestone-dolomite sequence of the Shan State continues towards the S into the Kayah and Karen States (HOBSON 1941) and further S, in the area of Salween, it links up with the **Moulmein Limestone** (Moulmein Beds, OLDHAM 1856, pp. 32, 33). The latter had long been classified as Carboniferous or Permo-Carboniferous on the basis of a few megafossils. The new fusulinid faunas, however, are taken as positive evidence of the Permian age of the Moulmein Limestone (WHITTINGTON in CLEGG 1953, p. 194, BRUNNSCHWEILER, 1970).

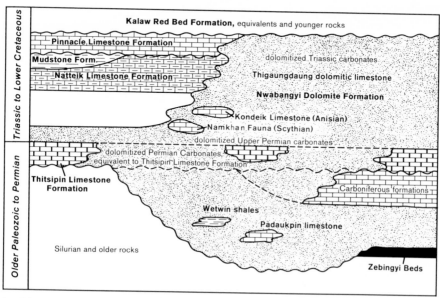

Fig. 41. Diagrammatic representation of stratigraphic relationships of the Shan Dolomite Group (after GRAMANN et al. 1972, AMOS 1975).

In the area of Moulmein – Pa-an, the Upper Carboniferous strata of the Taungnyo Series are overlain with an angular unconformity by reef limestones of the Moulmein Limestone (BRUNNSCHWEILER 1970, pp. 65, 66). They form rugged mountains characterized by karstic phenomena (CHHIBBER 1934, pp. 93 et seqq., Pl. IX) and rise as steep crags from the alluvial plains of the Salween River (Fig. 42). The lower section of the reef limestone is often built up of biostromal, well-bedded limestone which frequently fills an old relief in its substratum. It contains fusulinid faunas of the **Artinskian** (BRUNNSCHWEILER 1970, p. 66). Massive, unbedded bioherms emerge from the biostromal limestone, and in high elevations of the old relief they lie directly on top of older strata (BRUNNSCHWEILER 1970, Figs. 8, 9). A profile drawn through the Zwekabin Range reveals that the Moulmein Limestone is just under 1,000 m thick.

In the S Tenasserim Division as well, the Mergui Series are overlain by mostly coarsely crystalline thick limestone. RAU (1933, pp. 16–18, Pl. 9) described three morphological zones, the most westerly of which forms some islands in the Mergui Archipelago (Pinnacle Rock, West Rock, Mali Kaing, Mali Don, St. Matthes's Island). Identifiable fossils were found near Kyauktaung, NW Tharabwin (NOETLING 1893, PASCOE 1959, p. 716, RAU 1933, pp. 17, 18). These include *Schwagerina oldhami, Lonsdaleia salinaria, Lithostrotion* sp., *Productus* cf. *sumatrensis* (PASCOE 1959, pp. 716, 717). Waagenophylliid corals and a species of the Guadalupian productid brachiopod *Tyloplecta sumatrensis,* which indicate an **Upper Permian** date, were identified by GOBBETT (1973, p. 138).

A sequence of sandstone and shale, which may possibly overlie the Moulmein Limestone and which can be traced for more than 70 km to the N of Martaban, was given the name **Martaban Beds** by PASCOE (1959, p. 713). In addition to unidentifiable plant remains, the carbonate shale from a location near the Martaban Railway Station contains brackish water

Fig. 42. Massive limestone near Kwauktalon, ca. 16 km S of Moulmein; Permian. Photo: F. BENDER.

molluscs, i.e. *Palaeoanodonta okensis* and *P. subcastor,* on the basis of which the Martaban Beds are placed in the **Upper Permian** to possibly the lowermost **Triassic**.

The **Yinyaw Beds** are composed of a sequence of shales and argillites with thin intercalated layers of silicified sandstone. They are exposed W of Loilaw. HOBSON (1941, p. 118, Pl. 2) described them as being intercalated in the Permian (Plateau) Limestones. HOBSON (1943, p. 120) quoted a Permian fauna identified by REED, which consists principally of brachiopods of the genera *Marginifera, Productus, Spirifer* and *Martinia,* and which, according to GOBBETT (1973, p. 137), reveals similarities with the Tonbo Fauna (p. 73).

The Permian fusulinid and reef limestones can be correlated with other areas of the Tethyan Province, while the brachiopod and coral faunas correspond to the Middle and Upper Permian of the Salt Range. GOBBETT (1973) grouped the carbonates under the heading "posthercynian limestones" as a transgressive facies. With stratigraphic gaps and usually with angular unconformities these limestones overlie older rocks of various ages. The "Ratburi Limestone" in Thailand is assigned to the

Sakmarian and Artinskian. The "Chuping Limestone" of W Malaysia belongs to the Artinskian and Guadalupian, while Artinskian and Sakmarian faunas were discovered in the upper section of the "Kinta Limestone" (GOBBETT 1973, GOBBETT & HUTCHINSON 1973).

In all probability, the Martaban and Yinyaw Beds correspond to the clastic section of the Lower "Mae Moei Group" in the Mae Sot area of W Thailand. This alternating sequence of shale and sandstone overlies the Permian Ratburi Limestone and merges towards the top with the Upper Triassic to Lower Jurassic "Kamawkala Limestone" (v. BRAUN & JORDAN 1976, Fig. 4, p. 20).

Probably pre-Mesozoic epimetamorphic rocks are exposed in association with Triassic sediments at the E edge of the Arakan Yoma and the Naga Hills. They have not yet been dated by fossils. They include talc schist, mica schist, green chloritic schist and quartzite ("**Kanpetlet Schists**"; COTTER 1938, p. 31) in the area of Mt. Victoria, rubellite schist several hundred metres thick from Panmum Chaung S of Kalemyo (BRUNNSCHWEILER 1966, p. 166) as well as a sequence of very thick green and purple shale, containing large quantities of vein quartz in the Minbu and Kyaukpyu Districts (HAYDEN 1896, pp. 74−76). In addition to quartzite layers, the latter also contain quartzite pebbles and boulders several centimetres to several metres in diameter with longitudinal axes oriented parallel to the bedding planes. Further S, at 18° 94′ 40″ N, metamorphic clastic rocks (blackish phyllite with quartzite bands) were observed by CLEGG (1938, p. 187, quoted from BION's Field Note Book, 1915).

Metamorphic, probably pre-Mesozoic sequences, occur also N of Mt. Victoria from the Somra Tract to Lahe (26° 30′ N). BRUNNSCHWEILER (1966, pp. 174, 175) grouped these rocks together as the **Naga Metamorphic Complex.** It includes dark graphitic phyllite, biotite-muscovite and biotite-sillimanite schist, banded gneiss and sericitic quartzite.

In the W, the Kanpetlet Schists border on the Chin-Arakan flysch-type sediments (p. 24). In the E, they apparently grade into non-metamorphic Triassic sediments (D. BANNERT, verbal information 1982). It may be therefore assumed, that at least parts of the Kanpetlet Schists are of Triassic age. In Fig. 30, they are still placed all together into the pre-Mesozoic.

3.12 Mesozoic and Cenozoic

3.121 Triassic

Triassic strata were known only from a few locations, while in recent years it has been found that Triassic rocks are widespread in Burma (Fig. 31).

The Triassic of the **Indo-Burman Ranges,** on whose existence doubt was cast soon after its discovery by THEOBALD (1871), has been confirmed in the meantime. It occurs here as a shale-fine sandstone sequence of the "**Halobia Schists**". Outcrops are limited to a narrow intermittently exposed zone at the E edge of the Indo-Burman Ranges.

In the **East Kachin/Shan Unit** of the Sino-Burman Ranges, large portions of the "Upper Plateau Limestone", which was formerly regarded as Permo-Carboniferous, and part of the shattered dolomites, which were previously assigned to the Devonian, have now been identified as Triassic rocks. The scarcity of macrofossils and the predominance of the dolomitization of the platform carbonates, which extend from the Upper Paleozoic to the Triassic without any striking lithological break, are the reasons for the late discovery of Triassic fossils.

It is not certain where the boundary lies between the Triassic of the East Kachin/Shan Unit and the Permian. The uppermost Permian, Dzhulfian or Tatarian, was thought by BRÖNNIMANN et al. (1973, p. 6) and AMOS (1975, p. 65) to exist in miliolid dolomites with *Nankinella* and later confirmed by ZANINETTI et al. (1979) and WHITTAKER et al. (1979). These dolomites, together with carbonates of Triassic age, belong to a group of formations which apparently cannot be lithologically separated. The youngest Permian strata are located in the Ye-ngan area in the S of the Shan States and N of this area, to the E of Mandalay.

GRAMANN et al. (1972), AMOS (1975) and BRÖNNIMANN et al. (1975) anticipated that the carbonate sedimentation would extend beyond the Permian-Triassic boundary. Local gaps and transgressive contacts between Permian and Triassic strata seem to be exceptions rather than the rule. However, a regional discordance was observed in the Shan States at the base of the Permo-Triassic sequence known as the Nwabangyi Formation (AMOS 1975). A similar discordance was found also in the N in the Kyaukme-Langtawkno region N of Hsipaw (WHITTAKER in BRÖNNIMANN et al. 1975, Fig. 2) (Fig. 31).

In the extreme S, in the Martaban area, the Permian Moulmein Limestone is overlain by the Martaban Beds, which belong to the transition zone between the Permian and the Triassic (p. 75).

Scythian: An often quoted fauna, which is in need of revision, from Nankam N of Lashio and W of Hsenwi, indicates the occurrence of lowermost Triassic (Scythian, Griesbachian, Induan, *Otoceras-Ophiceras* strata). Provisional lists in CLEGG (1941b, p. XVI) and PASCOE (1959, pp. 902–903), contain *Ophiceras (Glyptophiceras)* and *Vishnuites,* i.e. cephalopods which are typical of the earliest Triassic, while *Owenites,* which is also mentioned, characterizes somewhat younger Scythian faunas. It can be assumed that these lists give the fossil contents of a larger rock complex. *Otoceras,* which at other sites is also much rarer than *Ophiceras,* seems to be missing, so that possibly only Upper Griesbachian as defined by TOZER (1973) is present. This fauna is not mentioned in the special literature on the distribution of the lowermost Triassic. If it should be possible to confirm the existence of the *Otoceras-Ophiceras* strata of Nankam, this would be proof that the region of the East Kachin/Shan Unit belongs to the central part of the Triassic Tethys. The lowermost Triassic of Malaysia and of the Chinese-Indochinese regions (Yunnan, Laos) adjoining Burma, exhibits a *Claraia* facies characterizing marginal parts of the Triassic Tethys.

The earliest Triassic cephalopod fauna from Nankam was found in thin-bedded, marly limestone and shale that are intercalated in dolomite. Because it was assumed that the dolomite in the area was Paleozoic or limited to the Devonian, it was believed that the Triassic strata were in tectonic contact to the dolomite (PASCOE 1959, pp. 902, 903). According to recent knowledge, the marly sequence with the *Ophiceras* fauna is of the same age as the surrounding dolomites (Fig. 41).

Anisian rocks were identified in the East Kachin/Shan Unit containing the following fauna (GRAMANN, FAY LAIN & STOPPEL 1972; the nomenclature has in part been revised):

Cephalopods:	*Acrochordiceras* cf. *enode* HAUER, 1892
	Paraceratites thuillieri (OPPEL, 1863)
	Flexoptychites mahendra (DIENER, 1895) (Fig. 43)
Conodonts:	*Gratognathodus hochi* (HUCKRIEDE)
	Cypridodella mediocris (TATGE)

Cypridodella venusta (HUCKRIEDE)
Diplododella sp.
Enantiognathus petraeviridis (HUCKRIEDE)
Enantiognathus ziegleri (DIESEL)
Gondolella mombergensis TATGE
Gladiogondolella tethydis (HUCKRIEDE)
Metapolygnathus pectiniformis (HUCKRIEDE)
Metapolygnathus spengleri (HUCKRIEDE)
Neogondolella constricta (MOSHER & CLARK)
Neogondolella navicula (HUCKRIEDE)
Prinoidella ctenoides TATGE.

This fauna is evidence of **Upper Anisian** (Illyrian, *Trinodosus*-Zone) and was found in dark, flaky, micritic limestone with chert nodules which merge towards the top into bedded grey cephalopod-bearing limestone. This sequence, which resembles the "Gutensteiner Limestone" of the European Alps, was observed in the Southern Shan State near Kondeik, S of Aungban and the Kalaw-Heho Road and therefore it is given the name "**Kondeik Limestone**". In the European Alps, comparable deep-water carbonates were interpreted as having been deposited in stagnant parts of the sedimentary basin (BECHSTÄDT & MOSTLER 1976, p. 274). Another occurrence is located near Nammekon, E of Loikaw in the Kayah State (19° 37′ N, 97° 0,6′ E).

The Kondeik Limestone is intercalated in light-coloured dolomitic limestone. Near Nammekon it forms the youngest rock sequence in a syncline. The light-coloured dolomitic limestone is named "Thigaungdaung Limestone" (mountain chain W of Heho; GRAMANN, FAY LAIN & STOPPEL 1972, p. 13). It dates at least in part from the **Anisian** as is indicated by the occurrence of an *Acrochordiceras* sp. and of complex-structured *Glomospira* resembling *Glomospira densa* (PANTIC). The division between these and the similarly light-coloured carbonate rocks that represent the Middle to Upper Permian, and also their full extent within the Triassic are still unclear. Upper Triassic might be indicated by *Kerocythere raibleriana* and *Clypeina*. In the European Alps and in Anatolia, the calcareous alga *Clypeina,* which is also present in this limestone, does not occur before the Karnian (OTT 1972).

AMOS (1975) and WHITTAKER (1975, in BRÖNNIMANN et al. 1975) regarded the Thigaungdaung Limestone as being identical in age to the Nwabangyi Dolomite Formation of the area around Ye-ngan. Here, the formation covers the entire biostratigraphic range from the Permian through to the Middle Triassic-Upper Triassic boundary. *Glomospirella irregularis* (MOELLER) of **Middle Triassic** (Scythian?) age was found in the uppermost part of the Thigaungdaung Limestone near Kyaukme.

According to AMOS (1975, pp. 58–63), the Thigaungdaung Formation consists of dolomitic limestone or dolomite (microsparites). The brecciated carbonates (diagenetic breccia) are called "shattered dolomites". Millimetre-thick bedded or cross-bedded microsparite and coprolitic microsparite occur quite frequently. Along the road from Aungban to Heho, at the steep drop to the Heho Basin, chert stringers are intercalated with millimetre-thick bedded carbonates. According to GARSON et al. (1976) and AMOS (1975), the Nwabangyi Dolomite Formation is here represented laterally by limestone sequences which have in part escaped dolomitization. The sequences in question are the **Natteik Limestone Formation,** which consists of thin-bedded calcarenite and which contains siliceous concretions and layers in the middle and upper parts. Thus, instead of corresponding to the micritic Kondeik Limestones, as AMOS (1975) assumed, they more closely resemble

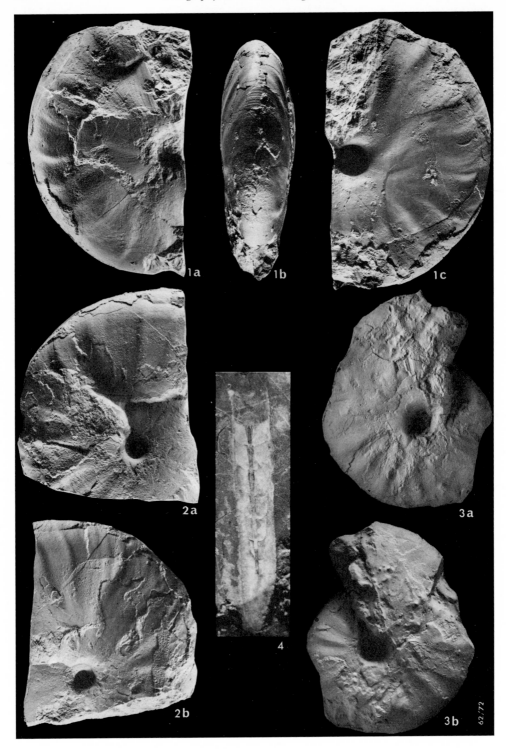

rhythmically deposited dolomitic calcarenite (microsparite) with siliceous layers, such as are observed in the Thigaungdaung Range.

The so-called Mudstone Formation in this area forms a lenticular intercalation that wedges out laterally.

There are doubts about the age of the "Pinnacle Limestone Formation" (GARSON et al. 1975). The carbonates assigned to it presumably belong to the latest Triassic portion of the Nwabangyi Formation.

The calcareous algae *Permocalculus* and *Clypeina,* which are regarded as bathymetric indicators of shallow water sedimentation, are found in the Thigaungdaung carbonates. The great abundance of holothurian sclerites is evidence of still bottom water. The *Favreina* coprolites described by AMOS (1975, p. 64) in the Natteik Limestone Formation are regarded in the European Alps, too, as characteristic of shelf sedimentation (BECHSTÄDT 1976, p. 6).

GRAMANN et al. (1972) assumed that the Permo-Triassic rock series in the S part of the Shan States are >1,000 m thick. The thickness of 2,500–5,000 m given by AMOS (1975) for the Nwabangyi Dolomite Formation is similar to the total thickness of the Triassic sequence of the Northern "Kalkalpen" in Europe. The lithological and biofacies characteristics are very similar to the "Ramsau Dolomite" and "Wetterstein Limestone" of the European alpine Triassic carbonates.

Upper Triassic: Fossils from the Upper Triassic (Karnian, Norian, Rhaetian) have been known for a long time in Burma. Their occurrences are widely scattered. From the N of the Shan States, the Kayah State (Karenni), to the Amherst District, they belong to the East Kachin/Shan Unit of the Sino-Burman Ranges (p. 115). The occurrences W of Kalemyo, at the foot of Mt. Victoria in the Pakokku District and from Thabyegaing in the Thayetmyo District are located along the E edge of the Indo-Burman Ranges ("*Halobia* Schists").

THEOBALD (1871) discovered *Halobia*-containing beds of **Karnian** age at the E edge of the Indo-Burman Ranges (Chin Hills, Arakan Yoma), near Thabyegaing, 19° 2′ N; 94° 57′ E, in the W of the Thayetmyo District. The Halobiidae were identified as *Halobia* sp. (TIPPER 1906), which led to the assumption that the Arakan Yoma had a stratigraphic sequence that in general extended downwards at least to the Triassic, and also gave rise to the much-discussed "Axial" concept (p. 22).

MYINT LWIN THEIN (1973) found Halobiidae at the E edge of the Indo-Burman Ranges close W of the town of Kalemyo, at the E foot of the Chin Hills. Further finds were made in the extreme W of the Pakokku District, on the E slope of the Mt. Victoria Massif, particularly in the Te-Chaung Stream NW of Kyauktu (GRAMANN 1974, p. 280).

The Halobiidae of Kalemyo and those of Te-Chaung in the Pakokku District seem to be identical; they were found in the same dark grey shale. The accompanying fossils at both places are flattened, tubular, conical-pointed structures which MYINT LWIN THEIN (1973) referred to as *Titahia birmanica.* These tubular structures, which are also known to exist in

Fig. 43. 1–3: *Flexoptychites mahendra* (DIENER 1895). 1: Specimen, partly with shell, Kondeik Limestone near Kondeik village, S of Kalaw-Heho Road. Diam. 134 mm, thickness 41 mm; Anisian, 2: Crushed specimen from same block of limestone as 1, greatest length 81.75 mm, 3: Specimen from Nammekon, E of Loikaw (lat. 19° 37′, long. 97° 06′), diam. 70.25 mm, thickness 22.5 mm. – 4: *Michelinoceras* sp. – locality as in 1, length 38 mm; Anisian (from GRAMANN, FAY LAIN & STOPPEL 1972, Pl. 6).

Halobia schists in other countries, may be large primitive arenaceous-test foraminifers of the genus *Bathysiphon*. The Halobiidae themselves are pseudoplanktonic and indicate conditions hostile to life at the sea floor.

The finds from Kalemyo were identified by MYINT LWIN THEIN (1973) as *Daonella lomelli*, but they exhibit characteristics of the genus *Halobia*. The specimens from Te-Chaung can be identified as *Halobia* cf. *comata* BITTNER 1899. The rock series from which they come consists of dark grey shale to slate and of quartzitic fine sandstone which are frequently interbedded with the latter and exhibit "flow-cast" sole marks. Dark micritic limestone with "filaments" and conodonts also occurs. The entire rock sequence is known meanwhile as **"Thanbaya Formation"** or **"Pane Chaung Formation"**.

STOPPEL (in GRAMANN 1974, p. 280) identified *Gondolella polygnathiformis* BUDUROV & STEPANOV in this sequence, thereby confirming its age as Karnian.

The Thanbaya Formation of the Pakokku District grades into the metamorphosed Kanpetlet Schists. The contact is usually accompanied by basaltic pillow lavas or dyke-like occurrences of serpentinized rocks. Schists with *Halobia* of Kalemyo are also located in close proximity to serpentinized basic rocks.

The Thanbaya Formation of the Pakokku District is discordantly overlain by the Albian-Cenomanian Kabaw Group (p. 89). Their position above epimetamorphic sediments is similar to the *Halobia* schists of Sumatra, which also lie on top of epimetamorphic Permo-Carboniferous sequences.

Dark, filamentous limestones, which can be compared with those of the Thanbaya Formation, occur also at the S end of the Arakan Yoma, W of Bassein. The Triassic Thanbaya Formation is very similar to the laminated flysch series of the Lower Tertiary which constitutes the mass of the rocks in the Indo-Burman Ranges.

The Thanbaya Formation is probably about 1,000 m thick.

Triassic olistoliths can also be expected to occur in the Tertiary flysch series (BRUNNSCHWEILER 1966; dolomitized limestone block containing presumably Triassic ammonites and foraminifers on the road from Falam to Lake Hri). An occurrence of Halobiidae in Kayah State (Karenni) has an obscure origin. TIPPER (1906, p. 134) mentioned a specimen containing *Halobia* and *Monotis* sp. that was labelled as follows: "Sent to Mr. Theobald by Mr. D. O'Riley from Karenee, Aug. 1864." This "occurrence" was never confirmed by new observations. Nevertheless, for regional-geological reasons Halobiidae-bearing Triassic sequences can be expected to occur in the East Kachin/Shan Unit of Burma (p. 38), as is shown by finds in Yunnan (China), in neighbouring Thailand and on the Malayan Peninsula.

The Upper Triassic of the Northern Shan State has also been known for a long time. The **Napeng Formation** (LA TOUCHE 1913) yielded a fauna that was classified in the Rhaetian.

The area around Hsipaw (type location of Napeng Formation) has been re-investigated (BRUNNSCHWEILER 1970). In the process, the Mesozoic stratigraphic sequence was expanded to include the Pagno Evaporites underlying the Napeng Formation.

MITCHELL et al. (1977), extracts of whose work are published in BRÖNNIMANN, WHITTAKER & ZANINETTI (1975), have in turn added to and modified the stratigraphic scheme proposed by BRUNNSCHWEILER (1970). They regarded the Upper Triassic-Jurassic stratigraphic sequence as a formation group and gave it the name of **"Namyau Group"** after the Jurassic Namyau Formation, which is part of it and which contains an Upper Bathonian fauna. The Namyau Group is usually diachronously superjacent on the Nwabangyi Dolomite

Formation (Permo-Triassic). The Namyau Group commences with bedded limestone and differs from the Nwabangyi Dolomite Formation through the reddish colours, the occurrence of evaporites and the transition to limestone conglomerates that takes place in its uppermost part. From the limestone in the lowest section of the Namyau Group, BRÖNNIMANN et al. (1975) described a foraminiferal fauna that is characterized by Involutinidae. Most species are limited to the **Ladinian-Norian.** Altogether, the fauna from Burma corresponds to the fauna from the lowest Bidestan member of the Nayband Formation in SE Central Iran, which is assumed to date from the Norian. However, the possibility cannot be excluded that the fauna extends down into the Karnian.

KEMPER, MARONDE & STOPPEL (1976) described similar Upper Triassic limestone with Involutinidae from the Kanchanaburi area of W Thailand.

BRÖNNIMANN et al. (1975, p. 7) also observed Charophytes from the lower portions of the Namyau Group. The joint occurrence of carbonates, clastic material, evaporites and the microfossils mentioned indicate that the Namyau Group is built up of sediments deposited in a shallow shelf surrounded by arid regions (BRÖNNIMANN et al. 1975). A fauna from the Napeng area has been described by HEALEY (1908). It was obtained from shales and consists almost exclusively of Lamellibranchiata and gastropods; apart from a large number of "new" species, it contains *Pteria contorta* PORTLOCK, which is known as the type fossil found in **Rhaetian** strata.

Similar faunas containing *Pteria contorta* were reported from neighbouring Indochina (PASCOE 1959, pp. 1196–1197).

In the Karen State, in the border region between Thailand and Burma, limestone with sparse faunas ("**Kamawkale Limestone**") is also regarded as Upper Triassic in age (DE COTTER 1923a, GREGORY, WEIR, TRAUTH & PIA 1930). According to V. BRAUN & JORDAN (1976), the Kamawkale Limestone in Thailand ranges from the Norian to the Rhaetian. It overlies shaly-sandy Middle Triassic sequences which are separated from the underlying Middle Permian limestone by red clastic sediments. The sedimentation of the Kamawkale Limestone continued through into the Jurassic and was followed in the Kamawkale region by Jurassic clastic sediments. VON BRAUN & JORDAN (1976) took the entire Upper Triassic-Jurassic sequence together as the "Mae Moi Group". The upper portions of the Jurassic sequence (Upper Mae Moi Group) are formed by the **Kalaw Red Beds** or the **Khorat Group** (Upper Triassic to (?) Tertiary) and thus clear analogies exist with the Mesozoic sequence in the region S of the Shan States but also with the Namyau Group in the N.

3.122 Jurassic

The following sedimentation pattern can be deduced from the observations made so far:

– The Triassic platform carbonates continued to form in the lowest Jurassic as well (Kachin – Shan Ranges)
– More often a change from marine calcareous to dominantly continental, mainly clastic sedimentation occurred in the Lower and Middle Jurassic
– Marine fossiliferous intercalations were deposited in this clastic-terrigenous series in the Toarcian-Aalenian and Callovian-Oxfordian
– Continental red beds formed during the uppermost Jurassic to Cretaceous.

The area in which this scheme applies extends from the Northern Shan State to the Karen State. In addition, the red beds occur in the Mergui District (Fig. 31). They are comparable with the Khorat Series of the other continental parts of SE Asia. Jurassic sediments are not known in other areas of Burma.

Northern Shan State

LA TOUCHE (1913) observed Jurassic strata in the Northern Shan State in the vicinity of the road from Namtu to Lashio. The brachiopod fauna of the "Namyau Limestones" was described by BUCKMANN (1917). REED (1936a) discussed the Lamellibranchiata. He also (1936) mentioned a similar fauna from the area around Hsenwi. In the Jurassic sequence which he called the **Namyau Group,** BRUNNSCHWEILER (1970, p. 72) distinguished between the **Tati Limestone** (Bathonian-Callovian) containing the brachiopod fauna in the lower section (about 36 m) and the **Hsipaw Red Beds** (Callovian to Upper Jurassic) in the upper section. The red sediments with calcareous conglomerates at their base were estimated by BRUNNSCHWEILER (1970, p. 74) to be 1,100–1,200 m thick. In its uppermost 100 metres this sequence gradually loses its red bed character by a transition into yellow weathered sediments.

GARSON, AMOS & MITCHELL (1976) arrived at different conclusions. They included in the Namyau Group the **Pangno Evaporites** which BRUNNSCHWEILER (1970) assigned to the Triassic. These authors could not delimit the Napeng Beds in which a Rhaetian fauna was found (p. 82). They did confirm the occurrence of thick red beds which merge upwards into grey, brown-weathered shale with thin sandstone beds and grey conglomeratic limestone lenses.

Southern Shan State

Despite the discovery of *Velebitella simplex* KOCHANSKY (normally Upper Carboniferous to Permian), GARSON, AMOS & MITCHELL (1976, p. 37) classified as Lower Jurassic the "Pinnacle Limestone Formation" which merges with calcareous conglomerates into the Kalaw Red Beds (p. 83). According to previous authors, the Jurassic of the area between Kalaw and Heho commences with the clastic, coal-bearing **Loi An Series.** The flora of the Loi An Series, from just above the coal seams, justifies classifying this series in the Jurassic (SAHNI 1937b, p. 387).

According to SAHNI (1937b), the connections with Far East floras of the same age are clearer than those with the floras of the Gondwana Province. Thin layers of limestone containing *Alectryonia* shells and crinoids in the Loi An Series indicate marine intercalations.

In 1968, ammonites were found in the vicinity of Thigyit-Hsi Khip, 96° 44′ E, 20° 20′ N, and near Pinlaung, 96° 46′ E, 20° 07′ N. They are preserved in the form of impressions and casts in a yellow weathered shale. The most likely genus in question is *Tmetoceras* (Fig. 44) which V. BRAUN & JORDAN (1976) obtained also from fine sandy shale of the Upper Mae Moi Group in the Mae Sot area of Thailand and which is widely distributed in the **Aalenian.** In the area of Thigyit a coal-bearing fine sandstone/shale sequence crops out similar to the Loi An Series.

Fig. 44. *Tmetoceras?* sp.; Thigyit-Hsi Khip, 96° 44′ N/ 20° 20′ E, Southern Shan State; Middle Jurassic (Aalenian?). Photo: F. GRAMANN.

MIDDLEMISS (1900) and SAHNI (1936b) regarded the Kalaw Red Beds as equivalent in age to the red beds of the upper Namyau sequence in the Northern Shan State. The "Cretaceous ammonites" which FOX (1930) reported he had found in Kalaw Red Beds have been identified by SAHNI (1936b) as inorganic concretions. GARSON, AMOS & MITCHELL (1976, p. 53) pointed out the possibilities of correlating these red beds with the red beds sequence overlying the Kamawkale Limestone in the Karen State.

The same authors separated out a **"Pan Laung Formation"** from the area W of the Pan Laung Overthrust. On the basis of the fossils found they placed it in the range Upper Jurassic to Cretaceous. The mainly clastic series with terrigenous beds contains individual layers of limestone whose thickness fluctuates rapidly and attains a maximum of 200 m. The total thickness of the Pan Laung Formation is about 2,600 m.

Karen State

The geological investigations carried out in Thailand close to the Burmese border (V. BRAUN & JORDAN 1976) bear importance for the neighbouring Burmese territory as well, because no more recent data are available from the Karen State.

The Upper Triassic Kamawkale Limestone in Burma is overlain by clastic, chiefly red series. HEIM & HIRSCHI (1939) drew attention to the presence of uppermost Jurassic or of Lower Cretaceous limestone

S of Mae Sot in bordering Thailand. BROWN & BURAVAS et al. (1951) and KOMALARJUN & SATO (1964) also mentioned Jurassic marine faunas from the Mae Sot Region. VON BRAUN & JORDAN (1976) were able to define several Jurassic stages in the Upper Mae Moei Group:

Upper Toarcian with	*Pseudolioceras* sp., *Lytoceras (Alocolytoceras) ophioneum* cf. *toarcense* GECZY, *Haughia* sp.
Aalenian with	*Tmetoceras* cf. *dhanarajatai* SATO
Upper Aalenian to Lower Bajocian with	*Eudmetoceras (Planammatoceras)* sp.
Lower Bajocian, Sowerby Zone with	*Docidoceras* (D.) *longalvum* (VACER) *Sonninia?* sp.
Middle to Upper Oxfordian with	*Epimayaites* cf. *falcoides* SPATH *Phylloceras* sp. *Glochiceras?* sp. "*Posidonomya alpina* GRAS."

Rapid facies changes can be observed in the Upper Mae Moei Group. Near Kamawkale in Burma, the group consists mostly of biostromal limestone, whereas at Mae Sot in Thailand it takes the form of a mainly arenaceous-argillaceous sequence.

A stratigraphic sequence comparable to the Kalaw Red Beds or the Khorat Red Series overlies the sediments of the Mae Moei Group. Because of the superposition on strata containing Oxfordian fauna, V. BRAUN & JORDAN (1976, p. 16) thought it very likely that this sequence is Cretaceous in age. This is quite possible in Burma, too, for at least parts of the clastic red sequence on top of the thick Tethys carbonates.

Equivalents of the Kalaw Red Beds were mentioned by RAU (1933, Pl. 9) from the area N and E of Mergui, too.

3.123 Cretaceous

Cretaceous sedimentary rocks occur in W and N Burma. The view that fossiliferous marine Cretaceous exists also in the Shan-Tenasserim Highlands is based on a misinterpretation of concretions from the Kalaw Red Beds (p. 83).

Since CLEGG's (1941 a) synoptic presentation of the distribution of the Cretaceous in Burma, further paleontological evidence has been accumulating. Many questions, however, are unanswered so that any attempt to draft a picture of the paleogeographical evolution during the Cretaceous is still very vague. One important problem is the source of the allochthonous Cretaceous rocks in the Indo-Burman Ranges and of their W foreland. After BRUNNSCHWEI-LER (1966), the view is held that the region of origin is a geotectonic unit which in the meantime has by and large disappeared and which would have to be sought at the boundary between the Indo-Burman Ranges and the Inner-Burman Tertiary Basin adjoining to the E.

It is still not certain whether Cretaceous sedimentary rocks can be assumed everywhere below the Tertiary sequences in the Inner-Burman Basin, and if so, how they are constituted.

Pre-Albian

A stratigraphic gap appears to exist between the Triassic and the Cenomanian in W Burma. There is no paleontological evidence of older Cretaceous rocks. The *Orbitolina*-bearing beds of the second defile of the Irrawaddy (p. 89) are younger than originally thought and can therefore be dealt with under the "Cenomanian".

Albian-Cenomanian-Turonian

Cenomanian rocks were identified by THEOBALD (1873 d), following the discovery of a specimen of *Inflaticeras inflatus* (SOWERBY) near the village of Maie (=Mai-I) in the Sandoway District. This fossil might have come, however, from an exotic limestone block in the Arakan flysch sediments. Later, a *Mammites daviesi* SPATH 1936 was found near the Konbwe Bay of Ramree, an area which also contains Cretaceous only in the form of exotic rocks. In the Arakan Coastal Area, allochthonous Cretaceous rocks occur as *Globotruncana*-bearing limestone of the Campanian-Maestrichtian, which is amply represented, and as mottled cherty limestone containing both Foraminifera and radiolarians of the Albian to Cenomanian.

Autochthonous Cretaceous rocks seem to be limited to the E foothills of the Indo-Burman Ranges and to the steep-walled valleys in the upper reaches of the Irrawaddy. From S to N the areas in question are as follows:

According to unpublished geological surveys, late Upper Cretaceous *Globotruncana* limestone is extensively exposed in the area **SW of Bassein.** Further to the W, and S of Ywahaungyi, a fossil-bearing dark limestone was collected in 1972. The limestone bears *Turrilites* cf. *circumtaeniatus* KOSSMAT 1895, *Placenticeras* sp. and *Acanthoceras* (?) sp. and therefore dates from the Cenomanian. The fossils from Ywahaungyi are very similar to those of the Middle Utatur Stage (Cenomanian) of the Cretaceous from the Trichinopoly District of S India.

Another area with autochthonous Cenomanian rocks was found by GRAMANN 1973 at the E edge of the Indo-Burman Ranges in the **vicinity of Mt. Victoria.** In the valleys of streams draining into the Yaw River, dark limestone, cherty limestone and bituminous platy limestone of varying thickness are exposed ("Kabaw Group", Figs. 31, 45 a). They have yielded the following fossils in Kyi Chaung:

Vertebrata, Actinopterygia:

> *Neopachycormus birmanicus* TAVERNE 1977
> *Benthesykime* sp.

Cephalopoda, Ammonoidea:
> *Turrilites* cf. *costatus* (LAMARCK, 1801)
> *Puzosia* cf. *denisoniana* (STOLICZKA, 1865)
> *Acanthoceras* cf. *turneri* WHITE 1889

Foraminifera:
> *Hedbergella amabilis* LOEBLICH & TAPPAN, 1961
> *Hedbergella* cf. *planispira* (TAPPAN, 1940)
> *Heterohelix globulosa* (EHRENBERG, 1840)
> *Rotalipora evoluta* (SIGAL, 1948)
> *Schackoina cenomana* (SCHACKO, 1896)

The megafossils come from a higher level, while the microfauna comes from near the base of the exposed section. Below it, the Upper Triassic *Halobia*-bearing beds appear (GRAMANN 1974, pp. 284–285).

Fig. 45 a. Limestone of Kabaw Group; Kyi Chaung, upstream of Kyi Village, NW of Kyauktu, Pakokku District; Cenomanian. Photo: F. GRAMANN. – b, Thin section with *Siderolites* sp. in limestone of Kabaw Group; Cretaceous; Ché-Chaung near Yeshin Village, SW of Kyauktu, Pakokku District; Maestrichtian. Photo: F. GRAMANN.

Indications of pre-Maestrichtian rocks of the Cretaceous in this area were *Nerinea* sp. and *Trochalia* sp. from the confluence of the Kyaw Chaung with the Kyauksit Chaung just upstream from Kachaung (COTTER 1938, p. 32). This marine, calcareous-siliceous sequence, which commenced possibly in the Albian and extends perhaps into the Turonian, is overlain by clastic sedimentary rocks with fossil leaves. They are in turn overlain by limestone containing Maestrichtian marine fossils (Fig. 45 b).

In the **area W of Gangaw,** so far only Campanian-Maestrichtian rocks have been observed (p. 89).

Orbitolina birmanica SAHNI (1937 a) (= evidence of Cenomanian, according to SAHNI & SHASTRY 1957) in the second defile of the Irrawaddy near Zinbon, downstream from Bhamo, is important for interpreting the regional geology. According to CLEGG (1936, p. 353), below the *Orbitolina birmanica*-bearing beds is a sequence of chocolate-coloured mudstone of unknown age which in turn is underlain by the Chaung Magyi Group (p. 51). The *Orbitolina* are found in calcareous sandstone which belongs to a sequence of sandstone, limestone and shale with a total thickness of more than 300 m.

This rock sequence is intruded by serpentinites. They are both discordantly overlain by clastic, presumably terrestrial Tertiary rocks.

CLEGG (1941) described a sequence of rocks containing macroforaminifera and *Lithothamnia* from the first defile of the Irrawaddy below Tonbo (24° 24′ N, 97° 07′ E). In his opinion it is of the same age as the *Orbitolina*-bearing strata of the second defile. However, the reference to *Dictyoconus* and *Lithothamnia* would indicate that the age is more likely to be Lower Tertiary. Reworked *Orbitolina* limestone was described from the Hukawng Valley (SAHNI & SHASTRY 1957).

Campanian-Maestrichtian

THEOBALD (1873 d) already knew about the widely distributed *Globotruncana*-bearing limestone in W Burma. He described it in an unmistakable way as "flea-bitten" limestone. NAGAPPA (1959, p. 165) recognized it as *Globotruncana* limestone of the late Upper Cretaceous from an occurrence on the island of Ramree. Allochthonous *Globotruncana* limestone has also been found in the Arakan Coastal Area and in the area of the Indo-Burman Ranges. The pelagic, micritic limestone occasionally contains red or green marl layers and appears to have been originally associated with red shale. The size of the exotic material varies greatly from rocks of the size of nuts to large blocks, some of which are being quarried, while pagodas have been built on top of others. The transition of the limestone to red shale is apparent in a large occurrence 1 km S of "mile 70" on the road leading to the Taungup Pass. Other exotic rocks are in close contact with ophiolite blocks. The pelagic limestones W of the Taungup Pass and those of the Pagoda Mountain of Bawlaba on Ramree contain an early Maestrichtian foraminiferal fauna with *Globotruncana arca*. Along with rare younger faunas, Foraminifera from the late Upper Cretaceous are contained as reworked fauna in many samples of microfossils from the Arakan Coastal Area.

Since BRUNNSCHWEILER (1966) described autochthonous *Globotruncana* limestone in the area of Gangaw, this region, together with its extension in the direction of the strike, has been regarded as the area from which the allochthonous *Globotruncana* limestone originated. The *Globotruncana* limestone of Gangaw is a part of the **Rangfi Formation** of Campanian-Maestrichtian age. COTTER (1938) described also *Orbitoides*-bearing limestone from this area. This leads over to the late Upper Cretaceous rocks of the Mt. Victoria foothill zone above the village of Kyauktu (Fig. 45 b). They contain limestone with *Siderolites* and *Orbitoides,* and the **"Kabaw Shale"** in which one finds *Trigonia* cf. *scabra* (SOWERBY) together with *Lepidorbitoides (Asterorbis)* sp., *Globotruncana arca* (CUSHMAN) and *Pseudotextularia elegans* RZEHAK, 1895 (GRAMANN 1974, p. 285). These sediments were deposited in a marine environment closer to the coast than the *Globotruncana*-bearing limestone.

The age of the probably exotic radiolarite from Kyauknimaw on Ramree has still not been precisely determined. Radiolarites were also observed together with *Globotruncana* limestone as exotic blocks in the southernmost outcrops of the Arakan Flysch Sediments.

3.124 Tertiary

THEOBALD (1869) proposed a general classification of the Tertiary rock sequences into the nummulite-bearing Eocene, the marine "Pegu Group", and the non-marine Plio-Pleistocene

"Fossil Wood Group" of the "Irrawaddian". In the following years, research concentrated on the oilprovinces of the Inner-Burman Tertiary Basin.

In the Arakan Coastal Area, as well as in the Inner-Burman Tertiary Basin, the dominantly clastic sediments of the Tertiary attain thicknesses of more than 9,000 m (Fig. 31). Calcareous sediments developed only occasionally. While turbidites formed in the Early Tertiary in the Arakan Coastal Area and in the area of the present Indo-Burman Ranges, changes from shelf sedimentation to delta sedimentation took place during the period from the Paleocene to the Upper Miocene in the Inner-Burman Tertiary Basin. During the Upper Miocene to Pliocene, non-marine clastic sediments were laid down.

So far, only Early Tertiary sequences are known from the Indo-Burman Ranges. The Tertiary sequence is more or less complete in the Arakan Coastal Area and in the W part of the Inner-Burman Tertiary Basin, while the Tertiary of the Sino-Burman Ranges is apparently limited to the Late Tertiary fillings of intramontane basins.

Attempts at correlations run into difficulties because a regionally applicable Tertiary biostratigraphy is lacking or contains large gaps.

At present, the biostratigraphy is based almost exclusively on microfossils. The investigations were first concentrated on the benthonic Foraminifera, but the results have not been published. There is no definition of a zonal scheme, and no attempt to link the open nomenclature with documented Foraminifera of the region has been published. Some rock units have meanwhile been correlated with the zonal scheme of the planktonic Foraminifera. However, good planktonic foraminiferal faunas are rare, and very little nannoplankton is contained in the marine sediments of the Inner-Burman Tertiary Basin. On the other hand, palynological investigations, in particular on non-marine strata in the Chindwin Basin, have yielded good results. Vertebrate faunas of the Eocene Pondaung Formation and of the Miocene to Pleistocene Irrawaddy Group are known from old publications, but they should be revised and checked against material that has been collected from known geological horizons. This would be particularly desirable for the Irrawaddy Group because there are indications that vertebrate faunas of different ages exist within this group, which in its lower portion might be of Upper Miocene age.

A detailed lithostratigraphic scheme of the Tertiary rock sequences exists for the central parts of the Inner-Burman Tertiary Basin, although the lithologic breaks are frequently inconspicuous. Geological interpretation of aerial photographs and of satellite imagery have permitted progress to be made in applying this lithostratigraphic scheme over most parts of the basin.

Attempts to classify the rock sequences of the Indo-Burman Ranges on the basis of the lithostratigraphy have so far failed to yield satisfactory results. The sequences E of the Inner Volcanic Arc have not so far been studied to any great extent; Early Tertiary rocks seem to be missing there. An attempt was furthermore made to distinguish lithostratigraphically the Tertiary sedimentary rocks of the Pegu Yoma, without much progress so far. The Tertiary rocks of the intramontane basins in the Sino-Burman Ranges and in N Burma (Hukawng Basin, Putao Basin) are to a great extent unknown.

Paleocene

THEOBALD (1873) mentioned *"Cardita beaumonti"* and thus provided the first evidence of Paleocene rocks in Burma. Strata containing *Venericardia beaumonti,* which since NAGAPPA's work (1959) must be regarded as Lower Paleocene, are known to occur in the "Western Outcrops" of the Thayetmyo and Minbu districts (Fig. 17 a, folder). For a long time they were regarded as Cretaceous. The Cretaceous "Kabaw Shales" of the Pakokku District presumably also include some Paleocene strata in their uppermost portion.

The **Paunggyi Conglomerate (Paunggyi Formation)** plays a special role in separating the Tertiary of the Inner-Burman Tertiary Basin from the Upper Cretaceous. However, it is not clear whether it is the same conglomerate of the same age each time that is involved. NAGAPPA (1969) stated that he found *"Discocyclina* and mollusca of Ranikot affinities" in the Paunggyi Conglomerate, and he regarded the latter as Paleocene. The type area is the Minbu District. To the N of it, sediments which have been mapped as Paunggyi Conglomerate contain Eocene fauna. Like COTTER (1938/40), NAGAPPA (1969) believed it possible that the "Laungshe Shales" contained a Middle to Upper Paleocene component. The type fossil in the Laungshe Shale is, however, the Lower to Middle Eocene *Nummulites atacicus.*

In the Kyauktu-Saw area of the Central Basin (Fig. 27, folder), so far the only rocks that could belong to the Paleocene are the volcanoclastics containing graded tuff, soapy clay and subordinately siliceous rocks, which have been observed in the Upper Yaw Chaung.

In the Arakan Coastal Area, Paleocene has been found only in the form of an olistolith of shallow-water limestone. The olistolith contains Maestrichtian *Globotruncana* limestone and Paleocene algal limestone with *Ranikothalia,* so that a hiatus seems to exist between the Maestrichtian and Upper Paleocene (HELMCKE & RITZKOWSKI, in press).

So far no evidence of Paleocene rocks has been found in the Indo-Burman Ranges.

Eocene

Eocene nummulites have long been known from the Arakan Coastal Area. Studies by HELMCKE & RITZKOWSKI (in press) on the islands of Ramree and Cheduba have shown that olistoliths of Lower Eocene nummulite limestone occurred together with blocks of nummulite-bearing sandstone. Middle Eocene turbidites with layers of chert are regarded as autochthonous.

The **Ngapali Formation** (Fig. 47) of the Arakan Coastal Area – an argillaceous-clastic sequence containing olistoliths – yields planktonic Foraminifera of the Eocene. In addition, apart from Cretaceous olistoliths, Eocene limestone also occurs as an allochthonous element in the coastal region. Nummulitic sandstone is part of the rock sequence on the islands of Ramree and Cheduba and is referred to as the **Kyaukale Formation.**

The turbidites of the Arakan Yoma along the Padaung-Taungup Road very rarely contain large Foraminifera of Eocene age (Fig. 46). It can be assumed that a major portion of the turbidite sequences of the Indo-Burman Ranges are Early Tertiary in age and contain a considerable amount of Eocene.

The Eocene of the Inner-Burman Tertiary Basin is better known. From base to top, it can be subdivided into the Laungshe, Tilin, Tabyin, Pondaung and Yaw Formations. The **Laungshe Shale** (L. Formation) contains layers of sandstone and conglomerate and very occasionally macroforaminifera-*Lithothamnia* limestone. Together with *Nummulites atacicus* LEYMERIE 1846 (**Lower to Middle Eocene**) the following planktonic Foraminifera were found above the village of Kyauktu:

7 *

Fig. 46. *Lenticulina* sp. Nummulitic sandstone, Eocene. Indo-Burman Ranges, roadcut near summit Pass Road Padaung-Taungup. Photo: F. GRAMANN.

Fig. 47. Steep dipping sandstone, siltstone and mudstone with somewhat silicified concretions; at beach approx. 3 km N of Ngapali; Ngapali Formation, Eocene. Photo: F. BENDER.

Globorotalia aequa CUSHMAN & RENZ
Globorotalia aequa simulatilis (SCHWAGER)
Globorotalia pseudotopilensis (SUBBOTINA)
Globorotalia spinuloinflata (BANDY).

The **Tilin Sandstone** has been described from the region of the Pakokku-Tilin Road where it is at least 1,500 m thick and forms an extensive range of hills. S of Sidoktaya the Tilin Sandstone is only occasionally found. The upper portion of this marine sandstone sequence shows terrestrial influence: silicified wood, red beds and gravel. The rare fauna belongs to the **Eocene.**

The **Tabyin Clay** was described by COTTER (1914) from the village of the same name about 20 km W of Pauk, as dark, frequently spherically weathered clay containing particles of coal. E of Gangaw (Kunze Chaung), the upper 200 m of the Tabyin Clay contain platy limestone beds and lignite seams within dark-bluish clay. Thin layers of sandstone lead over to the Tilin Sandstone below and to the Pondaung Sandstone above. *Nummulites acutus* SOWERBY is regarded as the type fossil of these strata, which contain very few fossils. A correlation with the Indian Khirtar Stage (Middle Eocene) is assumed. COTTER (1938) observed tuffaceous material in fine sandstone layers from Myaukmakyin. The content of fine sandstone layers increases towards the N. The thickness of the Tabyin Clay varies around 1,200 m.

Since the Tilin Sandstone is missing N of the Myittha River, it is difficult to classify a rock sequence known as the **Tabyin-Laungshe Formation** which exists here. The sequence was deposited in a shallow, mainly marine environment and bears Eocene macroforaminifera. Layers of conglomerate and small seams of coal are evidence of continental influences on the depositional environment. The separation of this formation from the Cretaceous-Paleocene Kabaw Formation is not always certain because the Paunggyi Conglomerates may be missing. A Paleocene component is also probably present.

The **Pondaung Sandstone (Pondaung Formation)** forms frequently impressive ranges of hills. The Nwamataung Range in the N Minbu District, Yeyodaung and Dudawtaung in the area of the border with the Pakokku District, the Pondaung and Ponyadaung Ranges in the Pakokku District and in the S of the Chindwin Basin are mentioned by COTTER (1938, p. 52) as Pondaung Sandstone Ranges. Also the Mahudaung Range is made up by Pondaung Sandstone.

Apart from sandstone with layers of conglomerate, the lower portion of the Pondaung Formation contains clay layers which lead over to the Tabyin Clay. In its upper portion it contains argillaceous-earthy red beds from which a large number of species of vertebrate fossils have been collected. Despite the presence of silicified woods, the portion of the formation below the red beds appears to be marine and contains occasional marine molluscs.

The vertebrate faunas come from the area around the village of Kyawdaw (22° 01′ 40″, 94° 29′ 20″). Most of the fossils are Anthracotheriidae (PILGRIM 1928). Very important among this fauna are the finds of *Amphipithecus mogaungensis* COLBERT and *Pondaungia cotteri* PILGRIM, as they have been regarded as earliest primates. Whereas *Amphipithecus mogaungensis* has been recently regarded as a lemuroid (SZALAY 1970), a newly collected well preserved fragment of a lower jaw with second and third molar teeth confirms PILGRIM's view that *Pondaungia cotteri* is the earliest known anthropoid primate (BA MAW et al. 1979). Since the fauna contains many species, comparisons with other Early Tertiary faunas are indicating a **Middle to Upper Eocene** age.

The **Yaw Shales (Yaw Formation)** (COTTER 1914) from the Yaw River area commence with a distinct lithological break above the Pondaung Sandstone. They consist of argillaceous strata with occasional limestone beds and with increasing sandstone beds towards the top, deposited in a marine environment. The macroforaminifera *Nummulites yawensis* COTTER 1914 (from the group of *Nummulites striatus* (BRUGUIÈRE)), *Discocyclina sella* (D'ARCHIAC), *Operculina* cf. *canalifera* D'ARCHIAC & HAIME and rich molluscan faunas with *Velates perversus* (GMELIN) indicate an **Upper Eocene** age.

In the Chindwin Sub-basin, the bituminous coal (p. 170) W of Kalewa is regarded as a member of the Yaw Formation. POTONIÉ (1960) gave an inventory of its pollen content. From the upper seam KOTAKA & UOZUMI (1962) described *Pachymelania*, a genus of gastropods that lives in lagoons and estuaries in Africa.

The non-marine **Eocene sedimentary rocks at the E side of the Chindwin Basin** are divided into numerous lithostratigraphic units. Reliable correlations have been possible since (from 1977 onwards) pollen analysis has been used not only on the marine strata on the W side of the Basin but also on the limnic-fluviatile sequences in the E.

Oligocene

Oligocene marine sediments were first identified in the Minbu area. For a long time, it was expected that the Oligocene would be lacking in the other sedimentation regions. In the meantime, Oligocene strata have also been identified in the Arakan Coastal Area (HELMCKE & RITZKOWSKI, in press) and in the Chindwin Basin.

From base to top, the stratigraphic sequence in the Salin Syncline is divided up lithostratigraphically into the Shwezetaw, Padaung and Okmintaung Formations. They form the lower portion of THEOBALD's (1869, p. 80) "Pegu System", which also includes Miocene formations.

The boundary between the Eocene and the Oligocene may be drawn within the **Shwezetaw Sandstone** (named after the Shwezetaw Pagoda on the Man Chaung, 20° 7' N, 93° 34' E). *Globularia (Ampullinopsis) birmanica* (VREDENBURG 1922) is regarded as the type fossil, although it also occurs in the Yaw Formation. The lower portion of the sequence is calcareous and contains Foraminifera. A microfauna from the zone of *Globorotalia cerroazulensis* was observed here, which is evidence of an Upper Eocene component. The upper portion of the formation usually contains no calcareous matter, few Foraminifera and some small coal seams.

The **Padaung Formation** commences with up to 1500 m thick shale. In its upper part, thin sandstone beds (Fig. 48) with rich molluscan faunas are intercalated. Corals also occur as stenohaline-marine organisms. Towards the top, the sandstone layers become thicker, leading on to the overlying Okhmintaung Formation which makes the definition of the contact between both formations uncertain. In addition, the molluscs and the microfossils are to a large extent identical in both formations. Thus it is clear that the boundary is purely lithostratigraphic and heterochronous. According to EAMES (1951, p. 382), the gastropods *Lyria varicosa* VREDENBURG, *Neoathleta theobaldi* (VREDENBURG) and *Turricula (Orthosurcula) birmanica* (VREDENBURG) are restricted to the Padaung Clay. According to KYI MAUNG (1970, p. 76) the *Globorotalia opima* zone (P 21/N 2) of the Upper Oligocene is contained in the Padaung Clay of the Minbu area.

Fig. 48. Millstones of Padaung Sandstone; Gwegyo E of Chauk; Middle Oligocene. Photo: D. BANNERT.

Fig. 49. Chauk oil field; view from N along crest of structure; sandstone and shale; Upper Oligocene. Photo: F. BENDER.

Foraminifera faunas rich in species and containing planktonic Foraminifera, as assemblages of the outer shelf, occur mainly in the argillaceous, lower section of these sequences and are replaced in the upper sections by shallow-shelf faunas with *Ammonia, Florilus* and aggluti-nating Foraminifera.

The "Tiyo Formation" is a particularly argillaceous equivalent of the Padaung and Okhmintaung formations in the Minbu and Thayetmyo districts. It is possible that it also contains a component that is the same age as the Shwezetaw Sandstone.

The separation of the **Okhmintaung Formation** (Fig. 49) from the underlying Padaung Formation is rather questionable. Because of the similarity between the faunas, paleontologi-cal criteria cannot be used to define the boundary (EAMES 1951, p. 382). This goes not only for the molluscan faunas, but also for the microfaunas including the ostracods (GRAMANN 1975 a). Satellite imagery interpretation in the area of the Western Outcrops (Fig. 27, fol-der) indicates, however, that such contacts can be defined on the basis of lithological criteria. Locally, the large Foraminifera *Heterostegina* and *Lepidocyclina* occur in the Okhmintaung For-mation, and in some places they may accumulate to form *"Lepidocyclina* limestone". The best known occurrence is the *Lepidocyclina* limestone of Myinmagyitaung Mountain below Thayetmyo on the W bank of the Irrawaddy. It is quarried for cement raw material (p. 206).

Shale of the Pyawbwe Formation (Miocene) overlies the Okhmintaung Formation in the Minbu area.

In the Arakan Coastal Area, Oligocene sediments were observed for the first time in 1974. Turbiditic shale sequences on the island of Ramree contain Oligocene planktonic Foraminife-ra and nannoplankton in addition to varying contents of allochthonous Cretaceous and Eoce-ne faunal elements. The planktonic Foraminifera indicate an age in the range from P 21 to N 2–3 = *Globigerina suturalis* Zone to *Globigerina angulosuturalis/Globorotalia opima* Zones. The nannoplankton yields mainly NP 25 = *Sphenolithus ciperoensis* Zone. The paleobathymetric evaluation points to a sedimentation at great depths (HELMCKE & RITZKOWSKI, in press).

Pollen analysis of sediments of the "Western Outcrops" of the **Chindwin Basin** com-pared with those of the Minbu Basin reveals that the strata which were mapped as the Yaw, Tonhe and Letkat formations contain pollen associations that occur in the S in the Shweze-taw Formation and in the Padaung Formation. Thus, for the first time, evidence was obtained that Oligocene strata also exist in the Chindwin Basin.

Miocene

Marine Miocene sediments occur in the Arakan Coastal Area and in the Inner-Burman Ter-tiary Basin.

In the Salin Sub-basin the marine Miocene is divided into the Pyawbwe, Kyaukkok and Obogon formations. In the Irrawaddy-Delta-Sub-basin these formations have marine equiva-lents, while strata of the same age in the Chindwin Basin have been for the most part formed under non-marine conditions. The fossil content of the Pyawbwe Formation permits a correla-tion with the Telisa Formation of Sumatra.

The lower portion of the Miocene sequence of the islands in the Kyaukpyu Archipelago and in the **Arakan Coastal Area** is developed in flysch-type facies. The contact between flysch-type sediments and sediments which were deposited in marine shallow water is formed by the **Nga-Ok Conglomerate** of the island of Ramree. In the Nga-Ok area the conglomerate, which mainly consists of flysch-type sandstone, is overlain by clay with Foraminifera of the (N 8) – N 9-Zone = *Orbulina suturalis/ Globorotalia peripheronda* Zone of the **Lower Miocene** (HELMCKE & RITZKOWSKI, in press). The foraminiferal faunas are rich in benthonic species with calcareous tests and planktonic Foraminifera. They contain only a small quantity of allochthonous elements from the Cretaceous, Eocene and Oligocene. In the area of Sane the Lower Miocene is represented by sediments of the *Globigerinoides primordius – Globorotalia kugleri* Zones. The **Upper Miocene** which is exposed on Saddle Island, on the W coast of Ramree and on Cheduba, is sublittoral. The youngest member is the Leikhamaw Sandstone containing a *Spongeliomorpha* ichnofauna that is characteristic of coastal sands. All the fossiliferous strata of the Upper Miocene yield Zone N 16– N 17 of the BLOW (1969) scheme of planktonic Foraminifera or NN (10) 11 of nannoplankton, and belong to the **Early Upper Miocene.**

Upper Miocene sediments are exposed further to the N in the same lithofacies as on Ramree and extend into the Akyab region (Boronga Anticlines) as well as into Bangladesh.

Miocene sedimentary rocks in the **Minbu area** commence with the argillaceous **Pyawbwe Formation.** The shale contains *Calliostoma singuense* (NOETLING). From the sandstone of Monatkon, EAMES (1951) mentioned the Lamellibranchiata *Anadara peethensis* (D'ARCHIAC), *Dosinia protojuvenilis* NOETLING, *Indoplacuna birmanica* (VREDENBURG), *Paphia protolirata* (NOETLING). The faunally indicated hiatus between Oligocene and Miocene is expressed in local discordances and layers of conglomerate. In the Mann oil field the Oligocene/ Miocene contact has been revealed by drilling to be a concretionary conglomerate a few centimetres thick. It consists for the most part of detrital sediments. Within the clay of the Pyawbwe Formation, several concretionary layers are developed in the Chauk oil field. These hardground layers contain holes of stoneboring organisms which are evidence of interruptions in the sedimentary process within this formation. Individual corals which point to conditions of normal salinity occur relatively frequently. In the Prome-Embayment-Irrawaddy-Delta-Basin the *Miogypsina* limestone of the Kyanging Tondaung W of Myanaung is an equivalent of the Pyawbwe Formation. The "Kama Clay" from the W bank of the Irrawaddy at 19°02′N also belongs to the Pyawbwe Formation.

The ostracod fauna found in this formation is clearly Miocene in character. In the Western Outcrops the Miocene, commencing with the Pyawbwe Formation, contains presumably allochthonous, corroded radiolarians which might be evidence of an uplifting of the Arakan Yoma.

The **Kyaukkok Formation** is more sandy than the Pyawbwe Formation and is characterized by molluscan faunas rich in gastropods with *Conus (Leptoconus) bonneti* COSSMANN and the bivalves *Acropsis bataviana* (MARTIN), *Loxocardium minbuense* (NOETLING) and *Trachycardium subrugosa* (NOETLING). Claws of *Protocalianassa,* which indicate a shallow-water environment, are found relatively frequently and were illustrated by NOETLING (1901, Pl. 24) as *Calianassa birmanica.* A considerable portion of the molluscan fauna that occurs in the Pyawbwe Formation is missing in the Kyaukkok Formation (EAMES 1951, p. 383).

The **Obogon Formation** is an alternating sequence of fine sands with fine sandy clays. According to EAMES (1951), the molluscan faunas are not very characteristic. Stratification phenomena and trace fossils indicate that the fine sand layers represent beach sands. The occurrence of this formation is restricted to the area S of 21° 30′.

The Obogon Formation belongs to the *mayeri* zone (BOLLI) of the **Middle Miocene** (KYI MAUNG 1970).

Fig. 50. "Cut and fill" structure in alternating clay and sandstone near Tamanthi, N Chindwin Basin; Miocene. Photo: D. BANNERT.

Marine Miocene has also been found in the area of the **Prome Embayment** and of the **Ir-rawaddy Delta.** Here, the limestone of Kyanging Tondaung W of Myanaung in the Henza-da District is developed and it contains *Miogypsina, Lepidocyclina, Cycloclypeus, Heterostegina* and *Spiroclypeus.* This limestone was previously mistakenly thought to be of the same age as the *Lepidocyclina* limestone of Thayetmyo. According to SAING (1969), shale intercalations in the limestone from the Henzada District contained planktonic Foraminifera from the region of the *Catapsydrax stainforthi* zone of the Early Miocene to the *Globorotalia fohsi fohsi* zone of the early Middle Miocene (=N6 to N10 according to BLOW 1969). The *Miogypsina* limestone is regarded as an equivalent of the Pyawbwe Clay and of the "Local Zone No. 4" of KYI MAUNG & MAUNG MAUNG THA (1968) in the Myanaung area. This is also confirmed by the correlation with the Telisa Formation of Sumatra (=upper "Te" of the Indonesian "Letter-Classification").

The clays of Sitsayan (18° 55′ N, 95° 8′ 30″ E) on the bank of the Irrawaddy, which have been recognized as the equivalent of the Pyawbwe (EAMES 1951, p. 386) are important for the molluscan stratigraphy.

In the Irrawaddy Delta Area, marine sediments of Miocene age are interbedded with vol-canoclastic rocks which were encountered in several exploratory oil wells.

The lithological differences which make it possible to subdivide the Miocene into forma-tions in the Salin Sub-basin are not very pronounced in the delta region. The youngest marine stratigraphic member of the Tertiary is here, too, composed of sediments of the Middle Miocene *mayeri* zone (BOLLI 1966) = N14 after BLOW 1969. The "Local Zone No. 4" of the delta region apparently correlates with the Obogon Formation (GRAMANN 1975a, p. 6).

In the **Chindwin Basin,** Miocene strata have with the exception of the Natma Formation so far been assumed to exist only on the basis of palynological studies.

The predominantly argillaceous **Natma Formation** in the S part of the Chindwin Basin becomes increasingly arenaceous towards the N (from Kalewa onwards). From 23° 10′ N onwards the formation is predominantly sandy; at 23° 40′ N it contains conglomerate and finally, towards the "24° N Uplift" it wedges out as clay with conglomerate (Fig. 50).

Palynological studies have shown that the clay penetrated by drilling ("Indaw Clay") in the Indaw oil field can be correlated with the Natma Formation. It contains pollen from mangrove plants as does the Natma Formation, along with Foraminifera and Hystrichosphaeriidae.

The **Shwethamin Formation** is characterized by a thick sandstone series with clay layers in its lower portion. The remains of leaves, fossilized wood and thin red earth horizons emphasize its terrestrial depositional environment. Presumably a multiple sedimentation rhythm, commencing in each case with a layer of conglomerate and ending with fossil soil, is present in the Swhethamin Formation.

At the E edge of the Chindwin Basin, there are thick Miocene sedimentary sequences that have been given the names "Nandawbee", "Shauknan" and "Kaungton Formation". Palynological investigation has revealed that they contain no marine fossils either.

The **Pegu Yoma** is composed of clastic, folded Tertiary sediments. Apparently, only Miocene is exposed; it contains marine fossils. Poor accessibility, unfavourable exposure conditions and tectonic deformation have so far hampered attempts to work out a reliable biostratigraphical subdivision of this sedimentary sequence.

3.125 Late Tertiary and Quaternary

During this period, exclusively terrestrial, chiefly fluviatile but – in intramontane basins – also lacustrine sediments were deposited in what is now the continental area of Burma.

The **Irrawaddy Group,** with its fluviatile sands and gravels, attains particularly great thicknesses of up to more than 2,000 m and it is widely distributed in the Inner-Burman Tertiary Basin.

The change in sedimentation from the marine Pegu Group to the terrestrial Irrawaddy Group was often equated with the Miocene/Pleistocene boundary. COTTER (1938) demonstrated that this facies change is heterochronous.

When using modern stratigraphic criteria, the Irrawaddy Group covers the entire period from the Middle Miocene to the Lower Quaternary (Fig. 51). In the type region, the Yenangyaung oil field, the time interval between the uppermost Obogon Formation (p. 97) and the lowermost strata of the Irrawaddy Group is small.

In the Irrawaddy Group, as in the sedimentologically similar Siwalik Group of the Indian Subcontinent, vertebrate faunas are the only reliable time markers. The faunas of the Irrawaddy Group can be linked with the biostratigraphic sequence of the Siwalik Group, even though the faunas are not always completely identical. It is unfortunate that no comprehensive revision of the Burmese mammalian fossils has yet been carried out and the density of the finds is low (Fig. 51).

The **Mingin Gravels** of the S Chindwin Basin are lithologically identical to the gravels of the Irrawaddy Group. If they are included in the Irrawaddy Group, it can be assumed that

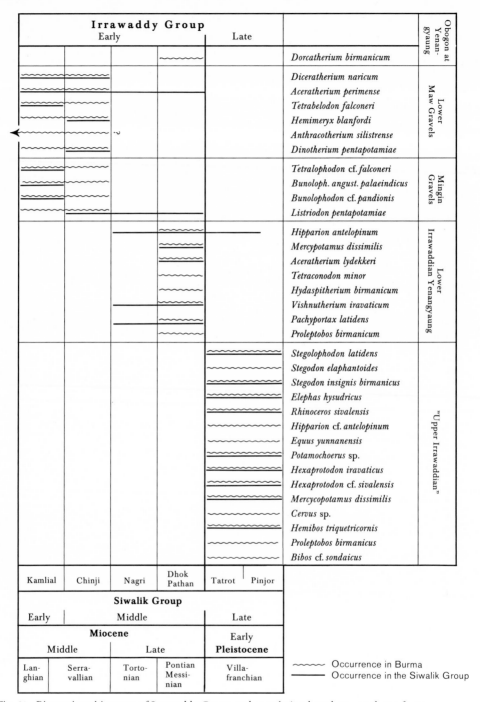

Fig. 51. Biostratigraphic range of Irrawaddy Group and correlation based on vertebrate fauna.

Fig. 52. Fossil Red Soil horizon at base of Irrawaddy Group, overlying with angular unconformity cross-bedded sandstone of Upper Miocene Kyaukkok Formation, dipping SE; Upper Miocene-Quaternary. Approx. 3 km SE of Chauk, roadcut. Photo: F. BENDER.

they are time-equivalent with the upper Pegu Group, as is indicated by the mammalian fauna of Taungbyinnge (22° 50′ 35″ N/94° 49′ 08″ E; COTTER 1938, p. 105) which dates at least from the Chinji Period and thus belongs to the **Middle Miocene.**

The **Maw Gravels** of the Kabaw Valley at the E edge of the Arakan Chin Hills are located outside the coherent distribution area of the Irrawaddy Group, presumably as structurally deformed valley fill material. According to COTTER (1938, p. 108), faunas found near Gangaw date from the Chinji Period (Middle Miocene). Otherwise the Maw Gravels contain faunas of the Dhok Pathan (uppermost Miocene).

If the faunas of Taungbyinnge and of Gangaw are classified in the Chinji Stage of the Si-walik Group (Middle Miocene, approximately Serravallian), then the Mingin Gravels and the Lower Maw Gravels are significantly older than the Obogon Formation of Yenangyaung, where the cranial remains of the specimen of *Dorcatherium* was found, which according to THENIUS (1959, p. 139) is the same age as the Dhok Pathan of the Middle Siwalik Group (uppermost Miocene?).

The Irrawaddy Group of the type region is usually defined in such a way that it should be-gin at the base with a red residual soil horizon 1.5–2 m thick (Fig. 52), which can be traced over large areas of the steppe region of the Yenangyaung oil field. This red soil horizon is un-derlain in places by up to 20 m thick non-marine clastics with a small coal seam, signs of bleaching, and a wedge shaped clay bed containing *Batissa crawfordi*, a brackish-water bivalve related to Cyrenae. It can be assumed that a gradual transition occurs from the marine sediments of the Pegu Group to the fluviatile sediments of the Irrawaddy Group.

The red soil horizon frequently contains pebbles which have been ferritized or silicified to the point where the original lithology is no longer recognizable. It seems to contain fragments of rhyolite. Remains of reptiles, in particular the dermal bones of crocodiles, carapace fragments of turtles – *Colossochelys atlas* as well as *Trionyx* sp. – are found more frequently than mammalian bones. The mammalian fauna from the basal portion of the Irrawaddy Group of Yenangyaung has been interpreted in a variety of ways. While mention is mostly made of one single fauna that is supposed to date from the Dhok Pathan (uppermost Miocene), STAMP (1922 c) assumed that *Stegolophodon latidens* and *Hexaprotodon irravaticus* came from a more recent horizon. He therefore regarded the basal red beds of Yenangyaung as an equivalent of the Indian Nagri Stage. However, according to PILGRIM (1927, p. 160), the faunas that do not contain *St. latidens* and *H. irravaticus* are not older than Dhok Pathan. COLBERT (1938, p. 277) also agreed with this view. Dhok Pathan is in turn regarded as being of the same age as the Turolian or the Pontian of the Paratethys. According to modern classification, these units belong to the uppermost Miocene or the Messinian. As far as is known, faunas of definitely Pliocene age are lacking not only in the Irrawaddy Group but also in the Siwalik Group of the Indian Subcontinent.

The faunas that contain *Stegolophodon latidens* and *Hexaprotodon irravaticus* are known from many localities (COLBERT in DE TERRA & MOVIUS 1943, p. 424). They constitute the youngest faunal complex of the Irrawaddy Group and are usually correlated with the **Lower Pleistocene Tatrot stage** within the Upper Siwaliks.

The entire Irrawaddy Group is known for its wealth of silicified wood ("Ingyin Kyauk" in Burmese). *Dipterocarpus,* which confirms this Burmese name is very frequent. Fossil palmwoods are rarer.

N of the Yenangyat structure, the Irrawaddy Group contains at least 10 red soil horizons which can be traced in aerial photographs; one of the horizons is more than 10 m thick. Most of these layers are earthy and contain light-coloured carbonate concretions. In addition to cross-bedded sand, gravel and red soil horizons, the lower portion of the Irrawaddy Group also contains thick light-coloured silt to fine sand layers which are frequently solidified into concretions or sandstone beds. Hollow hematitic or limonitic iron concretions also occur in layers, often in combination with red horizons.

The upper Irrawaddy Group contains fewer red soil horizons; it is more gravelly and seldom consolidated. Eolian components appear to be lacking in both groups of strata, nor are there any wind-shaped pebbles and desert varnishes, which are sedimentological criteria that point to eolian erosion and exposure to weathering in arid regions.

Evidence of structural deformation during the Quaternary is shown by the minor tilting which also affected the Upper Irrawaddy Group in the Inner-Burman Tertiary Basin.

Stratigraphic gaps, erosional discordances and minor angular unconformities are frequently observed at the base of the Irrawaddy Group (Fig. 52). In the Ondwe structure, the Upper Irrawaddy Group containing *Stegolophodon latidens* overlies directly the Kyaukkok-Obogon Formation. A tooth of *Stegolophodon latidens* was found at the base of a gravelly sequence that comes directly on top of the Miocene Shwethamin Formation (p. 99) in the Wethaw-Kadeik region of the SE Chindwin Basin and presumably corresponds to the upper part of the Mingin Gravels.

The **Upper Tertiary of the Sino-Burman Ranges** with coal seams in the N Shan region (p. 172) and the **oil shales of the Tenasserim** region (p. 169) are regarded as continental Upper Miocene to Pliocene. The only datable fossils are some leaves and gastropods that are comparable with the recent fauna of the area. These sequences are also tectonically deformed. The stratigraphic sequences of the intramontane basins in the area of Lake Inle and further S in the Shan Highlands have not yet been studied.

Apart from the information provided by DE TERRA & MOVIUS (1943), little is known about Quaternary strata above the Upper Irrawaddy Group. Since the catchment area of the Irrawaddy extends into the snow-covered heights in the N, it is to be expected that the Himalayan glaciations left their marks as terrace sequences or loesses in the more southerly located parts of Burma. The "Pagan Silt" in the Dry Zone is supposed to contain a loess component. DE TERRA (1943, p. 330) correlated it with the fourth terrace of the Irrawaddy and a fourth glaciation.

From the section of the Irrawaddy between Magwe and Mandalay, DE TERRA & MOVIUS (1943) described a terrace system that shows up clearly, particularly in the area between Sale and Chauk.

The lowest "Terrace 5" is located in the area of Chauk 12 m above the (dry season?) level of the Irrawaddy, and it consists of sand and fine sand, similar to the present-day sediment load carried by the Irrawaddy, without any red beds.

"Terrace 4" at 17–20 m commences with red gravels and red sand. It contains stone implements of the Late Paleolithic, and it is followed by eolian sand, the loess-like "Pagan Silt".

"Terrace 3" at 27–33 m above the river is clearly observable. Normally it consists of a pebble-bearing gravel at the base, above which come fluviatile and finally eolian sands. Worn bones, possibly from the Irrawaddian Group, and implements from the Early Paleolithic are characteristic. The base section is often strongly solidified into an ironstone hardpan.

"Terrace 2" can be observed only as highly eroded cuestas. Its presence is marked by a soil horizon known as "Nyaung-U-Red Earth", which covers it. This horizon is several metres thick and causes a miniature badland topography.

"Terrace 1", 30–60 m above "Terrace 2", is somewhat problematical. It lies with coarse red gravels on hill tops. DE TERRA & MOVIUS (1943, p. 303) correlated these gravels with CHHIBBER's "Uyu Boulder Conglomerate" from the Myitkyina District. Similar five-element systems of terraces were described from the Namtu River and the Salween by DE TERRA & MOVIUS (1943).

The discovery of hominids in the karst caves in China encouraged DE TERRA, MOVIUS & TEILHARD DE CHARDIN to carry out a largely unsuccessful investigation of karst caves in the "Shan Plateau" in 1939 to 1940. According to DE TERRA & MOVIUS (1943), cave fillings in the area of Mogok contain remains of the fossil bamboo bear *Ailuropoda baconi* (WOODWARD), the elephant-like *Stegodon orientalis* and *Palaeoloxodon namadicus*. Therefore, a Middle Pleistocene age was assumed. They are regarded as occurring during pluvial periods, and they are correlated with the Choukoutien of China and Terrace 2 of the Irrawaddy. Caves in the area of the Lashio coalfield contained Holocene faunas. BEQUAERT (1943) believed that the gastropods found there are the remains of human meals.

From the N Myitkyina District and the Hukawng Valley, CHHIBBER (1929, 1932) described the Uyu Boulder Conglomerate, which consists of a series of more than 300 m thick, poorly sorted sediments of boulders (occasionally of jadeite, p. 210) and blocks in a matrix of red earth or sand, parts of which he believed to be at least of Pleistocene age.

A sediment similar in age to the Uyu Boulder Conglomerate is described by CHHIBBER (1934a, p. 276) as "Tanai Hka Boulder Conglomerate" from the Hukawng Valley N of 26° 11'.

Nothing comprehensive on raised beaches has been reported since CHHIBBER (1934a, pp. 66–70). THEOBALD (1871/1873) had already observed raised beaches on the S Arakan coast. CHHIBBER assumed the existence of at least three successive raised beaches in the Arakan Coastal Area, particularly on Ramree. BRUNNSCHWEILER (1966, p. 159) mentioned marine terraces on Ramree and Cheduba. According to him, the highest raised beach is 60 m above the W coast of Cheduba, and the highest on the island of Ramree, directly S of Konbwe Bay, is at an elevation of 30 m, with reference in each case to the low-water line. Towards the interior, the littoral sediments of these marine terraces interfinger with terrestrial deposits.

As far as the Tenasserim coast is concerned, it was mostly assumed that it has undergone subsidence. A critical revision of the entire question of coastal uplift and subsidence, and also of Quaternary eustatic sea level fluctuations, should be of interest for placer tin ore exploration in the Tenasserim region.

3.2 Tectonics

The data available on the tectonics are as incomplete as the data on the stratigraphy. Detailed structural analyses are so far limited to only a few regions. Geological and structural evaluation of satellite imagery, which has been used here on a large scale, has contributed a great deal to the understanding of the structural features. Since the Upper Cretaceous, structural development has been very distinctly influenced by the effects of the converging plates to the W of the present-day Indo-Burman Ranges. The structural map (Fig. 17b, folder) gives an overview.

3.21 Indo-Burman Ranges and their W foreland

The tectonics of the Bay of Bengal, the Arakan Coastal Area and the Indo-Burman Ranges is a result of the collision between the Indo-Australian Plate and the Asian Plate. The collisions have affected the area since the Tertiary, and since the Miocene they have resulted in the uplift of the Indo-Burman Ranges (p. 22).

From W to E this area can be divided into three tectonic units:

– the Indian continental shelf and continental slope
– the Bay of Bengal with oceanic crust, which formed in the course of the northwards drift of the Indian Plate and was overlain by thick clastic sediments of the Bengal Deep Sea Fan
– on the E side of the Bay of Bengal the present continental slope and continental shelf of the Asian Plate.

DU TOIT (1937) showed already that the two continental slopes are basically different. While the Indian shelf is bordered by a passive continental slope, the continental slope on the E side of the Bay is active.

The Indian shelf region and the adjoining continental slope are dominated by germanotype block faulting. The fractures run parallel to the Indian block. Graben systems also formed;

some of them are filled with Gondwana sediments (PAUL & LIAN 1975) and covered by the Rajmahal trap (CHAUDHURY 1973). According to radiometric age measurements (K/Ar), these basalts are 100–105 m.y. old (MCDOUGALL & MCELHINNY 1970) and thus correlate with the drifting apart of India and Antarctica (SCLATER & FISHER 1974). The younger sediments – from the Upper Cretaceous onwards – of the shelf and continental slope have not been affected by major structural movements (PAUL & LIAN 1975, Fig. 3).

Most of the Bay of Bengal is underlain by oceanic crust which formed since the lower Early Cretaceous following the separation of India from Antarctica (MCKENZIE & SCLATER 1971, CURRAY & MOORE 1974). According to JOHNSON et al. (1976), the W section of the oceanic crust of the N part of the Bay of Bengal is 130 m.y. old. These authors quote an age of 80–64 m.y. for the E section. The oceanic crust is overlain by an extremely thick sedimentary sequence (Upper Cretaceous to Recent), which according to CURRAY & MOORE (1974) is interrupted by a widespread unconformity. The profiles obtained at the Deep Sea Drilling Project (DSDP) site 217 at the N end of the Ninetyeast Ridge (MOORE et al. 1974), the interpretation of the observations made by KARUNAKARAN et al. (1964) on the Andamans and by BRUNNSCHWEILER (1966) in the Indo-Burman Ranges, and also the data reported from Bengal and Assam (SENGUPTA 1966, EVANS 1964), led CURRAY & MOORE (1974) to assign an age of 55 m.y. to this unconformity. This correlates with a change that took place in the seafloor spreading in the Indian Ocean (MCKENZIE & SCLATER 1971, SCLATER & FISHER 1974). According to CURRAY & MOORE (1974) the stratigraphic gap covers the period from the Late Paleocene to the Middle Eocene.

The collision between the Indo-Australian Plate and the Asian Plate about 35 m.y. ago (MCKENZIE & SCLATER 1971) resulted in a change in the sedimentation. CURRAY & MOORE (1974) assumed that the formation of the modern Bengal Deep Sea Fan commenced during this time (Oligocene to Early Miocene). Two lobes of the fan were active up until the Middle Pleistocene, namely W of the Ninetyeast Ridge the Bengal Fan sensu stricto, and to the E the Nicobar Fan (CURRAY & MOORE 1974). The supply of sediments to the Nicobar Fan was then cut off by the convergence of the Ninetyeast Ridge and the Sunda-Andaman Trench. At present, only the axial area of the Sunda-Andaman Trench is still receiving sediments from the fan in the Bay of Bengal.

While the sediments in the open area of the Bay of Bengal are unfolded, intensive folding and imbrication commences towards the E continental slope of the Bay (PAUL & LIAN 1975) – this structure is typical of an active continental slope (SEELY et al. 1974). These conditions are shown by a seismic reflection profile off the E coast of the Bay of Bengal at Cox's Bazaar (PAUL & LIAN 1975, Fig. 3). The profile shows that shallow, open folding commenced W of the Cox's Bazaar borehole. The borehole had been drilled into the first clearly marked anticline; no vergency could be observed. The anticlines further to the E are narrower and exhibit a slight W-vergency.

On shore, these groups of folds are exposed in the hill chains of the Boronga Archipelago and the Mayu Range (=S extension of the Chittagong Hills of Bangladesh). The folds are crossed by large, apparently W-vergent overthrusts and they are composed of alternating sequences of Miocene sandstone and siltstone. To the S, near the Arakan Coastal Area, Eocene rock series occur (p. 91) which exhibit W-vergent folding and imbrication. On the islands of Ramree and Cheduba, the main portion of the rocks is composed of Oligocene flysch sequences with conspicuous circular synclines (Figs. 25, folder, 26). HELMCKE & RITZKOWSKI (in press) have shown that the synclines are located in a network of faults and anticlines striking 160° and 70°. E. LEHNER (unpubl. report 1939) regarded the circular structures as being connected with differing tectonics in the upper and in the lower subsurface, and with sedimentary volcanism, evidence of which exists in the form of the mud volcanoes in the upper subsurface. He therefore interpreted them as subsidence areas resulting from losses of material in the deeper strata. The distribution of mud volcanoes and circular structures, however, does not support this opinion. Mud diapirism also occurred in the surrounding anticlines and affected the shale and melanges of the Eocene to Miocene flysch sediments.

In the Middle Miocene folding and subsequent transgression of Upper Miocene clay-sandstone sequences in a brackish milieu occurred. According to BRUNNSCHWEILER (1966) at Taungup and Sandoway these sequences extend as far as the W foothills of the Arakan Yoma.

S of the islands of Ramree and Cheduba the mud volcano of Foul Island is an indication that similar geological conditions continue to the SE. At 18° N the course of the coastline changes from SSE to SSW and the Arakan Yoma extends with Eocene rocks to the coast.

The Indo-Burman Ranges continue to the S in the Andaman and Nicobar Islands. W-vergent folding predominates on the Andaman Islands as well. GEE (1926) and KARUNAKARAN et al. (1968) described a sequence of Mesozoic sediments with andesitic volcanism and ophiolites in the E section of Middle Andaman Island. Eocene to Oligocene flysch rocks adjoin to the W. Once they had been folded, Lower Miocene limestones and sandstones were deposited and accumulated between the islands of the Outer Island Arc. Another unconformity followed in the Middle Miocene, which can also be observed in the Arakan Coastal Area.

The Arakan Yoma forms the S part of the Indo-Burman Ranges. In the S, the mostly W-vergent folds strike NNE, parallel to the W coast (Fig. 16). The transition of the Eocene flysch rocks to the Mio-Pliocene sediments of the Western Outcrops of the Irrawaddy Delta area (p. 98) seems to be concordant, but the possibility of a stratigraphic gap cannot be excluded. At 19° N the strike of the Arakan Yoma changes from NNE to NNW. A major fault zone occurs at the E side of the Arakan Yoma, which separates here the sediments of the Indo-Burman flysch from the sediments of the Inner-Burman Tertiary Basin.

In the area where the strike of the Indo-Burman Ranges changes from NNE to NNW (at approximately 19° N), an SE–NW-striking lineament approaches from the SE, out of the S Shan Plateau (p. 118), in the extension of the Moei-U Thai-Thani Fault (Fig. 29); the SW-vergent folds of the 20° uplift lie along the course of this lineament (p. 36). It can also be traced on satellite images within the Arakan Yoma. BANNERT & HELMCKE (1981) assumed that it is caused by a transcurrent fault. It may perhaps represent the N limit of the movement mechanism of the Andaman Sea spreading centre which has been opening gradually since the Miocene (EGUCHI et al. 1979). So far, however, the character of this lineament as a transcurrent fault has remained unexplained, because no corresponding lateral movements have been observed in the Sino-Burman Ranges at the Moei-U Thai-Thani Fault Zone or at the parallel-striking Three Pagoda-Ratchaburi Fault Zone (KOCH 1973, 1978).

Mainly Upper Cretaceous flysch-type sediments with intercalated basic volcanic rocks occur in the area of 19° N at the E edge of the Arakan Yoma. The tectonics is marked by W-vergent folds and imbricate structures (Fig. 20).

Along the E boundary of the Indo-Burman Ranges, in the area of the Chin Hills, partially serpentinized ultrabasic rocks occur in the contact with Triassic flysch rocks (p. 81). N of

25° N imbricate structures of the Naga Metamorphic Complex (BRUNNSCHWEILER 1966) submerge towards the E under Eocene sandstone sequences of the Inner-Burman Tertiary Basin (Chindwin Basin). The Naga Metamorphic Complex is thrust far westwards over Upper Cretaceous flysch-type sediments and older rocks. Near Pansat, BRUNNSCHWEILER (1966) described limestones and metamorphic rocks beneath Upper Cretaceous flysch sediments (Fig. 19). In this area, the course of the E boundary fault of the Indo-Burman Ranges is not clear, because major faults were observed at several places between the Eocene sandstones of the Chindwin Basin and the Naga Metamorphic Complex. The overthrust plane of the imbricate structures of metamorphic rocks could be the separating fault between the Indo-Burman Ranges (Naga Hills) and the Inner-Burman Tertiary Basin (Chindwin Basin). Thus, the Naga Metamorphic Complex would have to be regarded as the basement for the sediments of the Chindwin Basin. This view is favoured also by the occurrence of the low-grade metamorphic rocks of the 26° uplift, which can be regarded as a horst-like structure with basement rocks of the Chindwin and Hukawng Basin (pp. 113, 114).

At about 25° N, the strike of the Indo-Burman Ranges (Naga Hills) changes gradually from the S–N direction to NNE, and at 26° 30' N it turns to the E in the Patkai Ranges. Only flysch rocks occur in the area of the Ledo Road between the Hukawng Valley and Assam in India. According to STUART (1923) these flysch sediments are comparable to the Eocene Disang Series, the Eocene to Oligocene Barail Series and the Mio-Pliocene Tipam Sandstones of Assam. These rocks are compressed together into eight NW-vergent recumbent folds (EVANS 1964, p. 84); they are overthrust over a width from 20 to 40 km. The Kohima-Patkai Syncline, with the Barail Series at the centre and the Disang Series at the flanks, is located in the interior of the Indo-Burman Ranges. It is divided into several anticlines and synclines. On satellite images it can be seen that the intensity of the folding increases greatly towards the E. The imbricate structures of the Naga Metamorphic Complex probably represent the structurally highest unit. Serpentinite occurs occasionally at the E edge and in the flysch sediments.

In the area of the N frame of the Hukawng Basin, the folds of the Patkai Ranges strike E–W towards the Mishmi Fault (Fig. 22). According to KARUNAKARAN (1974), these folds are overthrust from the NE along the Mishmi Fault by the metamorphic and crystalline rocks of the E Himalayas. BANNERT & HELMCKE (1981) regarded the compressed rocks of the Indus-Tsangpo Suture Line (GANSSER 1964, HUANG CHI-CHING 1978) as the continuation of the Cretaceous to Lower Tertiary flysch sediments of the Indo-Burman Geosyncline.

Paleogeographically, the Indo-Burman Ranges can be interpreted as the E section of the Indo-Burman Geosyncline (BANNERT & HELMCKE, 1981). Since the Eocene, this geosyncline has been compressed and folded during the collision of the Indian-Australian Plate with the Asian Plate. Considerable subduction must have occurred in the area of the present-day Indo-Burman Ranges and the oceanic floor of the former Tethys.

3.22 Eastern Himalayas

Because of inadequate knowledge of the regional geology of the E Himalayas and the mountain ranges in their SE extension in NE India, N Burma and SW China, the interpretation of the tectonic relationships in the area of the "eastern syntaxis" of the Himalayas (see, inter al., GANSSER 1964) is only sketchy and questionable.

It is safe to state that the tectonic macro-units which have been established for the W and Central Himalayas, namely the Sub-Himalaya, Lower Himalaya, Higher Himalaya and Tethys-Tibetian Himalaya, as well as the overthrusts that separate them can be traced to the E only about as far as to the gorge of the Dihang (=Tsangpo=Brahmaputra) in the Indian State of Arunachal Pradesh, where they abruptly end. In the Lohit Himalaya adjoining to the SE, completely different lithotectonic units occur and then strike SE towards N Burma.

The course of the Indus Suture Line, the geotectonically most important lineament in the Himalayas, which is supposed to correspond to the course of the S suture of the former Tethys (Fig. 13; STONELY 1974), is also only adequately verified further to the W (GANSSER 1964, p. 254). Ophiolites have been described from the valley of the Tsangpo, and typical melanges containing exotic blocks of Triassic and Jurassic limestones, which probably mark the E extension of the Indus Suture Line (HUANG CHI-CHING 1978, pp. 631 et seqq.), have been observed from the area E of Shigatze. However, in the area E of the gorge of the Dihang (=Tsangpo), there is no evidence of the Indus Suture Line, and its position is often quite hypothetically plotted (STONELY 1974).

The tectonics of the Lohit Himalaya is marked by two large, on the whole NW–SE-striking structural elements in the Indian State of Arunachal Pradesh: the Mishmi Thrust and the Lohit Thrust (p. 22). According to VALDIYA (1976, pp. 367 et seq., Photograph 8), the Mishmi (=Miju) Thrust between the Dihang and Dibang Valley in Arunachal Pradesh cuts off the four above-mentioned macrotectonic units of the Himalayas. In this area the Mishmi Thrust is probably characterized by northwestward strike-slip movements.

According to VALDIYA (1976), the Mishmi Thrust then strikes further SE into N Burma where it borders on the NNE–SSW-striking Patkai Synclinorium (=N Indo-Burman Ranges) and the Naga-Disang Thrusts. In this area, the Mishmi metamorphics of the Lohit Himalayas are overthrust over the Tertiary sediments of the northernmost Indo-Burman Ranges. In N Burma the overthrust plane of the Mishmi Thrust can be traced on satellite imagery SW of Putao, where it is exposed in the gneiss of the Kumon Range (Fig. 53).

The Mishmi metamorphics, which strike approximately NW–SE, are bounded towards the NE by the Lohit Thrust. In general, the Lohit Thrust runs parallel to the Mishmi Thrust. In Arunachal Pradesh, a broad and approximately 100 km long zone of ultrabasic rocks was observed between the Tidding and Twelang valleys along the Lohit Thrust (KARUNAKARAN 1974). This fact may be significant in geotectonic terms. However, it is not possible to link this belt of ophiolites with the Indus Suture Line. The ultrabasic rocks which are shown W of Putao on the 1 : 1 million Geological Map of Burma (1977) can probably not be regarded as the extension of the belt of ophiolites along the Lohit Thrust, because they are situated in the extension of the strike of the Mishmi metamorphics.

Mainly granite and granodiorite, which are referred to as Lohit Granodiorites, crop out NE of the Lohit Thrust in Arunachal Pradesh. Towards the SE, they strike into the Burman section of the E Himalayas, to the N and E of the basin of Putao, which is characterized by a relatively uniform NW–SE-strike of the structural elements (STUART 1919). Between Putao and the valley of the Nmai Hka, some more or less wide zones of schist, phyllite and quartzite as well as highly siliceous slaty limestone are intercalated in the locally slightly schistose granite which composes most of this region. In some areas "lit-par-lit injections" of granite occur in the metamorphic rocks. The belts of calc-silicates, which could only be observed locally by STUART (1919) and JURCOVIC & ZALOKAR (1961), are particularly easy

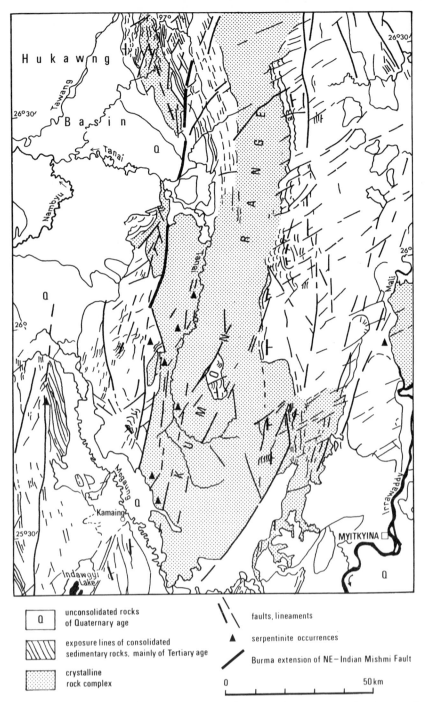

Fig. 53. Geological sketch map of Kumon Range and surrounding area; Landsat-imagery interpretation.

to pick out on Landsat images, and they can be traced as continuous zones extending to the S into the area of Myitkyina. Here, STUART (1923) found a belt of fine crystalline white limestones, striking N–S and dipping almost vertically, in the valley of the Hpyepat. These parts of the E Himalayas can be linked structurally with the Mogok Crystalline Rock Complex (p. 40) and are linked with the East Kachin/Shan Unit of the Sino-Burman Ranges (p. 38). SEARLE & BA THAN HAQ (1964) assigned these rock series to the "Mogok Belt" (p. 40), which runs from Putao to Bhamo in an arc that is slightly convex to the E, as well as further to the S, and crops out over a total width of more than 100 km. According to these authors, the metamorphic grade in the Mogok Belt seems to increase from W to E.

The Lohit Thrust strikes from Arunachal Pradesh to the SE towards the Putao Basin; so far it has not been observed in Burma. The frequent occurrence of ultrabasic rocks in the S extension of the Lohit Thrust is remarkable (STUART 1923).

3.23 Inner-Burman Tertiary Basin and Gulf of Martaban

The **Andaman Sea** contains a spreading centre which, according to EGUCHI et al. (1979, p. 47), may have been active for the past 10 m.y. CURRAY et al. (1979) assumed that the E rise in the Andaman Sea (=Mergui Terrace) is built up of the continental crust of the Asian Plate. The spreading centre is accompanied in the W by a number of seamounts and faults which continue to the N in the volcanic islands of Barren Island and Narcondam. CURRAY et al. (1979, Fig. 2) linked the most strongly marked "West Andaman Fault" with the Inner-Volcanic Arc in Burma (p. 19). Between the Andaman-Nicobar Ridge in the W and the Malayan Peninsula in the E, the Andaman Sea Basin is divided up by this arc into a W and an E basin. The E basin corresponds to the Back-Arc Basin and is characterized by thick Mio-Pliocene sediments (CURRAY et al. 1979). In the W basin (=Inter-Arc Trough) the sediments are not as thick (p. 160).

The structural units can be traced from the Andaman Sea to the onshore region in the N. In the W, the Arakan Yoma (p. 106) forms the extension of the Andaman-Nicobar Ridge. The Irrawaddy Delta and the Prome Embayment correspond to the Inter-Arc Trough to which is adjoined the Inner-Volcanic Arc in the E (Fig. 27, folder). The Pegu Yoma and the Sittang Valley are located in the area of the Back-Arc Basin which is bordered in the E by the Shan Boundary Fault (p. 38).

In the **Irrawaddy Delta,** only a few shallow anticlines of Miocene rocks are visible SW of Rangoon because this area is essentially occupied by Recent to sub-Recent fluviatile sediments. The W boundary is formed by Tertiary sediments of the "Western Outcrops" (Fig. 27, folder), which can be traced to the N for 1,400 km along the E side of the Indo-Burman Ranges.

The **Pegu Yoma** consists of an anticlinorium of Mio-Pliocene sediments that is uplifted like a horst and accompanied by ± N–S-striking faults. It plunges to the S beneath the delta deposits of the Irrawaddy and of the Sittang. To the W, the Pegu Yoma border with a fault escarpment on the Irrawaddy Delta. In the E, they are separated by the Pegu Fault of the Sittang Valley. The Pegu Fault belongs to the Shan Boundary Fault System as well as the Sittang fault of the Sittang Valley where it runs in the foothills of the Shan Escarpment.

At 19° N, a series of anticlinal structures emerges to the N from the Irrawaddy Delta. As they approach the "**20° uplift**" (p. 34), these structures swing to an NNW direction. They link up with a complex of mainly WSW-vergent anticlines, which split off in the SE from the Pegu Yoma (Fig. 27, folder). The WSW-vergency and the NNW–SSE-strike of the structures of the 20° uplift do not coincide with the normal directions of strike that are observed within the Inner-Burman Tertiary Basin. This could point to a right-lateral displacement along the E boundary fault of the Indo-Burman Ranges.

N of the 20° uplift, the axis of the **Salin Syncline** plunges northwards. Most of the oilfields of Burma are located at its E edge (Fig. 27, folder). The Salin Syncline runs for more than 200 km in the N–S direction. W of the Salin Syncline, the Western Outcrops broaden to 40 km. It is here that the Pondaung and Ngahlaingdwin Anticlines occur; their axes strike SE in the N and then change to an SSE-strike in the S. On their W flank they are accompanied by steep normal faults which separate them from broad synclines (from the S: Man, Thanwin and Handauk Syncline). Between the Ngahlaingdwin Anticline and the Handauk Syncline one of the anticlines is cut off by a main fault zone, but further to the NW this anticline is clearly evident in the rocks of the Tabyin, Tilin and Laungshe formations. The Lower Tertiary and Cretaceous rocks (Kabaw Formation, p. 89) are more intensively folded than the younger rocks above the Pandaung Formation. Between 21° and 21° 30′ N, sediments of the Kabaw Group contain Cenomanian limestone resting unconformably on Triassic Halobia beds and the metamorphic Kanpetlet Schists (GRAMANN 1974; p. 87). Morphologically, this area around Mt. Victoria and Kanpetlet belongs to the Indo-Burman Ranges. However, geologically the rocks represent the basement of the Inner-Burman Tertiary Basin together with the Triassic flysch sediments and the Kabaw Formation.

No structural subdivision has yet been observed within the area of the Salin Syncline. At 21° N, a culmination of the approximately N–S-striking synclinal axis occurs which was also revealed by gravity measurements. Seismic investigations showed that the axis of the syncline in the pre-Irrawaddian sediments runs E of the axis that can be observed at the surface. The E edge of the Salin Syncline is formed by a series of narrow elongated anticlines (Fig. 27, folder, p. 162). From S to N these are:

- Mann-Minbu Anticline
- Yenangyaung Anticline
- Yenangyat-Chauk Anticline
- Letpanto Anticline.

As the Salin Syncline increases in depth and width towards the N, the anticlines at its E edge are offset "en-echelon" to the E. The locally strong E-vergency of the structures is remarkable, and in the case of the Yenangyat-Chauk Anticline it led to overturned bedding and major overthrusting along its E flank. The Mann-Minbu Anticline is also E-vergent. In addition to a W-dipping E boundary fault, a steep E-dipping upthrust fault was found on the W side.

To the E of these anticlines a further belt of anticlines is exposed in the N extension of the W part of the Pegu Yoma, in the plains to the N of Taungdwingyi. From S to N they are:

- Ondwe Anticline
- Yedwet Anticline
- Gwegyo-Ngashandaung Anticline
- Pagan Hills
- Myaing Anticline.

To the E of these structures, the E part of the Pegu Yoma Fold Belt turns to the N. It is built up by individual anticlines and synclines and, at about 22°N, it plunges beneath the unconsolidated Quaternary sediments of the Shwebo Plain. At about 21°N the Mt. Popa Volcano (1,518 m), which consists mainly of andesitic lavas and tuffs, has been building up since the Miocene (p. 131). It belongs to the Inner-Volcanic Arc. The Shinmataung Range, where volcanic activity occurred during the Oligocene and Miocene (p. 135), lies to the N of this volcano.

The E edge of the Pegu Yoma is formed by the Pegu Fault which corresponds to the Shan Boundary Fault. It cuts the NNW–SSE-striking folds of the Pegu Yoma. Mio-Pliocene rocks occur E of this fault in front of the Shan Escarpment. The Sittang Valley narrows here to a few kilometres.

The "22° uplift area" commences with the Letpanto Anticline, Kyaukwet Structure and Myaing Anticline. It is generally characterized by the emerging Salin Syncline and the structures that accompany it to the N. The 22° uplift separates the Salin Syncline in the S from the Chindwin Basin in the N. Evidence of this uplift exists as early as in the Upper Eocene and Lower Oligocene (p. 93). There are no sediments younger than the Upper Eocene Yaw Formation. Since the Oligocene, the 22° uplift has influenced the clastic sedimentation in the N (Chindwin Basin) and in the S (Salin Syncline). A number of N–S-striking anticlines and synclines are found in the area of the 22° uplift. From W to E these are:

– Kyaw Syncline (with the Upper Eocene Yaw Formation in its core)
– Patolon Anticline (with the Middle Eocene Tabyin Formation in the core, overturned towards the E)
– Patolon Syncline (with the Oligo (?) – Miocene Letkat Formation in the core)
– Mahudaung Anticline (with the Middle Eocene Tabyin Formation in the core, E-vergent).

The age of the folding is unknown. The wedging out of the Upper Miocene Mingin Gravels at the W flank of the Mahudaung Anticline indicates an uplift movement in this area since the Upper Miocene. E of the Mahudaung Anticline and the Myaing Anticline, an area with smaller anticlines rising out of Quaternary deposits leads over to the **Monywa-Salingyi Volcanic and Crystalline Complex** 30 km to the E (Fig. 61). This rock complex, which belongs to the Inner-Volcanic Arc (p. 131), provided some of the detrital components in the Middle Eocene Pondaung Formation 30 km further to the W.

To the E of Monywa, basalt dykes intruded Miocene sandstone which represents the northernmost spurs of the Pegu Yoma. They dip gently to the E underneath the Quaternary unconsolidated sediments of the **Shwebo Basin.** The Shwebo Basin belongs to the Back-Arc Basin. According to aeromagnetic surveys, the magnetic basement in this area is up to 8,000 m deep (p. 168).

The E flank of the approximately 100 km wide Shwebo Basin is occupied by Miocene sediments which are cut by N–S-running faults belonging to the fault system of the Shan Boundary Fault. To the N of Shwebo, there is an N–S-striking anticline with Oligocene sediments exposed in its crestal area.

A more than 200 km long zone of Miocene rocks strikes N along the Shan Boundary Fault at the E edge of the Shwebo Basin. It is located on the E flank of the volcano-magmatic **Wuntho Massif.** The latter strikes approximately NNE. Parts of it belong to the Inner Volcanic Arc. Since it strikes at an acute angle to the Shan Boundary Fault, the Back-Arc Basin becomes narrower towards the N. At 25°N the faults of the Shan Boundary Fault transect the Wuntho Massif E of the largest volcano in Burma, Taungthorlon (1,708 m, p. 136). Miocene sediments accompany the area of this fault and link up with the sediments of the N Chindwin Basin.

The **Chindwin Basin** is also a large syncline whose axis plunges towards the N at about 22° N (Fig. 27, folder). Flanked in the W by the rocks of the Western Outcrops, and in the E by a monocline with individual, locally complex, anticlinal structures, the Chindwin Basin can be traced as far as the presumably pre-Cretaceous rocks of the "26° uplift". Since it is bounded in the E by the Inner Volcanic Arc (part of the Wuntho Massif), it belongs, in accordance with plate tectonic criteria, to the Inter-Arc Trough. The Western outcrops border on the Myittha and Kabaw Valley, a graben-like N–S-striking zone in which the Upper Miocene Maw Gravels were deposited; these can be correlated with the Mingin Gravels of the Chindwin Basin (p. 97). Holocene river sediments, which mask the contact to the E Indo-Burman Ranges, are widespread in the Myittha-Kabaw Valley. This contact is visible N of 25° N. The sediments of the Chindwin Basin overlie the Naga Metamorphic Complex with a pronounced unconformity (BRUNNSCHWEILER 1966, p. 176). The sediments of the Western Outcrops are folded. The age of the folding is questionable; as far as is known, only pre-Oligocene rocks have been folded. Mention should be made of the Yenan Anticline at 24° N on whose E flank Upper Eocene and Oligocene sediments wedge out. This permits the conclusion that uplifting took place during this period in the Yenan area. The thickness of the post-Eocene sediments increases continuously towards the N. No reduction in thickness of the strata appears to have occurred as the 26° uplift is approached. The approximately 75 km long N–S oriented, shallow, hydrocarbon-bearing Indaw Anticline occurs in the centre of the Chindwin Basin S of 24° N (p. 125). Its E flank is overthrust along a W-dipping fault. Towards the N, at 25° N, near the Western Outcrops, the Uyu Anticline rises up from the Quaternary sediments of the Chindwin and Uyu rivers. It can be traced for more than 70 km to the NE. Presumably Oligocene or Miocene rocks crop out in its crestal area. The SE-vergent Uyu Anticline encounters the Northern Chindwin Fold Belt at right angles. The latter is an NW–SE-striking complex of folds in which two anticlines contain *Orbitolina*-bearing Cenomanian limestones in their cores as they approach the 26° uplift (SAHNI & SHASTRY 1937). Two shallow synclines, the Tamanthi Syncline in the S and the Hkamti Syncline in the N, occur to the W of the Uyu Anticline and Northern Chindwin Fold Belt. They contain Mio- to Pliocene sediments.

The E boundary of the Chindwin Basin is formed N of the 22° uplift by the elongated Mahudaung Anticline which submerges towards the N under Mio-Pliocene sediments. Towards the NE, the E boundary of the Chindwin Basin is formed by a W-dipping monocline, which represents the northwards extension of the Medin Anticline from the region between the Mahudaung Anticline and the Monywa-Salingyi Volcanic and Crystalline Complex. At about 23° 25′ N it encounters the double anticline of the Wethaw-Kadeik Structure rising from the SW out of the Chindwin Basin to the NE. This double anticline in turn appears to be connected to the SW via the shallow and buried Makkadaw Anticline with the submerged Mahudaung Anticline. At about 24° N, an E-vergent anticlinal belt occurs basinwards of the monocline. The anticlines contain Miocene rocks in their crests, while the monocline is composed of W-dipping, alternating sequences of Early Tertiary sandstone and conglomerate. The conglomerate contains mainly rocks (diorite, granodiorite, meta-andesite, trachyte) of the Wuntho Massif in the E (p. 136), which formed the E boundary of the Inter-Arc Basin (AUNG KHIN & KYAW WIN 1969). Presumably Cretaceous limestone and alternating sequences of limestone and greywacke are found beneath the Lower Tertiary sediments which further to the S overlie the Wuntho Massif. The fault system of the Shan Boundary Fault runs S–N to the E of the Wuntho Massif.

In terms of structural geology, the area to the E of the Shan Boundary Fault N of Mandalay is referred to as the West Kachin Unit (Fig. 29). In accordance with plate tectonic criteria, parts of the West Kachin Unit can be assigned to the Back-Arc Basin. Miocene rocks, individual occurrences of which are linked with the northernmost of the Tertiary basins, the Hukawng Basin, were observed SW and N of Myitkyina.

The **26° uplift**, striking ±N–S occurs as a separating uplift in the Inter-Arc Trough between the Chindwin Basin in the S and the Hukawng Basin in the N. In the N part of the 26° uplift, the upper reaches of the Chindwin River flow through low-grade metamorphic rocks occupying the crest of an anticlinal structure. Further N, these rocks then submerge beneath Mio-Pliocene sediments. The latter continue from the Chindwin Basin into the Hukawng Basin. The S part of the 26° uplift in the area of the Jade Mines District is made up of gneiss, serpentinite, peridotite, gabbro and granite (SOE WIN 1968, CHHIBBER 1934b, p. 315, STUART 1923, Pl. 29). To the S, these rocks continue into the area around Indawgyi Lake where they are covered by Cretaceous and Tertiary sediments. With the exception of its N part, the rocks of the 26° uplift are all in structural contact with the rocks of the Tertiary basins.

The **Hukawng Basin** is a syncline approximately 100 km in diameter (Figs. 7, 27, folders). The sediments dip on all sides at approximately 5° to 20° towards the centre of the basin. STUART (1923) drew attention to the similarity between the sediments on the W and N flanks and the Eocene Disang Series and the Oligocene Barail Series of Assam.

The youngest strata correlate with the Pliocene Tipam Sandstone. On the E flank, crystalline rocks of the Kumon Ranges form the basement of the basin sediments (Fig. 53). According to satellite imagery, the older Tertiary sediments seem to be missing. The S and SW flank of the Hukawng Basin is formed by rocks of the 26° uplift, against which the basin sediments appear to border along a fault system. The centre of the Hukawng Basin is covered by river sediments of the Tanai-Hka (Chindwin River) and its tributaries. So far, satellite images have revealed three anticlines in the centre of the basin. The largest is located to the E of Taro and has a NW-plunging axis. STUART(1923, p. 404) describes a moderate dip of ± 10° for the NE flank. Blue, amber-bearing (p. 211) Eocene clay is exposed in the crestal area. In the E the anticline is cut off by an N–S-striking fault. It is followed by two small N–S-striking anticlines with Cenomanian limestone at their crests, which are separated by a syncline. In the W flank of the basin, NW of Taro, aerial photographs revealed a further anticline plunging to the S towards the Tanai Hka, within a thick sandstone sequence.

3.24 Sino-Burman Ranges

Large areas of the Sino-Burman Ranges are geologically almost unknown. Nevertheless, recent research results have led to a better understanding of their tectonic style and of their structural setting within the Indo-Chinese Peninsula. As discussed in Chapters 2.2 and 2.5, the Sino-Burman Ranges can be subdivided into the **East Kachin/Shan Unit**, the **Karen/Tenasserim Unit**, which is separated from the former by the Pan Laung Fault, and into the **West Kachin Unit** (Fig. 29). The major structural events in these three tectonical domains were accompanied by various phases of magmatic activity both in the Paleozoic (p. 127), and in the Mesozoic and Cenozoic (p. 139).

3.241 East Kachin/Shan Unit in the Northern Shan State

In the Northern Shan State, the East Kachin/Shan Unit is composed of the Mogok-Namhkam Belt of crystalline rocks (p. 43). It is followed in the E by a belt of mainly Paleozoic and Mesozoic sedimentary rocks of the Burmese-Malayan Geosyncline, which extends to the Thai border. Tertiary and Quaternary unconsolidated sediments form the fillings of large graben-like intramontane basins, whose subsidence occurred after the consolidation and uplift of the Sino-Burman Ranges during the Upper Mesozoic and Tertiary.

The metasediments (gneiss, marble, calc-silicate and quartzite) and migmatite of the Mogok Belt generally strike ENE and dip steeply to the S. SEARLE & BA THAN HAQ (1964) mentioned three major directions of the fold axes: 15° E, 50° SE and 50° SSW. The first of these is superimposed on the two latter orientations. The structures are of the "chevron" folding, overfolding and asymmetrical synclinal and anticlinal fold types. The amplitudes of the folds are in the cm- to m-range. The folds are disturbed by numerous overthrusts; the morphologically most prominent of these is the "Mogok Thrust", which extends from the head of the Mogok Valley to Kabaing, where it is cut off by the intrusion of the "Kabaing Granite" (p. 142). SEARLE & BA THAN HAQ (1964, p. 144) observed an intense degree and amount of brecciation, slickensiding, mullion structures and mylonitization in this fault zone that dips at 15−20° to the S. LA TOUCHE (1913), HERON (1937) and IYER (1953) placed the folding and the metamorphism of the metasediments in the Precambrian. SEARLE & BA THAN HAQ (1964) believed that the Mogok Series are part of the Mogok Belt and represent the metamorphic Chaung Magyi Series, and they regarded the metamorphism as being linked with the Himalayan orogeny. GARSON et al. (1976) saw considerable structural differences between the various formations that make up the Mogok Belt and therefore refused to accept that a uniformly Alpidic folded belt exists at the W edge of the Shan Highlands. LA TOUCHE's view (1913) appears to be more likely. According to him, the Mogok crystalline rocks form the basement of the Paleozoic and Mesozoic sediments at a contact extending as far as Yunnan, China. Here, a distinct unconformity was observed between the crystalline basement rocks and the Paleozoic/ Mesozoic rock sequences (BROWN 1923 d, p. 49).

The Paleozoic-Mesozoic sediments that adjoin the Mogok-Namhkam Belt in the SE form the NW flank of the Hsipaw Syncline which contains interfolded Jurassic sediments at its core. Its E or SE flank is characterized by an upthrust of Permo-Triassic carbonates (Fig. 54, folder).

The Chaung Magyi Series is separated from the Mong Long Mica Schists by an E−W-trending fault (MITCHELL et al. 1977, p. 20). The epimetamorphic slate-greywacke sequence is deformed in two fold axis directions: WNW-trending fold axes belong to the narrow to "chevron"-style folds, and the N-trending axes are moderately tight cross folds. The folding and the low-grade metamorphism are older than the discordantly overlying Upper Cambrian to Middle Ordovician Pangyun Formation.

The Paleozoic sediments with the Permo-Triassic carbonates dip at a shallow to moderate angle towards the E or SE. The tight fold structure with steep E-dipping overthrusts, which was assumed by BROWN (1918a) and LA TOUCHE (1913), has not been confirmed by more recent authors. BRINCKMANN & HINZE (1976) were able to show that the Bawdwin Volcanics do not form an anticlinal "low dome structure" within the Pangyun Formation (BROWN 1918a) but instead dip uniformly to the SE, and are intercalated in the Pangyun sediments (Fig. 54, folder). They saw the special folding of the Pangyun sediments and of the Bawdwin Volcanoclastics as being connected with the emplacement of the Bawdwin Rhyolite Massif, whose longitudinal extension runs parallel to the fold axes that plunge at moderate to steep angles to the SE.

In the area of Kyaukme-Longtawkno, MITCHELL et al. (1977) were able to show that the anticlinal structure traced by LA TOUCHE (1913, Pl. 23, sect. 4) at the Lilu Ferry/Namtu River forms together with the Lilu Thrust a normal E-dipping section, as can be reconstructed from Map 2 in MITCHELL et al. (1977) (Fig. 54, folder).

Many normal faults in the area of the East Kachin/Shan Unit run, often with considerable throws, N–S, parallel and sub-parallel to the large boundary faults, which further to the W mark the escarpment of the Shan Plateau. These faults include the N–S-trending Nangkashwe Fault Zone (MITCHELL et al. 1977, Map 1), where parts of the sequence of the Chaung Magyi Series and Pangyun sediments are repeated several times. In the Yadanatheingyi area the width of the exposed Pangyun and Naungkangyi formations is considerably reduced by NW–SE-trending normal faults (MITCHELL et al. 1977). The latter form the SW-dipping limb, which has been steeply tilted by block movements, of the Kalagwe Syncline, which contains Permo-Triassic carbonates of the Shan Dolomite Group at its core.

According to MITCHELL (1977, Map 2), in the Kyaukme-Longtawkno area the N-dipping Lilu Fault W of Manglang has throws of 2,000 m. Further S, at the Lilu Ferry, the vertical displacement is only 200 m at the outcrops of the Pangsha-pye Formation, which occur repeatedly on both sides of the Namtu Valley (Fig. 54, folder).

Together with the Permo-Triassic carbonates, the Paleozoic sequence forms the NW flank of the SW–NE-trending **Hsipaw Syncline** which runs from approximately 22° 10′ N to approximately 24° N. The tectonics of this syncline are characterized by complex micro- and macro-fold structures with overturned folds and overthrusts, as well as complicated imbricate structures. BRUNNSCHWEILER (1970) regarded the 10 to 100 m thick Upper Triassic Pangno Evaporites of the Bawgyo Group (p. 82), which are exposed in a narrow belt between Bawgyo and Tai, as being responsible for this structural style. They acted as a slip horizon along which the overlying Napeng Formation (p. 82) and the Jurassic Namyau Group (p. 82) started to slide when tilting occurred and ultimately became deformed into interlaced folds when they came up against abutments of horst-like uplifts. The competent Permo-Triassic carbonates remained largely unaffected during this process of disharmonic folding (BRUNNSCHWEILER 1970, Fig. 9).

In addition, diapiric movements occurred during and after the pre-Alpidic gravity-sliding events and these affected mainly the Pangno Evaporites themselves. Through a combination of both types of stress, evaporites became overlain on top of Jurassic Red Beds, and also profiles were formed in which Mesozoic strata are contained in juxtaposition and superposition to each other (LA TOUCHE 1913, p. 288). The age of the movements is younger than the Jurassic Hsipaw Red Beds, which were incorporated into the slip folds (BRUNNSCHWEILER 1970; Fig. 33). The SE boundary of the Hsipaw Syncline is formed by the steeply E-dipping "Onghkok Overthrust" where Permo-Triassic carbonates are overthrust onto the Jurassic Hsipaw Red Beds. Further to the E adjoin the slightly folded Paleozoic sequences in which, NE of Lashio, a further syncline was formed with Jurassic sediments at its core.

3.242 East Kachin/Shan Unit in the Southern Shan State

In the area of Ye-ngan, the Chaung Magyi Series occurs as the oldest rock sequence in two wide anticlinal structures (Fig. 54, folder). GARSON et al. (1976) assumed that this several thousand metres thick flysch sequence had undergone low-grade metamorphism and two periods of deformation. The discordantly overlying Lower Paleozoic sediments form moderately steep E- or W-dipping flanks of these anticlinal structures. The Permo-Triassic carbonates, which in turn overlie discordantly the Lower Paleozoic, dip at a shallower angle ($15° - 20°$ less) on the W flank of the anticline. This flank, which dips to the W towards the Shan Escarpment, is broken up by extensive N–S- and NW–SE-trending, mostly W-dipping synthetic normal faults, into individual blocks (in the Shan Highlands the "Karani" and the "Ingyi-Ingaung Fault"). The latter controls the fault scarp which, N of Ny-aing, forms the E boundary of the several kilometres wide plateau of Ye-ngan (Fig. 55).

Along the E flank of this anticline, another synthetic E-facing fault scarp occurs at which the Silurian rock sequence is reduced (BROWN & SONDHI 1933b, p. 235). To the E follows a narrow syncline containing Thitsipin and Nwabangyi Dolomite and the wide Mawson Anticline, whose S-dipping axis GARSON et al. (1976, p. 46) were able to trace through to the area immediately to the E of Heho (Fig. 55). The E flank of this anticline submerges beneath the Pleistocene lacustrine sediments of the intramontane Inle Lake Basin. The E boundary of this graben-like structure is formed by the imposing N–S-trending Taunggyi Fault Scarp (Fig. 55). Further to the N, in the area of Lawksawk, the Taunggyi Fault, which here trends NNW, cuts off the previously mentioned fold structures (SONDHI in HERON 1937, GARSON et al. 1976).

LA TOUCHE (1913) and BROWN & SONDHI (1933b) observed changes in thickness of the Ordovician and Silurian sediments in the areas of island-like uplifts of the Chaung Magyi Formation throughout the East Kachin/Shan Unit. In the course of further geosynclinal development, this relief was broken up even more. Uplift movements are evident in the varying degrees of erosion of Devonian and Silurian sediments (synsedimentary fold and fault movements). In the Kyaukme area, the Shan Dolomite Group follows on top of the partially missing Konghsa Marl Members or above the Namhsim Formation. Further to the W, in the Yadanatheingyi area, the Permo-Triassic carbonates are locally superjacent on the Lower Silurian Panghsa-Pye Formation. In this connection, MITCHELL et al. (1977, p. 19) noted that the low angle of the unconformity between the dolomite and the underlying rocks indicates that no major orogeny occurred in the region either in the Devonian or Early Permian. GARSON et al. (1976, p. 65) described wide folds in the Southern Shan State with NNW-trending axes in Cambro-Ordovician sediments. According to these authors, the folds were caused by vertical displacements of blocks of the underlying Chaung Magyi Group. In post-Permian times folding was continued about the same axes as a result of continued upward movements of the Chaung Magyi horsts. Consequently the pre-existing folds in the pre-Devonian rocks were accentuated, and the unconformably overlying Permian rocks acquired dips in the same direction but of lesser magnitude.

Locally, however, more intensive folding of pre-Permian sediments has been observed in the area of Bawsaing, from which GOOSSENS (1978a, Fig. 17) described an isoclinal fold structure characterized by overturned synclines and anticlines with overthrusting in Cambro-Ordovician sediments.

In the area of Neyaungga, the Pan Laung Fault, which runs in the Shan Scarp (Fig. 55), separates the predominantly carbonate sequence of the Shan Plateau, which extends from the Ordovician into the Jurassic, from the Jurassic-Cretaceous Pan Laung Formation. The Nwalabo Fault Belt, which accompanies the Pan Laung Fault in the E, forms a zone of intensely faulted and shattered limestone, divided up by subsidiary sub-parallel faults into numerous narrow wedges of widely different thicknesses and extent. In most places, shattering and brecciation of the limestones has obscured any original bedding and organic remains, and it is impossible to relate the rocks to any specific formation (GARSON et al. 1976, p. 59). However, rocks of individual block sections could be identified, i.e. Ordovician Doktoye Limestone, faunally dated Thitsipin Limestone and Nwabangyi Dolomite as well as Kalaw Red Beds. The latter are evidence of an age of at least Upper Jurassic for the movements along this fault complex. GARSON et al. (1976) assumed fault throws of altogether approximately 1,300 m, which is the amount by which the foreland in the W was lowered in relation to the Shan Plateau. To the S, the fault complex breaks up into an NNW–SSE-trending fault bundle which can be traced together with the Pan Laung Fault over several hundred kilometres (Fig. 55). Some of these faults control the fault scarp between the Shan Plateau, which is here composed mainly of Permo-Triassic carbonates, and a foreland containing chiefly clastic sediments and locally para-metamorphic rocks, where carboniferous fossils were recently discovered (Geol. Map of the Soc. Rep. of the Union of Burma, Explan. Brochure, 1977). The Pan Laung Fault therefore seems to represent a zone in which considerable structural movements were active already during the Paleozoic. Within the Burmese-Malayan Geosyncline it marks the facies change between the shelf region, with predominantly carbonate sedimentation, and a depositional environment adjoining in the W with chiefly flysch-type sediments. Late Mesozoic and Tertiary block movements have reactivated this zone.

Another section of the NNW-trending faults, which detach themselves from the Nwalabo Fault Belt, controls the elongated, graben-like structures which, near Kalaw, contain the tightly folded Jurassic Loi-an Series and the overlying Kalaw Red Beds (BROWN & SONDHI 1933b). GARSON et al. (1976) compared the disharmonic folding of the Loi-an Series with the Triassic-Jurassic Bawgyo and Namyau Group of the Hsipaw Syncline. They assumed that Triassic evaporites may also be present in the area of Kalaw. They may have played a role as slip horizons, thus contributing to the complex fold structure with overthrusts oriented in all directions.

The pronounced ancient relief of this region, with its long NNW-striking basin and uplift zones, was caused by synsedimentary block movements. This relief yielded the clastic debris of the Jurassic red sediments. BRUNNSCHWEILER (1970, p. 74) and GARSON et al. (1976, p. 37) stressed that the red beds concordantly overlie the older sediments in the basins. This observation points to continuous block-tectonic events which later are followed by the enormous uplift movements during the Cretaceous and the Tertiary. Tertiary fault movements also created the intramontane basins which received the Tertiary and Quaternary fresh-water sediments such as those in the Inle (Yawngwe) Lake Basin, the Thamakan and Heho basins (BROWN & SONDHI 1933b) and in the Lashio Basin (LA TOUCHE 1913).

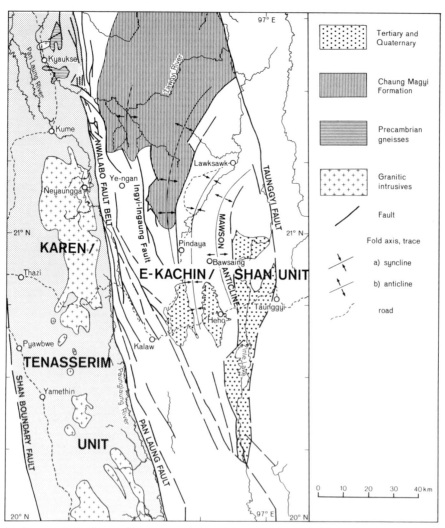

Fig. 55. Structural sketch map, Taunggyi area, Southern Shan State. Modified after GARSON et al. 1976.

3.243 Karen/Tenasserim Unit

The name Karen/Tenasserim Unit is used here for the area located W of the Pan Laung Fault and bounded in the W by the Shan Boundary Fault (Fig. 29). This unit extends from Mandalay via Mawchi to Taungnyo and continues via Tenasserim into Thailand and Malaysia (Fig. 56). The thick, clastic, low-grade metamorphic, tightly folded sediments were described by MIDDLEMISS (1900) and considered by SEARLE & BA THAN HAQ (1964) as part of the Mogok Belt.

The facies of these sediments and newly discovered fossils of Carboniferous age (Lebyin, Taungnyo Group and Mergui Series; Geol. Map of the Soc. Rep. of the Union of Burma, Explan. Brochure 1977, p. 8; BRUNNSCHWEILER 1970) reveal a separate flysch-type depositional environment located W and in front of the shelf-type depositional environment of the Shan Plateau. The clastic sequences, which possibly extend deeper into the Paleozoic (Fig. 30, folder), are underlain by Precambrian crystalline rocks. Permo-Triassic carbonates (Moulmein Limestone, Kamawkale Limestone) overlie in the N discordantly and, locally, in the S concordantly, the Lower Paleozoic strata. Marine Late Mesozoic sediments are represented by red clastics of the Jurassic, by greywacke and slate of the Ma-u-bin Formation and by sandstone, limestone and conglomerate of the Pan Laung and Yebok formations. Marine Tertiary sediments are observed in the N (Dahatpin Limestone Formation) and in the S in a number of basins of the Tenasserim region. They occur probably also beneath the Quaternary sediments in the area of the Shan Boundary Fault.

In the area of Mandalay-Kyaukse, crystalline rocks (Precambrian), occur in a number of uplifts that are surrounded by unconsolidated Quaternary sediments. The general direction of the strike of the gneiss, calc-silicate and granulite is NNW–SSE, and these rocks dip at 30–45° to the E. As in the Mogok area, the metasediments exhibit minor isoclinal folds and, for example, at the Mandalay Hill, the folds have amplitudes of 20 m, and their axes dip at 30° to the SE (SEARLE & BA THAN HAQ 1964, p. 146). According to these authors (p. 147), the general structure of the Sagaing Hills W of Mandalay is that of a NNW–SSE-striking synclinal structure complicated by overthrusts, faults and special folding. The calc-silicate gneiss and marble in the central parts of the Sagaing Hills have been deformed into tight anticlines and synclines with drag foldings in their limbs. The folds are intersected by overthrusts along the strike which dip ± 40° NE. The fold axes plunge gently NW and thus differ from the general dip of the axes of the Mogok Crystalline Rocks, which – as in the Minwin Ridges to the W – dip ± 45° SE. In the valley separating the two mountain chains, SEARLE & BA THAN HAQ (1964) observed a major tectonic feature which they regarded as a thrust plane along which the Sagaing Hills were overthrust onto the Minwin Hills. The next outcrops of crystalline rocks that follow in the S, in the area of Kyaukse and E of Meiktila, contain augengneiss and marble whose fold axes also dip 20°–25° SE.

The approximately 30 km wide belt between Pyinmana and Martaban, which is identified on the geological map (Fig. 17a, folder) as Precambrian gneiss, also strikes NNW–SSE (Fig. 56, folder). In the W it is accompanied by an equally extensive granitic intrusion, which is probably the largest of its kind in Burma. It extends into the Mawchi area, still striking NNW (HOBSON 1941, pp. 2 and 4). N of Papun, a narrow strip of Precambrian crystalline rocks branches off in the SE direction and extends as far as the Dawna Range where it is cut off by a granite body (COTTER 1922, 1924; Fig. 56, folder).

On satellite images, the contact to the low-grade metamorphic clastic sequences which adjoin the crystalline rock in the W is marked over wide areas by a fault zone (Fig. 56, folder), which HANCOX (in HOBSON 1941, sect. C–C[1], Pl. 2) represented as the fault-bound contact between the crystalline rocks and the Mawchi Series in the Mawchi area.

The flysch-type sediments that occur E of the zone of crystalline rock (MIDDLEMISS 1900, DUTT 1942), in which Carboniferous faunas were recently identified (p. 69), are tightly folded and strike generally NNW–SSE. The sediments dip at varying angles to the E. Isoclinal folds have frequently been observed. The Mesozoic clastic rocks of the Pan Laung and Yebok formations (GARSON et al. 1976), which have been described from the region of Neyaungga, also strike NNW and dip 30°–70° E. Flat-lying beds of the Dahatpin Limestone Formation, which is assigned to the Upper Cretaceous to lowermost Tertiary, overlie discordantly the above-mentioned formations.

In the E, the zone of the tightly folded clastic sediments of the Karen/Tenasserim Unit borders on the Pan Laung Fault or on the Nwalabo Fault Belt which separates this unit from the widely folded zone of the East Kachin/Shan Unit (Fig. 56, folder). Judging by the position of the block located to the W of these faults, uplift movements must have occurred relative to the shelf area adjoining to the E. In fact, from the distribution of the conglomerates in the Kalaw Red Beds, GARSON et al. (1976, p. 61) were able to deduce a rapidly uplifted block of limestones on the W of the Pan Laung Fault. Throughout the Cretaceous and the Tertiary, however, the W block was subsiding.

Towards the SSE, the Paleozoic clastic sediments W of the Pan Laung Fault lead over into the area of the Mawchi Series (Fig. 56, folder). These strike NNW–SSE with generally steep dips to the W and E. MITCHELL & HANCOX in HOBSON (1941, p. 149, Pls. 2–4) assumed that the fold structure is isoclinal, although no fold patterns can be observed in the monotonous sequence. The SE continuation of the Pan Laung Fault probably runs along the Nammekon Plateau, which is built up of Permo-Triassic carbonates and of the Yinyaw Beds. HOBSON (1941, Pl. 2, sect. A–A', B–B') reported that the carbonates are steeply tilted and the argillites of the Mawchi Series are tightly folded. Further to the SSE, the areas with sediments of the Taungnyo Group and Mergui Series of the Tenasserim Division follow on without any break in the lithofacies and in the structural setting (Fig. 56, folder). These clastic sediments strike generally NNW–SSE.

Strong structural deformation prevails in the E part of the Tenasserim region, while in the W part and on the islands of the Mergui Archipelago only moderate dips of strata were observed (RAU 1933, p. 14). Because of the monotony of the several thousand metres thick sediments, individual fold structures have rarely been identified. In the upper Tenasserim valley, BOSE (1893, p. 157) observed that the river flows in the core of an anticline with steeply W- and E-dipping limbs. Tight folding in these clastic sequences has also been observed in the neighbouring region in Thailand (KOCH 1973, 1978, Sheet Thong Pha Phum, and GARSON et al. 1975, Phuket area). Low-grade metamorphism, which is probably due to a dynamometamorphism, is common to the entire region.

A phase of pre-Permian folding events becomes evident by the discordant overlying of Permian Moulmein Limestone on strata of various ages of the Taungnyo Group in the area of Pa-an (BRUNNSCHWEILER 1970, p. 66). Further to the S, however, the change from clastic Devonian-Carboniferous sediments to the Permian shallow-water carbonates was obviously not interrupted by folding events (RAU in PASCOE 1924, p. 32, GARSON et al. 1975).

GARSON et al. (1975) placed another major episode of folding in the period following the deposition of the Triassic-Jurassic Ko-Yao Formation. KOCH (1973) was similarly able to show that Lower Jurassic limestone sequences were affected by major folding in the area of Thong Pha Phum in Thailand. This dating of an orogenic event holds true in the Tenasserim region as well. It is in particular verified by the flat-lying Upper Mesozoic red sandstone series of Mergui which rest with an angular unconformity on top of the folded Mergui Series.

Several major fault zones can be observed in aerial photographs and satellite images of the Tenasserim region. They extend into Thailand where some of them are used for subdividing tectonic units (CAMPBELL & NATALAYA in STOKES et al. 1975). From S to N, these are the Ranong Fault Zone, the Three Pagoda-Ratchaburi Fault Zone and the Moei-Uthai Thani Fault Zone (Fig. 56, folder).

The **Ranong Fault Zone** at the S border of Burma strikes NNE–SSE; this is the most easterly of the faults and, according to GARSON et al. (1975, p. 39), it forms part of a fault system tens of kilometres wide containing a graben structure with a vertical displacement of several hundred metres. In addition, sinistral movements of at least 150 km are assumed to have occurred at this fault system. Since the faults offset the Ko Yao Formation, they date later than the Late Mesozoic.

The **Three Pagoda-Ratchaburi Fault Zone** trends NNW – SSE. It extends from the Khwai Noi Valley in Thailand via the Three Pagoda Pass to the area S of Moulmein. On the basis of aerial photographs, BRUNNSCHWEILER (1970, p. 66) interpreted it as the S boundary of the distribution area of sediments of the Taungnyo Group. In the area of Thong Pha Phum, KOCH (1973, p. 183, 1977) observed that this fault system affected not only the Paleozoic clastic sediments but also the Permian and Mesozoic carbonates. He therefore placed it in the Late Mesozoic period of folding.

The fault movements remained active even after the consolidation and uplift of the Tenasserim region. They created the graben-like depression areas, which run mainly along the strike of the mountain ranges, and in which lacustrine and terrestrial sediments bearing lignite and oil shales were deposited during the Tertiary (Tenasserim Valley, Thein Kun Valley and Pakchan Valley deposits). RAU (1933, p. 35) described the Tertiary sediments as gently folded with maximum dips of 35° at the edges of the basins.

Tertiary basins containing oil shales are located E of the Dawna Range to the S of the Kamawkale Peak, and of these the Htichara Basin borders directly on the belt made up of gneiss of the Dawna Range to the SW (p. 169). COTTER (1924b, p. 289) described the bedding of the max. 38° dipping Tertiary sediments in the synclinal basins as undulated to horizontal.

The basins, which are filled with unconsolidated Quaternary sediments, provide evidence of vertical movements and erosion of high-lying areas right through to the present time. Relative sea-level changes resulted in marine ingressions in narrow depressions near the coast, which extend fjord-like into the mountain zone (e.g. Heinze Chaung).

3.244 Regional setting

Together with W Thailand and W Malaysia, the E parts of the Sino-Burman Ranges form a tectonic unit which STAUFFER (1973) referred to as the "West Malayan Block". This block contains essentially the Late Mesozoic consolidated sediments of the Burmese-Malayan (KOBAYASHI 1964b) or of the Yunnan-Malayan (BURTON 1967a) Geosyncline on the Precambrian basement. According to STAUFFER (1973), in the E there follows a Paleozoic island arc with andesitic and basaltic volcanics and slope-deposited clastics of the "East Malayan Block". This in turn is followed by the Precambrian-floored microcontinent of the Indochina Block, which extends into E Thailand. STAUFFER (1973) believed that these three blocks were welded together during the Permian as a result of a plate collision. This hypothesis has been modified several times in recent years (PONGPOR ASNACHINDA 1979, THEERAPONGS THANASUTHIPITAK 1979, CHANTARAMEE 1979, BUNOPAS & VELLA 1979, MACDONALD & BARR 1979).

The concept of subduction through convergent movements between the W continental section of the crust, consisting of E Burma-W Thailand-W Malaysia, and the E plate of the Indochina Block, as well as the concept of a W-dipping Benioff Zone beneath the first-mentioned section of plate, are common to all the hypotheses. MITCHELL (1977) also believed that the SE Indian Peninsula formed as a result of a plate collision in the Early

Mesozoic, but on the basis of structural criteria he thought that the Benioff Zone dipped towards the E beneath the relatively W-moving Indochina Block.

JONES (1968) and HUTCHISON (1973) postulated a Precambrian consolidated craton in the W of the Paleozoic, Precambrian-floored Burmese-Malayan sedimentation area. This craton is now located in the present-day Andaman Sea. These authors assumed that the craton was the source area for the Upper Precambrian and Lower Paleozoic clastic rocks (Machinchang Formation, Chaung Magyi Formation; p. 51). It is possible that the Devonian-Carboniferous flysch sediments of the Karen/Tenasserim Unit also derived from this land mass. According to HUTCHISON (1973), this craton, with its W-dipping Benioff Zone, was in subducting contact to the Burmese-Malayan crustal section. He believed that the basic volcanic rocks of this craton between Chiang Rai and Chiang Mai (Fig. 56, folder) constituted an ophiolitic belt paralleling this Paleozoic trench. He also believed that the separation of the land mass in the W from the oceanic crustal section in the E took place in the Paleozoic, whereas ASNACHINDA (1979) and MITCHELL (1977) thought that the separation took place during the Triassic, when it was caused by rifting or by sphenochasm (MAUNG THEIN 1973). According to this view, the present structural W-edge of the Sino-Burman Ranges would be located at the Shan Boundary Fault (Fig. 56).

Many structural phenomena in the SE Indian Peninsula can be explained by vertical movements, for example the synsedimentary fault movements in the Paleozoic, which divided up the Burmese Malayan Geosyncline into areas of uplift and subsidence. The movements at these faults lasted beyond the Paleozoic sedimentation and were renewed in the Mesozoic and Cenozoic. The curved directions of strike of the Paleozoic and Cenozoic sediments around a Precambrian core, e.g. the structure of Chiang Mai (Fig. 56, folder), point to horst-like uplifts of the older basement.

The general strike of the faults caused by tectonic movements that have been active since the pre-Cambrian is NNW–SSE. A direction which is also followed by the main fault structures bordering the tectonical domains of the Sino-Burman Ranges (Figs. 55, 56, folder).

The W boundary of pre-Mesozoic outcrops runs very noticeably parallel to the Pan Laung Fault, which is where the E boundary of the Paleozoic flysch trough of the West Kachin/Tenasserim Unit is located. Accordingly, the W part of this flysch trough, whose Precambrian basement is known from the West Kachin/Tenasserim Unit, is overlain by the thick sediments of the Central Tertiary Basin. The subsidence of this basin, which together with the formation of the Inner Volcanic Arc and of the Back Arc Basin is regarded as connected with Tertiary and Quaternary plate tectonic movements, thus took place in a zone which was already an area of subsidence during the Paleozoic.

3.3 Magmatism

3.31 Pre-Mesozoic igneous rocks

3.311 Precambrian igneous rocks

The oldest magmatic rocks in Burma occur in the area of the Mogok Belt (p. 49). They are composed of a large number of rock types which SEARLE & BA THAN HAQ (1964) referred to as the "Alaskitic Suite". The suite includes syenite and nepheline-syenite (FERMOR 1932,

p. 81), also tourmaline granite, augite and hornblende granite and alaskite-like rocks which were probably originally of nordmarkitic composition. In the alaskites and nordmarkites the quartz content fluctuates from 50 to 5 vol.%, and the mafite content is less than 5%.

The Alaskite Suite, whose sheet-like intrusion bodies are interfolded in metasediments, is closely linked with the migmatization of the metasediments (SEARLE & BA THAN HAQ 1964, p. 143). The large number of rocks types mentioned above formed during the reactions of the leucocratic, alaskitic-nordmarkitic magma with the metasediments. As a result, nepheline-bearing and cancrinite-bearing syenite formed through assimilation of low-quartz carbonate sediments. Tourmaline granite as well as − when increasing amounts of mafic components were assimilated − hornblende-augite-granite, and ortho-syenite to hornblendite formed from quartz-rich contact rocks. The formation of the corundum minerals (ruby and sapphire) and of the spinels is also ascribed to reactions of the alaskitic and syenitic aluminium-rich magma. The corundum group formed in the low-magnesium milieu of the contact whereas the spinels formed in the magnesium-rich milieu. The rocks of the Alaskite Suite are rich in accessory minerals, which indicate that the magma contained a high percentage of gaseous components: allanite, fluorite, ilmenite, samarskite, titanite, topaz, tourmaline, zirconium and pitchblende.

In addition to the rocks of the Alaskite Suite, SEARLE & BA THAN HAQ (1964) identified mafic and ultramafic rocks, pegmatites and aplites and the extensive intrusive bodies of the Kabaing Granite in the Mogok area (p. 142).

The **mafic and ultramafic rocks** are not very widespread. They consist of dolerites and ultrabasic rocks such as picrite, hornblende-pyroxenite, pyroxenite and peridotite. The rocks are partly serpentinized and uralitized. They intrude the Alaskite Suite and are regarded as its derivatives.

The **pegmatites and aplites** of the Mogok area contain tourmaline and rare earths, including radioactive minerals (p. 172). They transect the metasediments and the Alaskite Suite in cm- to decimetre-thick veins. ADAMS (1926) described a large, over 30 m thick pegmatite vein from the area E of Sakangyi. This vein consits of kaolinitized orthoclase and quartz and was being mined for its content of semi-precious gemstones (topaz). The pegmatites and aplites, as well as the "Kabaing Granite", are probably of Tertiary age (p. 142).

3.312 Cambro-Ordovician igneous rocks

The rocks of the "Bawdwin Volcanic Stage" (LA TOUCHE 1913, p. 55) or of the "Bawdwin Rhyolite Series" (BROWN 1918a) were defined as true rhyolites, rhyolitic tuffs, true tuffs of various kinds and occasional bands of volcanic breccia and subordinate layers of feldspatic grit (BROWN 1918a, pp. 39 et seqq.). BRINCKMANN & HINZE (1976) divided these rocks into Bawdwin Volcanic Clastics and Bawdwin Rhyolites. The volcanoclastic sediments belong to an older phase of the rhyolitic volcanism and are intercalated in the Cambro-Ordovician Pangyun Formation (p. 53), which is positive proof of their Lower Paleozoic age. They originated from volcanic products which were deposited together with rhyolitic flows in a subaquatic milieu. The Bawdwin Rhyolites belong to a younger intrusive phase. They intruded not only the Pangyun sediments but also the volcanoclastics. The lead-zinc ore body of Bawdwin (Fig. 76) was emplaced during a subsequent hydrothermal-pneumatolytic phase of the volcanic acitivity. According to BRINCKMANN & HINZE (1976) the Bawdwin Volcanic Clastics are up to 1,000 m thick and consist of:

a) Agglomerate and conglomerate: pebbles and angular fragments of alkaline feldspars, rhyolite, quartzite and schist in a fine-grained quartz-alkaline feldspar matrix. The light grey to greenish-grey rock is massive and largely unbedded (facies close to eruption centre).

b) Conglomerate and greywacke: compared to Type a, the lithology is similar but the rock components are smaller in diameter, better sorted and well bedded. The matrix contains more detritic quartz. This rock type shows smooth transitions – laterally as well as vertically – to the massive Type a) (facies far from eruption centre).

c) Greywacke and arkose; fine-grained, well-bedded, dilution of volcanic rock content by detritic quartz; more frequent intercalations of quartzite and sandstone of the Pangyun Formation (facies far from eruption centre).

In general, the **Bawdwin Rhyolites** form a massive, brittle, rarely amygdaloid rock with phenocrysts of colourless, rounded quartz up to 1 cm in diameter, and mm-sized greenish-grey inclusions of slates; in addition to quartz, MÜLLER & WEISER (1975) also observed clouded alkali feldspar and biotite as the phenocryst generation. Zirconium and apatite occur as accessories; rutile belongs to the fine-grained secondary paragenesis (Figs. 57a, b).

The rhyolitic volcanic rocks of the Bawdwin area have undergone intensive alteration through sericitization and silicification. According to MÜLLER & WEISER (1975), the deep-reaching alteration influenced the primary rock chemism in such a way that when corresponding norm calculations are carried out, most of the rhyolite samples that were investigated fall in the alkali feldspar-syenite field. However, the normative cordierite contents show that the originally rhyolitic volcanic rocks did not have the character of alkaline rocks before the intensive sericitization took place.

The alteration of the volcanic rocks also influenced the Rb/Sr isotope ratio. Radiometric age determinations gave a Triassic age of 212 ± 4 m.y. (MÜLLER & LENZ 1981), an age which is certainly not in accordance with the age of the volcanic rocks that are intercalated in Cambro-Ordovician sediments. Furthermore, the high initial ratio of Sr^{87}/Sr^{86} indicates that this age cannot be related to the intrusion age.

Bawdwin Volcanics also occur in the Kyaukme-Longtawkno area about 20 km to the S of Bawdwin where they were described (MITCHELL et al. 1977, p. 15) as rhyolitic lavas, pyroclastic rocks and "tuffaceous sediments". They form lenticular volcanic masses, mainly at the base of the Pangyun Formation, with which they interfinger laterally and vertically.

STUART (1923, p. 407) reported a further occurrence of presumably Lower Paleozoic tuffs in N Burma from the Lagwi Pass approximately 80 km ENE of Myitkyina. The occurrence consists of dark grey, fine-grained siliceous tuff composed of highly angular quartz fragments surrounded by finer material containing much chlorite. The tuff is in contact to gneiss and metamorphic limestone. The occurrence strikes into China.

Sporadic occurrences of acid to intermediate volcanic rocks of the Lower Paleozoic are also known from Perak and Selangor in Malaysia (JONES 1970, YEW 1971). STAUFFER (1973, p. 103) believed these volcanic rocks were associated with a Benioff Zone which dips towards the W and submerges beneath a craton that is assumed to have existed to the W of the Burmese-Malayan Geosyncline (p. 36).

Fig. 57 a. "Fluidal structure" in rhyolite. Upper Pangyun Valley above Bawdwin mine. Photo: W. HANNAK. – b, Pyroclasts composed mainly of fragments of rhyolite. Piece of PbS/ZnS ore (in upper right corner of picture). Upper Pangyun Valley above Bawdwin mine. Photo: W. HANNAK.

3.313 Ordovician and Silurian igneous rocks

From the Pindaya Range N of Heho/Southern Shan State, BROWN & SONDHI (1933 b, p. 214) and MYINT LWIN THEIN (1973, p. 161) described rhyolites and rhyolitic tuffs which are interbedded in the limestone and slate of the Orodovician "Pindaya Beds". The light grey rock contains quartz phenocrysts and exhibits locally a banded flow structure.

Also from the Pindaya region, MYINT LWIN THEIN (1973, p. 161) reported volcanic tuffs and ashes which are intercalated in the Silurian Wabya Formation.

3.314 Carboniferous igneous rocks

The "Tawng Peng Granite" (LA TOUCHE 1913) in the Northern Shan State is one of the largest intrusive bodies in the East Kachin/Shan Unit (Fig. 17 a, folder). It is an elongated rock body that runs parallel to the NE–SW-strike of the mountain range from Namhsan via Namkhan to Yunnan/China. MITCHELL et al. (1977, p. 21) described it as a light-coloured porphyric rock with phenocrysts up to 3 cm in diameter consisting of quartz, microcline, plagioclase, orthoclase, muscovite and biotite. Titanite, apatite and zirconium were identified as the primary accessory minerals. The granite is frequently crushed, sheared and foliated; thin sections reveal sericitization, kaolinitization and saussuritization. With high initial Sr^{87}/Sr^{86} ratios, total rock analysis using the Rb/Sr method on samples taken from mile 71 along the Kyaukme-Namhsan Road gave a Carboniferous age of 340 ± 34 m.y., which is a date that must be treated with caution because of the divergencies in the Rb/Sr values (MITCHELL et al. 1977, p. 21). If this dating is correct, then the intrusion of the Tawng Peng Granite belongs to the generation of the Carboniferous intrusive rocks which is also known in Thailand (BURTON & BIGNELL 1969) and Malaysia (SNELLING et al. 1968).

Volcanic rocks occur in the Mergui Archipelago of S Burma. They are intercalated in the sediments of the Mergui Series (p. 70) and contain tuffs and agglomerates whose components consists of pumice, volcanic glass, porphyry, rhyolite, volcanic bombs, lapilli and fragments of sediments. Occurrences of these rocks, which are often associated with granitic and porphyric stocks, are also found on the islands of Elphinstone and Maingay as well as on several outer islands in the archipelago. Furthermore, on the W coast of Elphinstone, large masses of rhyolites and porphyries occur. There, as well as on King Island and Iron Island, the major eruption centres are assumed to have existed in the Carboniferous. Their volcanic eruption products reached the area of the present-day Tenasserim mainland, where they were intercalated in the sediments of the Mergui Series (RAU 1933, p. 10).

3.32 Mesozoic and Cenozoic igneous rocks

The distribution of post-Paleozoic intrusive and extrusive rocks reveals links with Cenozoic plate tectonic events. Their regional distribution is more or less parallel to the converging plates (Fig. 16).

From W to E one can differentiate between

– basic and intermediate sills, dykes, serpentinite stocks and submarine extrusions in the Arakan Coastal Area and in the W part of the Indo-Burman Ranges

– ophiolitic rocks mainly in the E part of the Indo-Burman Ranges and along the structural contact between the Indo-Burman Ranges and the Inner-Burman Tertiary Basin

– basic, intermediate and acid intrusive and extrusive rocks in the Inner-Burman Tertiary Basin, which belong to the Inner Volcanic Arc.

The distribution of the post-Paleozoic igneous rocks that are associated with the Shan Boundary Fault Zone (4) and are found in the Sino-Burman Ranges (5), however, point to intrusions and extrusions along zones of weakness which originated already in pre-Mesozoic times.

3.321 Igneous rocks in the Arakan Coastal Area and in the western parts of the Indo-Burman Ranges

Intermediate to basic dykes, sills and stocks of serpentinite occur at numerous points (e.g. Ngapali near Sandoway) in the Arakan Coastal Area and in the W part of the Indo-Burman Ranges. The diameters of these occurrences are in the order of a few tens of metres and rarely exceed 100 m. Sometimes they are associated with tuffite and agglomerate. According to BRUNNSCHWEILER (1966, p. 141), ophiolitic rocks are also concordantly interbedded in Upper Cretaceous strata as sills and as agglomerates and tuffites.

Basalt dykes and plugs are also frequent in this area. Near Sandoway, they penetrate Eocene flysch sediments. At the coast, approximately 4 km N of Ngapali, a steep hill measuring about 20 m × 30 m rises about 10 m high above the beach sands (Fig. 58 a). It consists of agglomerate, tuff and basaltic rock (Fig. 58 b). The latter is composed of plagioclase and chlorite-montmorillonite aggregates in a matrix of fine-grained devitrification and alteration products. Amygdale fillings consist of calcite (BGR-Lab. No. DS 25993, 1980). SSW of Taungup and on the island of Ramree, BRUNNSCHWEILER (1966, p. 156) found basalt vents which penetrated Lower Miocene strata so that a post-Lower Miocene age can be assumed for this basalt volcanism.

The occurrences of igneous rocks continue to the N and S from the Sandoway area.

3.322 Igenous rocks in the eastern parts of the Indo-Burman Ranges and along the fault system between the Indo-Burman Ranges and the Inner-Burman Tertiary Basin

Basic volcanic rocks occur sporadically in the central part of the Arakan Yoma. They seem to be largely lacking in the central (Chin Hills) Indo-Burman Ranges, but they occur again in the N (Naga Hills). Occurrences of these rocks are frequent in the E part of the Indo-Burman Ranges, e.g. about 65 km W of Thayetmyo and about 60–80 km WSW and W of Minbu (Fig. 17 a, folder). BRUNNSCHWEILER (1966, p. 161) mentioned thin feeder sills of ophiolitic rocks in Triassic sediments (p. 81) and submarine extrusions in the flysch sequence of the Kanpetlet area. GRAMANN (1974) observed spilitic basalts and dyke-like serpentinized volcanic rocks at the contact between the Kanpetlet Schists in the W and the

a

b

Fig. 58a. Basalt and tuff agglomerate at beach
approx. 4 km N of Ngapali. Photo: F. BENDER.
– b, Close-up photograph of same exposure.
Photo: F. BENDER.

Halobia-bearing Triassic sediments in the E. Ophiolites occur in belts up to 70 km long and up to 20 km wide, and they follow the general NE–ENE–W–E strike of this N section of the Indo-Burman Ranges (Fig. 14).

An approximately 1,200 km long chain of **basic to ultrabasic igneous rocks** (Fig. 17a, folder) runs, with a few interruptions, along the structural contact between the Indo-Burman Ranges and the Inner-Burman Tertiary Basin. In the S (W of Bassein and Henzada) the rocks are serpentinites (CHHIBBER 1934a, p. 239) which are either located along or in the immediate vicinity of the main fault zone, which strikes SW–NE here, where Eocene flysch-type sediments are in fault contact with Miocene sediments. In the section from W of Prome to W of Minbu, serpentinite masses are observed along the fault contact between the Cretaceous flysch sequences in the W and Eocene sediments in the E.

An approximately 100 km long and up to about 15 km wide chain of ultrabasites occurs along the general NNW–N–NNE strike of the main fault zone W of Kalemyo. The occurrences cover areas up to 70 km² with N–S lengths of up to 20 km and E–W widths of up to 8 km. They are located at the structural contact between the Chin Hill Flysch Series (Upper Cretaceous) in the W and mainly Triassic rocks in the E. The occurrences are mainly small ones of serpentinite, gabbro and diorite and large ones of peridotites. The latter consist of a dense serpentinized ground mass with up to cm-sized bastite-like pyroxenes. Transitions to saxonites (harzburgites) and dunites exist. Chromite and magnetite are widespread as accessory minerals. Chromite occurs in streaky and stock-like, ore-deposit-forming concentrations. Chalcedony, quartz, chrysotile, magnesite, chlorite, talc, antigorite and nickel-silicate are found in joints and crevices in the ultrabasites. Garnierite is widespread in the lateritic weathering zone (p. 198). The ultramafitite to mafitite rock bodies exposed here are "dismembered ophiolites" (basal portions of metamorphic tectonites and lowermost "cumulate complex"). This sequence dips to the SSW and WSW in the Mwetaung area, and in the Bhopi Vum area steeply towards the E.

Further ultrabasites are known from the W edge of the N Chindwin Basin and from the S border of the Hukawng Basin.

Views differ on the age and mode of emplacement of these basic magmas. Because the Eocene sediments have not been altered anywhere by contact metamorphism, THEOBALD (1873) and COTTER (1938) assumed a pre-Tertiary age for the intrusions. CLEGG (1938, 1941a) thought it was more likely that the magmatism was associated with the Tertiary folding of the Indo-Burman Ranges. BRUNNSCHWEILER (1966, pp. 164–167) found that the W edge of the Webula Taung (1,808 m; W of Kalemyo) basic rock complex is formed by an overthrust in which the basic rocks were thrust steeply over the E-dipping flysch sediments. From this he concluded that the W edge of the Inner-Burman Tertiary Basin is overthrust on the E edge of the Indo-Burman Flysch Ranges and that in the process parts of the Paleozoic or even Precambrian basement were uplifted tectonically to a higher level together with the ultrabasites. However, since these ultrabasites are clearly associated with the more than 1,200 km long structural contact between the Indo-Burman Ranges and the Inner-Burman Tertiary Basin, and this fault contact does not take the form of a very large-scale overthrust throughout its entire length, it is unlikely that such an upthrust of the basic rock complex occurred. It is more likely that an Upper Cretaceous-pre-Eocene emplacement occurred along a tectonic suture (zone of weakness) where overthrusting took place afterwards in the course of the Oligocene and Miocene Alpidic orogeny.

3.323 Igneous rocks along the Inner Volcanic Arc

In the **Inner-Burman Tertiary Basin** the **Inner Volcanic Arc** can be traced which runs from W Sumatra via the volcanic islands of Barren Island and Narcondam Island through the Gulf of Martaban as far as N Burma (Fig. 16). This arc is characterized by a discontinuous chain of basic, intermediate and acid, Cretaceous and Tertiary to sub-Recent intrusive and extrusive rocks which follows the S−N general strike of the Inner-Burman Tertiary Basin. Volcanic activity is nowadays only found on Barren Island. N of the area of Mt. Popa in Central Burma, basic and intermediate intrusives occur in more or less parallel zones together with igneous rocks of various chemical composition.

The Inner Volcanic Arc was identified by seismic reflection in the Gulf of Martaban, and in the Irrawaddy Delta area it was encountered in two exploratory oil wells ("volcanic rocks, basic tuffs and agglomerates"). Further to the N, from the W edge of the Pegu Yoma, CHHIBBER (1934a, pp. 301, 427−436) described dolerite dykes and intrusions in the Miocene Pegu Series:

− In the area SW of Kyaukpyu (Tharrawaddy District), 5 occurrences of intrusive olivine-dolerite, largely serpentinized, in baked Miocene mudstone, with the chemical composition given in Table 2.

Table 2. Chemical composition of olivine-dolerite SW of Kyaukpyu (after CHHIBBER 1934a).

		Molecular Proportions		Norm
SiO_2	49.70	0.828	Orthoclase	7.78
Al_2O_3	18.04	0.176	Albite	26.20
Fe_2O_3	2.10	0.013	Anorthite	27.52
FeO	4.92	0.068	Nepheline	3.69
MgO	7.65	0.191	Diopside	
CaO	9.50	0.170	$CaOSiO_2$	7.54
Na_2O	3.90	0.063	$MgOSiO_2$	5.10
K_2O	1.35	0.014	$FeOSiO_2$	1.85
			Olivine	
$H_2O+105°$	1.80	0.100	$2\ MgOSiO_2$	9.80
$H_2O-105°$	0.10	0.006	$2\ FeOSiO_2$	4.08
CO_2	−	−	Magnetite	3.02
TiO_2	0.45	0.005	Ilmenite	0.76
P_2O_5	0.29	0.002	Apatite	0.67
MnO	0.20	0.003		
S	−	−		98.01
			Water	1.90
Total	100.00		Total	99.91

− In the area of Myenettaung (E of Allenmyo, Prome District), three occurrences of intrusive olivine-dolerite and dolerite sills, highly serpentinized, in baked Miocene host rock, with the chemical composition given in Table 3.

Further N along the Inner Volcanic Arc follow the **volcanic rocks of the area of Mt. Popa** (approximately 40 km E of Chauk; Figs. 59, 60), which occupy an area of approximately 700 km² with an N−S extent of 35 km and an E−W width of approximately 20 km. The volcanic ruin of Mt. Popa, which still contains an eruption vent, is at the centre

Table 3. Chemical composition of olivine-dolerite of Myenettaung E of Allenmyo, Prome District (after CHHIBBER 1934 a).

		Molecular Proportions		Norm
SiO$_2$	48.00	0.800	Quartz	1.26
Al$_2$O$_3$	18.70	0.183	Orthoclase	3.89
Fe$_2$O$_3$	3.08	0.019	Albite	16.77
FeO	4.29	0.060	Anorthite	40.03
			Diopside	
MgO	8.30	0.207	CaOSiO$_2$	1.16
CaO	8.80	0.157	MgOSiO$_2$	0.80
Na$_2$O	2.00	0.032	FeOSiO$_2$	0.20
K$_2$O	0.72	0.007	Hypersthene	
H$_2$O+105°	3.70	0.206	MgOSiO$_2$	19.90
H$_2$O−105°	1.70	0.094	FeOSiO$_2$	5.02
CO$_2$	–	–	Magnetite	4.41
TiO$_2$	0.30	0.004	Ilmenite	0.61
P$_2$O$_5$	0.10	0.001	Apatite	0.34
MnO	0.20	0.003		
S	–	–		94.39
			Water	5.40
Total	99.89		Total	99.79

of this area. According to CHHIBBER (1934a, pp. 397–427), extrusions of the following rocks occurred here:

- pyroxene-andesite, acid andesite with spherulitic structure, augite-hornblende-andesite, biotite-pyroxene and basic andesite and hauyne-bearing pyroxene andesite, together with tuffs (S of latitude 20° 53′ N, presumably uppermost Tertiary)
- silicified andesite and silicified tuff from Gwegon and Konde (presumably uppermost Tertiary)
- rhyolite, andesite and silicified tuff from Kyaukpadaung and Taungnauk; the tuffs are intercalated in the Irrawaddy Group (p. 99) (Upper Miocene-Pleistocene)
- white, partially or completely silicified tuffs from Taurgni, Gwegon and Sebauk, also partially intercalated in the Irrawaddy Group
- black andesitic tuffs and ashes on the lower slopes of the Mt. Popa Plateau, partially intercalated in the Irrawaddy Group
- hornblende-augite-andesite, hornblende-augite-basalt, enstatite-basalt, olivine-basalt in countless lava flows from Mt. Popa, together with ashes and bombs (Late Pleistocene to sub-Recent, possibly through to early historic time).

The **Lower Chindwin Volcanics** are situated about 50 km further N along the NNW-striking section of the Inner Volcanic Arc. The occurrences of igneous rocks are here bound to two structurally pretraced zones. The W zone is the continuation of the "Pegu-Mt. Popa Line". It begins S of Kabauk in the Shinma Taung area and runs for about 110 km with a width of up to 18 km via Salingyi, Silaung Taung and Twin Taung to Natyin Taung in the N. The second line runs parallel to the first, E of Monywa, and can be followed for a distance of about 40 km (Figs. 61–64).

According to STAMP & CHHIBBER (1927b), CHHIBBER (1934a), BURRI (1931) and BARBER (1936), the major occurrences of igneous rocks along the Pegu-Mt. Popa Line are, from S to N (Fig. 61):

Fig. 59. Mt. Popa basalt volcano (background, left) and Taunggala neck of hornblende-augite-andesite (center). Photo: F. BENDER.

Fig. 60. Mt. Popa, view inside crater wall, looking N. Photo: D. BANNERT.

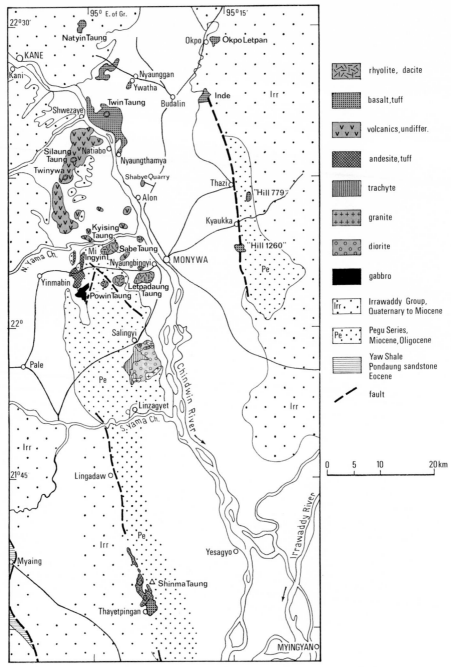

Fig. 61. Igneous rocks of lower Chindwin area, geological sketch map (modified after PINFOLD, STAMP & CHHIBBER 1927 and BARBER 1936).

- Shinmataung to Linzagyat: andesite, hornblende-andesite, olivine-basalt, picrite-basalt, dolerite, ashes and rhyolitic agglomerates and breccias
- Salingyi: quartz diorite, meta-andesite, trachyte, amphibolite, metagabbro, granodiorite, granite
- Sabe Taung, Letpadaung Taung, Kyising Taung approximately 10 km W of Monywa: biotite-dacite, rhyolite, rhyolitic tuff breccia, andesite, hornblende biotite porphyry, quartz porphyry, locally containing copper mineralization (p. 183).
- Powin Taung–Silaung-Kyaukmet area: granite, olivine-basalt (Silaung Taung), granite, hornblende-granite porphyry (S and W of Silaung Village)
- explosive craters S and N of the Chindwin River near Shwezaye: olivine-basalt, hornblende-pyroxene-andesite (in the Twin Taung crater with ejecta of quartz-diorite, augite-peridotite, biotite-hornblende-peridotite, hornblende, perknite, pyroxenite), rhyolite, quartz porphyry, liparite, ashes and coarse tuffs with lapilli
- Natyin Taung: olivine basalt.

From the Twin Taung crater, basaltic andesite was found of the chemical composition given in Table 4 (BGR Lab. No. 25844, 1980; in weight-%, trace elements in ppm).

Table 4. Chemical composition of basaltic andesite from Twin Taung.

	weight-%
SiO_2	51.19
TiO_2	0.77
Al_2O_3	17.72
Fe_2O_3	7.61
MnO	0.12
MgO	5.20
CaO	10.19
Na_2O	3.36
K_2O	2.54
P_2O_5	0.40
SO_3	0.01
LOI	0.60
Total	99.71

trace elements	ppm	trace elements	ppm
Ba	1517	Rb	40
Ce	142	Sc	28
Co	51	Sr	1567
Cr	61	Th	25
Cu	235	V	217
La	127	Y	20
Nb	0	Zn	53
Ni	52	Zr	145
Pb	30		

The basaltic andesite shows a porphyric texture (coarse pyroxene, hornblende, in a matrix of plagioclase, clinopyroxene, magnetite).

Fig. 62. Twin Taung explosive crater, oblique aerial photo from about 800 m above ground. Photo: F. BENDER.

The hornblende-biotite-porphyry from Kyising Taung and Sabe Taung W of Monywa intruded the rhyolitic tuff breccia and is therefore younger. A K/Ar dating of the porphyry gave an age of 5.8 million years, i.e. Middle Pliocene (Union of Burma, unpubl. rep. 1973).

The occurrences along the line to the E of the Pegu-Mt. Popa Zone in the "East Monywa Hill Belt" consist of (from S to N):

- picrite-basalt, olivine-basalt (Kyaukka Taung, "Hill 779", Thazi Hill)
- ashes with olivine-basalt blocks (Inde Hill)
- olivine-basalt, hypersthene-basalt, nepheline-tachylite (Okpo Letpan).

The basalts of Kyaukka Taung, from "Hill 779", from Thazi Hill and Okpo Letpan have intruded the sands and gravels of the Irrawaddy Group, and their ashes and tuffs are partly intercalated in this group, so that the age of these volcanic rocks can be assumed to be Upper Miocene-Early Pleistocene. The andesitic-dacitic volcanic rocks, like those in the Mt. Popa area, can probably be placed in the Plio-Pleistocene. The intrusive rocks are very likely hypabyssal representatives of the surface extrusive rocks, as CHHIBBER (1934a, p. 393) assumed.

The Inner Volcanic Arc continues for about 125 km NNE of the Lower Chindwin Volcanics in the **Wuntho Massif (Mingin Ranges)**. The Wuntho Massif is up to 40 km wide from E to W and is about 190 km long in the NNE direction. Results of geological reconnaissance carried out in this area have been published by, among others, NOETLING (1894), CHHIBBER (1934a), BROWN (1936) and SOMMERLATTE (1948). ZAW PE (unpubl. rep. 1976) reported on the results of geological mapping on the scale 1 : 100,000.

Fig. 63. Twin Taung explosive crater, view from E. Photo: F. BENDER.

Fig. 64. Silaung Taung explosive crater, aerial photograph from about 800 m above ground. Photo: F. BENDER.

Andesite stocks, flows, pillow lavas and volcanoclastic sediments are regarded as the oldest post-Paleozoic igneous rocks of the Wuntho Massif ("Namaw Chaung Andesite Group"). They were deposited in a marine environment and overlie with a distinct unconformity Upper Triassic (?) sediments. The Namaw Chaung Andesite Group also includes the Kyaukpasat Andesite with auriferous quartz dykes, silicified tuffs, andesitic lava and thin calcareous chert ("Shwetaung Formation").

More recent andesite and dacite dykes, biotite granodiorite, quartz diorite and granite intruded the Shwetaung Formation (ZAW PE unpubl. rep. 1976). Because of its position within the rock sequence, this igneous rock complex, which is up to 7,000 m thick, is thought to be of pre-Early Upper Cretaceous (pre-batholithic) age. It covers large areas of the E and central parts of the N Wuntho Massif.

According to K/Ar dating, the biotite-granodiorite, hornblende-diorite, granodiorite and quartz-diorite, garnetiferous two-mica granite with pegmatites and aplites and the widely distributed hornblende-granodiorite, which in turn intruded the Namaw Chaung Andesite Group, are probably Upper Cretaceous in age (93–97 m.y.). These igneous rocks are extensively exposed in the SW and central sections and also on the E and N edges of the Wuntho Massif. An Upper Cretaceous age might also be assigned to the dacite and quartz porphyry that are exposed for example approximately 10 km ENE of Pinlebu, also from Mongthon approximately 30 km N of Wuntho, from the S and N edges of the Wuntho Massif and from the Shangalon area. Here, they also contain Cu, Mo, and Au mineralization.

The "post-batholithic" sedimentation commenced with arkosic sandstone in which andesitic flows and sills are locally intercalated ("Mawlin Formation", village N of Pinlebu; ZAW PE, unpubl. rep. 1976). The Mawlin Formation is probably of Upper Cretaceous age and occurs on the W side of the Wuntho Massif.

The "Shangalon Granodiorites" in the SE part of the Wuntho Massif, tonalite porphyry, andesitic dykes and trachyte occurrences are assigned to the (?) Lower Oligocene.

Near Pinlebu, some basalt dykes penetrate cross-bedded sandstone, mottled mudstone and conglomerates which belong to the "Wabo Chaung Formation" (? Oligo-Miocene; ZAW PE, unpubl. rep. 1976). These basalts are probably equivalent in age to the Upper Miocene to Pleistocene basalt volcanism of the Lower Chindwin area. The volcano of Taungthorlon (1,708 m) in the N part of the Wuntho Massif with its basaltic and andesitic lava flows, tuffs, ashes and agglomerates, is also assigned to the Cenozoic.

The Inner Volcanic Arc continues to the N of the Wuntho Massif in the area of the Indawgyi Lake, in the Jade Mines Area (Myitkyina District) and along the W edge of the Kumon Ranges through to the W boundary of the Putao Basin (Figs. 17 a, folder, 53).

A complex of ultrabasic rocks extending approximately 60 km N–S and up to 40 km E–W, together with smaller occurrences, is exposed S of the Indawgyi Lake and continues to the W of it towards the N as far as the area W of Putao. The rocks consist of altered dunite, mica-, hornblende- and diallage-peridotite, pyroxenite and amphibolite. They are for the most part highly serpentinized; antigorite, marmolite, chrysotile and grains of chromite frequently occur. Jadeite occurs particularly in the serpentinite near Tawmaw, Meinmaw. Pangmaw and other places in the W Myitkyina District (Jade Mines Area; p. 211). These igneous rocks are partially surrounded by crystalline schists (graphite-, glaucophane-, hornblende-, chlorite-, kyanite-schists) and by altered diorite, gabbro, pyroxenite und perknite (CHHIBBER 1934a, p. 315). At other places they have intruded Cretaceous and Eocene sediments and contact metamorphism has been observed (Geol. Map of the Soc. Rep. of the Union of Burma,

Explan. Brochure 1977). Already CHHIBBER (1934a) assumed therefore that this intrusion phase was Cretaceous to Eocene in age.

From the area about 20 km NNE of the Indawgyi Lake, CHHIBBER (1934a, p. 40) reported a complex of granitic rocks: biotite-muscovite granite, hornblende granite, micro and graphic granite, granophyre, quartz-augite-monzonite. He assumed that these granite intrusions took place at the earliest during the Mesozoic era.

The chain of Cenozoic volcanic rocks continues in Mount Loi-Mye (1,562 m; about 70 km NNE of Indawgyi Lake). According to CHHIBBER (1934a, p. 295) this volcano started to develop with the eruption of greenish-black basaltic tuffs which were intercalated in lacustrine sediments of the Tertiary. Later, basaltic and andesitic lavas were produced, not only here but also in the volcanoes near Namyong in the N Jade Mines Area/Myitkyina District.

3.324 Igneous rocks along the Shan Boundary Fault System

Numerous intrusions and extrusions of magma are spatially and very probably also genetically associated with the Shan Boundary Fault System (Fig. 29) which separates the Inner-Burman Tertiary Basin from the Sino-Burman Ranges in the E (p. 36). They occur in the estuary of the Sittang River in the S and continue to the N to the region of Kabwet (Mandalay area) and further N and NNE to the W parts of the Kumon Ranges. Also, the fault system that branches away to the E from the Shan Boundary Fault N of Kabwet and runs in an arc via Bhamo to the NNE (Fig. 29) is characterized by numerous and extensive occurrences of intrusive rocks.

By far the largest (approximately 310 km S–N length and up to 40 km E–W width) coherent igneous rock complex in Burma extends from the area E of the Sittang Estuary to the N as far as the area NE of Pyinmana. It is located to the E of and parallel to the Shan Boundary Fault. In the S it consists of fine-grained soda-granite composed of quartz, idio-morphic albite with subsidiary allotriomorphic orthoclase and large laths of biotite, or it consists of medium-grained potash-granite. Quartz-, aplite-, dolerite-, and hornblende-granophyre dykes with pyrite and hematite are frequently encountered (CHHIBBER 1934a, p. 339). Judging by their position in the rock sequence and based on comparisons with better studied portions of these granites further to the N, an Upper Cretaceous to Lower Eocene age can be assumed for these intrusives. The rhyolites and rhyolitic tuffs, which CHHIBBER (1934a, p. 459) mentioned from the area about 12 km E of the Mokpalin Railway Station (Rangoon–Moulmein), are probably somewhat older.

The igneous rocks further N between Kalaw and Kyaukse in the region around Neyaungga and Ye-ngan are better known. Here, porphyritic rhyolites and dacites appear with andesitic tuffs, plugs, sills and dykes of dacite, rhyodacite and rhyolite-porphyry, which locally are intercalated between the Pan Laung Formation (p. 85) and the "Yebok Formation" (Upper Jurassic to Lower Cretaceous; GARSON et al. 1976, p. 44). These volcanic rocks are therefore in any event younger than Lower Cretaceous. To the W of the volcanic rocks appear their coarse-grained equivalents as large, N–S-extending batholithic masses of granodiorite, adamellite and biotite-granite with microgranite dykes, locally also with aplites and wolframite-bearing pegmatites. Further to the W, the Shan Boundary Fault runs in an N–S direction, parallel to the batholite. The intrusive rocks are younger than the extrusive rocks of this area and probably belong to the W belt of tin- and tungsten-bearing granites which runs to the S along the Shan Boundary Fault and extends into Indonesia via Peninsular Thailand. Petrographically comparable tin/tungsten-bearing granites W of Chiang Mai in N Thailand have an Upper Cretaceous to Lower Tertiary age (V. BRAUN et al. 1976).

In the Neyaungga area, dolerite and lamprophyre dykes penetrate not only the Pan Laung Formation but also the granodiorite, so that they probably date from the Tertiary.

Along the further extension of the belt of igneous rocks to the N, along the Shan Boundary Fault System, numerous smaller occurrences of intrusive and extrusive rocks are observed. They are similar in composition to those between Kalaw and Kyaukse.

IYER (1931, p. 95) and CHHIBBER (1934a, p. 355 and 449 et seqq.) described highly decomposed amygdaloidal basalt lavas from the area around Kabwet. These lavas are intercalated in the clastic rocks of the Lower Irrawaddy Group. They occupy extensive areas mainly N of Kabwet. These probably Late Tertiary lavas are followed by younger olivine-dolerite and "teschenite" intrusions. Their chemical composition is clearly of the alkaline (Atlantic) type, while the time-equivalent igneous rocks in the "East Monywa Hill Belt" (p. 134; part of the Inner Volcanic Arc), only about 80 km WSW of the Kabwet area, are pronouncedly calc-alkaline (Pacific) in their chemism.

S of Kabwet, "mugearite-basalt" lava (CHHIBBER 1934a, pp. 453 et seqq.) overlies, in an area of at least 100 km^2, Tertiary sandstone and, locally, sediments of the Upper Irrawaddy Group and older Irrawaddy river deposits. These most recent (Quaternary) lavas come from fissure eruptions along the Shan Boundary Fault System itself.

E of the Kabwet volcanic area lies a large N–S oriented occurrence of gabbroid and granitic intrusive rocks that presumably correlates with those that occur between Kalaw and Kyaukse.

Further occurrences of volcanic rocks that are associated with the Shan Boundary Fault are found about 80 km and about 110 km N of Kabwet. They are little known as yet, as are also the igneous rocks in the area of the Shan Boundary Fault where it begins to split up N of 24° N and bend towards the NW (Fig. 17a, folder).

N of Kabwet, a fault system branches off to the E from the Shan Boundary Fault and runs in an arc via Bhamo and E of Myitkyina to the NNE and along the E side of the Kumon Ranges to the N. It separates the West Kachin Unit from the East Kachin/Shan Unit (Fig. 29). It is characterized by a belt of ultrabasic, gabbroid and granitic rock complexes which attain lengths up to 100 km (W of Bhamo) and widths of up to 60 km (N of Myitkyina). These intrusions are always oriented along the strike of the fault zone (Fig. 17a, folder). So far, none of the intrusives in this region have been investigated in any detail. Their existence has been determined by photogeological interpretation coupled with some reconnaissance groundchecks. Their position partially in the Mesozoic sedimentary sequence, their shape and mode of emplacement, as well as their association with a major fault system allows the assumption, however, that these intrusions correlate with those along the Shan Boundary Fault System and belong therefore to the Cretaceous to Eocene generation of intrusive rocks.

3.325 Igneous rocks in the Sino-Burman Ranges

Numerous intrusive masses of granitic rocks occur from the southernmost islands (e.g. Davis Island) via the S tip of the Burmese mainland (Victoria Point) to the area N of

Mawchi (approximately 19° N). The intrusive rock bodies attain lengths of more than 100 km and widths of more than 50 km, and they are almost all oriented ± SSE–NNW. They are arranged in three belts in the Tenasserim, Mergui and Tavoy districts, and run parallel to each other at a spacing of about 20 to 50 km in the same direction and thus follow the general strike of the Paleozoic sediments into which the igneous rocks have intruded. As the "Three Pagoda-Ratchburi Fault Zone" approaches the Shan Boundary Fault, it is no longer possible to distinguish between the three belts. However, they are recognizable again N of this fault zone and then they finally disappear altogether as the "Moie-Uthai-Thani Fault Zone" approaches the Shan Boundary Fault (Fig. 56, folder).

Data on the Tenasserim Granite have been published by, among others, J. C. BROWN & HERON (1923), STAMP (1926), RAO (1930), CHHIBBER (1934 a) and MAUNG THEIN (1973). The granite is mineralogically very similar throughout the entire region: abundant quartz, orthoclase, acid plagioclase, sometimes microcline, biotite, muscovite, rarely hornblende; accessory minerals are remarkably scarce. It is coarse-grained, but towards the edges of the intrusions it becomes more fine-grained. Here the granite also contains numerous inclusions of rocks from the Mergui Series. Acid and basic pegmatites, tourmaline-pegmatite, quartz dykes and basic dyke rocks occur, but not frequently. In the Mergui District, RAO (1930) distinguished macroscopically between a coarse-grained, porphyric "old" granite, without any tin content, and a tin-bearing "young" biotite-granite. Microscopically, it is possible to distinguish between biotite-granite, hornblende-granite and tourmaline-muscovite-granite. According to CHHIBBER (1934 a, p. 338), the latter is the most interesting when prospecting for tin ore.

Views differ on the age of the Tenasserim granites. The E belt of granite occurrences in the Tenasserim region extends into Thailand and can be traced further N where physical age determinations (Rb/Sr and K/Ar) were conducted in the area of Chiang Mai (V. BRAUN et al. 1976). These datings revealed an Upper Triassic-Lower Triassic age (210–205 m.y.) of the granite of Mae Sariang (an N–S-running intrusion between 18° and 19° N at the W border of Thailand) and to the S of it, in a small isolated intrusion, an Upper Cretaceous age (70–80 m.y.). An Upper Cretaceous age was also determined on several occasions for granites in the "Central Crystalline Complex" E of the Mae Sariang Granite. In the granite belt E of Chiang Mai, a Lower Triassic age (232 ± 31 m.y., 235 ± 5 m.y., 236 ± 14 m.y.) was repeatedly found. Since these chains of acid intrusive rocks run from N Thailand to the S into the Tenasserim area of Burma, it is likely that here, too, a Lower Triassic, an Upper Triassic-Lower Jurassic and an Upper Cretaceous granite intrusion phase occurred. A further intrusive phase of acid granites took place at the turn of the Paleocene/Eocene: the tin/tungsten-bearing granite of Hermyingyi (p. 186) displayed an age of 55 m.y. (Rb/Sr-total rock; H. LENZ/BGR, in press).

RAO (1913) and CHHIBBER (1934 a) described olivine-basalt which overlies sediments of the Mergui Series on Medaw Island (Mergui Archipelago). It is the only occurrence so far known of Cenozoic volcanic rocks in the Tenasserim section of the Sino-Burman Ranges.

Apart from a few occurrences in the Mogok area (p. 142), so far only a few signs of post-Paleozoic volcanism have been observed in the central Sino-Burman Ranges between approximately 19° N and the Kabwet-Bhamo Intrusive Belt (p. 140). Some remnants of a deeply weathered rhyolitic rock were found in the iron ore deposit at Pang Pet (30° 44′ 30″ N, 97° 06′ 30″ E) approximately 11 km SE of Taunggyi (p. 194). As can be deduced from geological interpretations of aerial photographs, further extrusive rocks in all probability penetrated the Paleozoic-Mesozoic platform carbonates in the central Sino-Burman Ranges, probably during the Upper Tertiary. Presumably, also, the frequent thermal springs (p. 214) in this area are associated with a Late Tertiary-Quaternary volcanism.

Large complexes of granitic rocks are located to the E of the Salween River in the easternmost part of Burma. They correlate with the granites E of Fang in N Thailand to which a Lower Triassic age is assigned (V. BRAUN et al. 1976, p. 172). Pegmatites and aplites, to which SEARLE & BA THAN HAQ (1964) assigned a Miocene age (15 m.y.; U/Th/Pb), occur in the area of the Precambrian Mogok Crystalline Rocks. These authors also assumed a corresponding age for the "Kabaing Granite" in the same region.

The only basaltic volcano so far known in the Burmese section of the central Sino-Burman Ranges (Loi Han Hun, Loi Ling Range) is located about 60 km SE of Lashio. According to LA TOUCHE (1907) it is built up of columnar basalts and olivine-basalt lava masses and it is more a basalt plug than an eruption volcano, because no craters and no ashes or tuffs have been found. Since the lavas overlie Upper Tertiary sediments, and since the lava flows followed an already existing drainage pattern cut in these sediments, the volcanics of Loi Han Hun probably date from the Late or post-Tertiary. J. C. BROWN (1913) assigned the same age to the volcanoes of Lao-Kuei-po, Tay-in-Shan and Kung-po ("Teng-Yueh Volcanic Group") which are located about 290 km N of Loi Han Hun in Yunnan/China.

The post-Paleozoic igneous rocks in the N section of the Burmese sector of the central Sino-Burman Ranges are linked with the fault system that branches off from the Shan Boundary Fault N of Kabwet and runs E–NE before turning N.

4. Paleogeographic evolution

The geological knowledge gathered so far about Burma is not sufficient to reconstruct a reasonably complete history of the paleogeographic evolution. The synthesis that is attempted here can therefore only take account of the major events which were of regional-geological importance. The synthesis is based on own observations and on the evaluation of the literature on stratigraphy, structure and magmatism that have been referred to in the preceding chapters. The literature references and sources are given in these earlier chapters.

An attempt has been made in Fig. 65 (folder) to sketch the depositional environment, structural deformation, igneous activity and mineralization during the course of the paleogeographic evolution.

Precambrian

In the central and N Sino-Burman Ranges, metasediments, orthogneisses and migmatites (Mogok Series) form the basement beneath epimetamorphic or non-metamorphic rocks. Lithologically comparable basement crystalline rocks are also known from Yunnan/China and Thailand where they are overlain with a distinct unconformity by Lower Paleozoic strata. This basement rock, which is the oldest in Burma, underwent metamorphism, was folded and uplifted to varying degrees and then eroded to a varying extent, before it was overlain by the thick clastics of the Chaung Magyi Group.

Precambrian-Cambrian

The approximately 3,000 m thick epi- and mesometamorphic flysch-type sediments of the Chaung Magyi Group, consisting chiefly of pelitic rocks, subgreywacke and greywacke, display many sedimentological characteristics of having been deposited in a rapidly subsiding geosynclinal basin. In the stratigraphically higher (younger) sections of the Chaung Magyi Group there are indications that the sediments were deposited in a marine shallow water environment. Sediments of the Chaung Magyi Group are exposed in the central and N Sino-Burman Ranges and in Yunnan/China, which adjoins to the N. Lithologically comparable sediments are also found in the Indo-Burman Ranges (Kanpetlet Schists). According to recent observations, at least parts of them grade into non-metamorphic Triassic sediments, so that they do not correlate with the Chaung Magyi Group as shown in Fig. 65 (folder).

In general, the deposits of the Chaung Magyi Group have undergone epimetamorphism. Higher grades of metamorphism are found in the vicinity of intrusive rocks.

A distinct angular unconformity between the Chaung Magyi Group and the overlying sequences points to tectonic deformation and deep-reaching erosion prior to the deposition of Cambrian rocks.

In the Northern Shan State, the Chaung Magyi Group is overlain by an approximately 2,000 m thick sequence of sandstone, siltstone and mudstone with occasionally conglomeratic, dolomitic and quartzitic intercalations (Pangyun Formation). Locally rhyolitic volcanoclastics occur in the Pangyun Formation into which rhyolite has intruded (Bawdwin Volcanic Formation). The Pangyun Formation is comparable to the lithologically very similar clastics of the Molohein Group in the Southern Shan State, in which Upper Cambrian trilobites were found.

The lithofacies characteristics of the Pangyun Formation are indicative of deposition in a partially terrestrial-fluviatile, partially near-shore shelf environment.

The Cambrian sequences of the Molohein Group mark the early stage of the Burmese-Malayan Geosyncline, which extended from W Yunnan/China in the N via the Shan States, the Tenasserim area and the Thai-Malayan Peninsula to the S as far as Borneo.

Ordovician

In the central and N Sino-Burman Ranges, several 1,000 m thick fine clastic and carbonate sequences of the Ordovician overlie concordantly the Cambrian sediments (Naungkangyi Stage, Taungkyun Formation with overlying Li-Lu Formation). The content of carbonates increases from N to S (Mawson Group, Pindaya Group). In S Burma (Kayah, N Karen, Tenasserim region) the lower, mainly calcareous sequences of the Mawchi Series and of the Mergui Series probably belong to the Ordovician. The lithofacies and fossil content of the Ordovician rocks indicate that they were deposited in the shelf regions at the W edge of the Burmese-Malayan Geosyncline whose generally N–S oriented trough axis ran further to the E.

Acid volcanic rocks intruded in the Cambro-Ordovician sediments (rhyolite in the Pangyun Formation) and caused a Pb-Zn-Ag mineralization in the late phase of the volcanism. The widespread strata-bound impregnation of Ordovician carbonate rocks with Pb-Zn sulphides and baryte mineralizations (Bawetaung, Bawsaing, Kanchanaburi District/Thailand) may also be connected with the late phase of this volcanism.

Silurian

The sedimentation of fine clastics and carbonates in the shelf region of the Burmese-Malayan Geosyncline continued in the Silurian:

Shale together with phacoidal limestone (Linwe Formation, Lower Silurian, probably extending into the Ordovician), graptolite-bearing shale (Pangsha-pye Formation, Llandoverian), sandstone and marl (Namhsim Formation, Wenlockian, Upper Ludlowian) totalling more than 1,000 m in thickness, were formed during this period. These rocks occur in the N and central Sino-Burman Ranges and can probably be correlated with the middle portion of the Mawchi Series (Loikaw Beds).

Slight magmatic activity existed in the Llandoverian, as is indicated by rhyolitic tuff and volcanic ash, which are locally (Southern Shan State) intercalated in the sequences of the Silurian.

Viewed regionally, different $\pm$ S–N oriented facies types can be observed within the region of the geosynclinal sedimentation (from W to E):

- a fine-sandy-argillaceous-calcareous shelf facies with platform carbonates (W Malaysia, W Thailand, E Burma, Yunnan/China)
- a miogeosynclinal basin facies with mudstone and graptolite shale (Malaysia, Thailand, China, Vietnam)
- a „mixed facies" with siltstone, carbonate shale, limestone and rhyolitic volcanic rocks (Malaysia)
- a eugeosynclinal facies with flysch-type sediments and ophiolites (Malaysia).

Devonian

The pelitic-carbonate sedimentation in the shelf region of the Burmese-Malayan Geosyncline continued in the Devonian as well. About 60 m of limestone and mudstone (Zebingyi Beds, Lower Devonian) and presumably more than 1,000 m thick limestone, dolomitic limestone and dolomite with some mudstone layers (Maymyo Dolomite Formation with Middle and Upper Devonian faunas) were formed. In the Northern Shan State, the Upper Silurian Namhsim Formation, which occurs in a sandy, near-shore facies, was prior to the deposition of Devonian sediments affected by erosion which reached locally the sequences of the Ordovician. Therefore regionally limited structural uplifts probably took place in this region at the turn of the Silurian–Devonian.

Carboniferous

In S Burma more than 3,000 m thick sediments of the chronologically equivalent Mawchi, Taungnyo and Mergui Series and of the Lebyin Group were deposited: argillites, locally also with intercalations of carbonate, sandstone and greywacke with conglomerate. Upper Carboniferous fossils were found only in the upper part of the Taungnyo Series and of the Lebyin Group. The lower portions of the monotonous sequences very probably include also Lower Paleozoic. So far no sediments of definitely Carboniferous age have been identified in the central and N Sino-Burman Ranges.

Strong volcanism manifested itself in particular in the area of the Mergui Archipelago where tuffs and agglomerates are intercalated in the Mergui Series. Large masses of rhyolite and porphyry on the W coast of Elphinstone and on King and Iron islands mark the position of eruption centres from which volcanic products got into the Mergui Series.

The deposition of thick flysch-type sediments in S Burma points to significant tectonic events taking place during the Hercynian phase. Apparently, the Burmese-Malayan Geo-syncline became further differentiated. Uplift movements resulted in the raising of parts of the geosyncline in W Malaysia ("Malayan Geanticline"). The clastic erosion products from the uplifted areas accumulated mainly in the W part of the Lower Paleozoic Geosyncline: in NW Malaysia, SW Thailand, S Burma and NW Thailand.

In S Burma granitic magma intruded into the rocks of the Mawchi, Taungnyo and Mergui Series. Most of them probably belong to post-Carboniferous intrusive phases. The granite intrusion of Tawn Peng in the Northern Shan State, however, is possibly Carboniferous in age.

Permian

At the turn of the Carboniferous-Permian, the pre-Permian sequences were affected by regional tectonic movements which resulted in block faulting, tilting and local folding. In some regions of the Northern and Southern Shan States, the erosion of uplifted areas extended down to Silurian or even Ordovician rocks. The Permian carbonates of the Plateau Limestone of the Shan States (Dattawtaung Limestone, Thitsipin Limestone Formation, Thigaungdaung Limestone) therefore overlie older sequences with a regional erosion discordance and a local angular unconformity. Further to the S (Kayah and N Karen State) massive limestone formed, in places on top of red clastic rocks of the Mawchi Series and Lebyin Group. Still further to the S (S Karen and Mon State, Tenasserim Division) a thick series of biohermal limestones built up above the Carboniferous argillite and greywacke of the Taungnyo Group (Moulmein Limestone, up to 1,000 m).

In the Shan States the formation of carbonates continued with the > 2,500 m thick Nwabangyi Dolomite Formation through to the Upper Permian. The facies and the rare fossils are evidence of a marine shallow-water environment of deposition with local reef development. This type of sedimentation is also found in the S Sino-Burman Ranges. Here, however, with the occurrence of the Yinyaw Beds and the Martaban Beds, the lithofacies changed into pelite, argillite and sandstone in the Upper Permian. Apparently some of these fine clastic rocks were deposited in a brackish-water environment.

So far, no Permian rocks have been identified in the Indo-Burman Ranges, and consequently the extent of the Permian region of sedimentation to the W beneath the Inner-Burman Tertiary Basin has not been clarified.

Triassic

In the N and central parts of the Sino-Burman Ranges, the Permian carbonate deposition continued in the marine shallow-water environment in the Triassic as well (Skythian, Anisian, Karnian, Norian, Rhaetian): upper portions of the Nwabangyi Dolomite Formation, Natteik Limestone Formation, Kondeik Limestone, Thigaungdaung Limestone, Napeng Formation. The carbonates formed during the Permo-Triassic attained thicknesses between 2,500 m and 5,000 m, which are comparable with those of the northern "Kalkalpen" in Europe. In the Northern Shan State, evaporites formed during the Rhaetian-Lower Jurassic. They were probably deposited together with clastic sediments and carbonates (Namyau Group) in shallow marine-shelf basins partly surrounded by onshore areas with an arid climate. In the S regions of the Southern Shan State and in S Burma, the deposition of the Permian biohermal limestone was followed by mudstone, calcareous mudstone and sandstone (upper Yinyaw Beds, upper Martaban Beds) which in turn were overlain by calcareous sediments (Kamawkala Limestone). In addition, Triassic (Karnian) mudstone to fine sandstone with flysch characteristics and intercalated pillow lava were deposited in thicknesses up to 1,000 m in the area of the E Indo-Burman Ranges (*Halobia* Beds).

Apparently, during the Triassic the marine depositional environment included not only the region of the Sino-Burman Ranges, but also of the Inner-Burman Tertiary Basin and of the Indo-Burman Ranges. Near-shore shallow marine areas were located where mainly carbonates were formed in the region of the Sino-Burman Ranges and where also evaporites were deposited towards the end of the Triassic in the Northern Shan State. In the W, in the region of the Indo-Burman Ranges, the pelitic-fine clastic, thick Triassic sediments indicate a deeper marine depositional environment at a greater distance from the shore.

Intrusions of granitic magma appear to be limited to the area of the Sino-Burman Ranges. They took place at the turn of the Permian/Lower Triassic and at the turn of the Upper Triassic/Jurassic. They were associated with tin and tungsten as well as with gold, copper, antimony, bismuth and arsenic mineralizations.

Jurassic

Also in the earliest Jurassic, the formation of platform carbonates (Namyau Limestone) continued in parts of the sedimentation area in E Burma. Locally, in the Northern Shan State, evaporites continued to be deposited (Namyau Group, Upper Triassic-Jurassic).

A distinct change in the depositional environment was caused by uplift movements. This is evident in the change from carbonaceous to mainly clastic sedimentation during the Lower and Middle Jurassic. During the Toarcian to Oxfordian, clastic-terrigenous sedimentation with marine intercalations (Tati Limestone, Bathonian-Callovian) prevailed. The facies conditions in the area of the Shan States point to a general regression phase which was interrupted by minor transgressions. Turbidites and other clastic rocks were deposited in neritic and paralic environments. In the latter, widespread coal seams originated (Loi-An Series).

During the Upper Jurassic, the formation of red-coloured, coarse and fine clastics as well as of pelites (Hsipaw Red Beds, approximately 1,200 m) started, beginning with basal calcareous conglomerates (Northern Shan State). Red clastic rocks with pelites were discordantly laid down also on top of the Thigyit Beds (shale) of the upper Loi-An Series of the Southern Shan State (Kalaw Red Beds, approximately 2,600 m). Still further to the S, in the Karen-Tenasserim part of the Sino-Burman Ranges, similar clastic, mainly red sequences, formed after an hiatus above the Upper Triassic Kamawkale Limestone. As in neighbouring Thailand, the deposition of these formations lasted from the Upper Jurassic through to presumably the Lower Cretaceous, and they are thus comparable with the Khorat Series.

So far, no Jurassic sequences have been observed in Burma W of the Sino-Burman Ranges.

The facies and position of the Upper Jurassic sediments in the stratigraphic sequences, as well as their structural deformation, indicate that an uplifting-, faulting- and folding-phase (Kimmeridgean orogenic movements) took place in the area of the Sino-Burman Ranges prior to the deposition of Cretaceous sediments. Uplift movements presumably also affected the area of the Indo-Burman Ranges where apparently the entire Jurassic sequence is missing and Cenomanian sediments are found overlying Triassic rocks.

During the Upper Triassic-Lower Jurassic, granitic intrusions appear to have been limited to the SE part of the Sino-Burman Ranges.

Cretaceous

During the Cretaceous in the area of the Sino-Burman Ranges, the Kimmeridgean orogeny was apparently followed by a regional epirogenetic uplift, because neither terrestrial nor marine Cretaceous sediments have so far been observed in this area.

It is uncertain whether Cretaceous sediments underlie all of the Inner-Burman Tertiary Basin, although they occur in the area of the Wuntho Massiv and of Bhamo.

In the S and central parts of the E edge of the Indo-Burman Ranges, Cenomanian limestone, siliceous and bituminous limestone and calcareous sandstone formed above the Upper Triassic *Halobia* Beds or above chocolate-coloured shale of unknown age.

During the late Upper Cretaceous, micritic limestone, red and green marl and shale were deposited in W Burma, together with widespread *Globotruncana* limestone, occasionally associated with radiolarite.

The paleogeographical interpretation of the sedimentation during the Cretaceous is made difficult by the fact that strata of this age occur in a continuous stratigraphic sequence exclusively at the E edge of the Indo-Burman Ranges. Further to the W, Cretaceous sediments occur only as allochthonous rock masses and blocks in Tertiary flysch-type sediments. The allochthonous rocks originated in the area at the E edge of the Indo-Burman Ranges and of the Inner-Burman Tertiary Basin that adjoins to the E.

The period between the Upper Cretaceous and the earliest Tertiary, i.e. about 100 m.y. B.P. to approximately 60 m.y. B.P., was marked by increased magmatic activity.

Granitic magma intruded in the E, in the continental areas of the uplifted Sino-Burman Ranges, as well as at the E edge of the Inner-Burman Tertiary Basin. Here, they are associated with pre-existent zones of weakness (sutures) and particularly with the Shan Boundary Fault System. These magmas brought a local Sn–W-mineralization. Frequently, the intrusions were preceded by extrusions of acid to intermediate igneous rocks.

In the central and N Inner-Burman Tertiary Basin as well, the magmatic activity commenced in the Cretaceous with the extrusion of andesitic lavas and the formation of thick volcanoclastics (Wuntho Massif), which were deposited in parts in a marine environment. They were in turn intruded by batholithic masses of dioritic and granodioritic magmas as well as by dacite and quartz porphyries which were associated with Cu, Mo and Au mineralizations.

During the late Upper Cretaceous–Early Tertiary, ultrabasic magmas with Cr, Ni and Cu mineralizations intruded as large masses, stocks or dykes along presumably pre-existent fault systems at the E edge of the Indo-Burman Ranges. Ophiolitic igneous rocks emerged in the area of the Indo-Burman Ranges and in the W foreland of these ranges. Locally, pillow lavas, sills, tuffites and agglomerates were formed during the deposition of marine Upper Cretaceous sediments. Further intrusions of ultramafitic rocks took place along the central and N sections of the E border of the Inner-Burman Tertiary Basin. They were also associated with Cr, Ni and Cu mineralizations.

The facies of the autochthonous but also of the allochthonous Cretaceous sediments together with ophiolitic and basaltic rocks in the Indo-Burman Ranges and in the W foreland of these ranges permit an interpretation of the paleogeographical conditions. It is possible that the region from which the exotic Cretaceous rocks originated should be sought at the E edge of the Indo-Burman Ranges where an island chain existed. This area may have been built up partially of basaltic seamounts with overlying pelagic Upper Cretaceous limestone, and partly of sediments from the early Upper Cretaceous on top of Triassic *Halobia* Beds, into which the ophiolites or ultrabasites had intruded.

Tertiary

The post-Cretaceous paleogeographical evolution was dominantly influenced by plate tectonic events (Fig. 66a–g, 67):

- break-up of Gondwanaland with the separation of India during the Lower Cretaceous and with the northwards movement of the Indian Plate during the Upper Cretaceous (130–80 m.y. B.P.)
- further spreading and collision of the Indian Plate with the SE Asian Plate during the Late Paleocene (approximately 55 m.y. B.P.)
- spreading in the E Indian Ocean and thus a change in the direction of movement of the Indian Plate to the NE in the Upper Eocene and Oligocene (55–32 m.y. B.P.), associated with subduction at the NE side of the Indian Ocean
- continuation of the spreading-subduction mechanism through to the present.

The consolidation and uplift of the region of the Sino-Burman Ranges caused by the Kimmeridgean orogeny meant that during the Cretaceous, Tertiary and Quaternary, this region remained an area of continental erosion, and in intramontane basins it remained an area of accumulation of terrestrial deposits. During the Cretaceous, carbonates and clastics were deposited to the W of this area in shallow shelf regions. Pelagic Upper Cretaceous limestone at the E edge of the present-day Indo-Burman Ranges can be interpreted as evidence of the E edge of the Indo-Burman Geosyncline which stretched to the W into the Indian Ocean and in which flyschoid sediments were accumulated.

During the **Paleocene**, conglomerates (Paunggyi Conglomerate) and sandy and pelitic sediments were deposited above Upper Cretaceous rocks in the shelf W of the Sino-Burman Ranges in the Inner-Burman Tertiary Basin. Further to the W, shallow-water limestones formed, but they are only known from olistoliths in the Arakan Yoma and its W foreland. The formation of volcanoclastics, tuffites and andesitic rocks continued in the shelf area of the Inner-Burman Tertiary Basin.

During the **Eocene**, the supply of erosion products from the uplifted areas of E Burma increased. Very thick (> 5,000 m), mainly sandy and argillaceous sediments accumulated on the subsiding shelf platform to the W.

In the late Middle Eocene and early Upper Eocene, conglomeratic layers of reworked older rocks and shaly-earthy red beds (in the Pondaung Sandstone) point to a local continental erosion and re-deposition in uplifted areas at the W edge of the central and N Inner-Burman Tertiary Basin. Apparently, the rise of the Indo-Burman Ranges commenced with local uplift movements in their N and central parts.

During the Upper Eocene, the supply of sediments may have out-weighed the rate of subsidence in near-shore areas of the shelf in Central and N Burma. At the E edge of the Chindwin Basin, limnic-fluviatile sequences were laid down, and to the W these merge into sediments of an estuarine, paralic and marine shallow-water region. Coal seams also formed in this area (Tabyin Clay, Pondaung Formation, Middle and Upper Eocene).

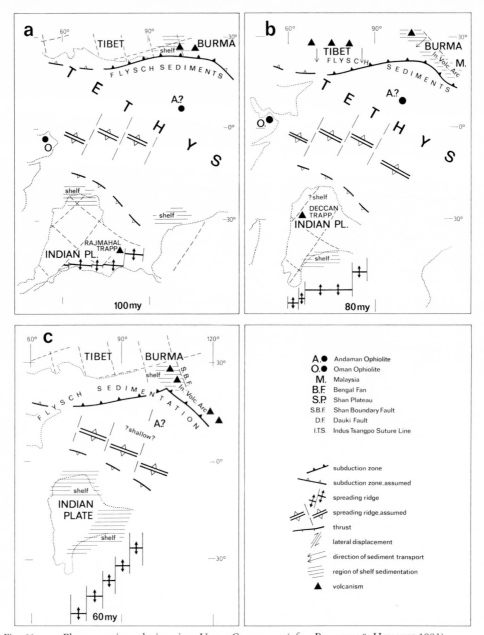

Fig. 66a–g. Plate tectonic evolution since Upper Cretaceous (after BANNERT & HELMCKE 1981).

Fig. 66a. Cenomanian (approx. 100 m.y. B.P.): Tethys occupies area between India/Antarctica/ Australia and Asia. Shelf sediments are deposited in front of paleo-continents. Subduction and seafloor spreading postulated N of Indian Plate. In Burma, the Inner Volcanic Arc begins to develop. – b, Santonian (approx. 80 m.y. B.P.): Indian Plate continues drifting N. Burma region starts rotation towards SW. In Tibet, granodioritic intrusions take place. – c, Paleocene (approx. 60 m.y. B.P.): Flysch sediments from Asian Plate cover parts of oceanic Tethys sea floor reaching to Andaman Islands area.

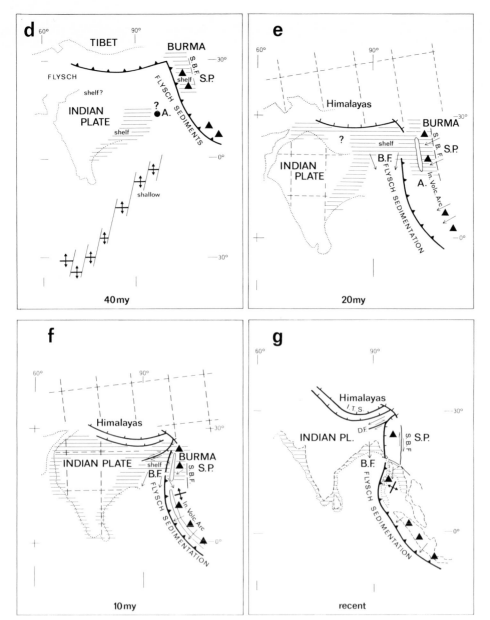

Fig. 66 d. Upper Eocene/Oligocene (approx. 40 m.y. B.P.): N drift of Indian Plate slows down due to collision with Asian Plate. Flysch sedimentation continues. Parts of Outer Island Arc begin to develop. – e, Lower Miocene (approx. 20 m.y. B.P.): Burma region approaches its present position. Subduction zone shifts to the W of Andaman Islands. Outer Island Arc further develops. Uplifting in the Himalayas and parts of Indo-Burman Ranges. – f, Upper Miocene (approx. 10 m.y. B.P.): Thrusting of Indian Plate under Asian Plate continues. Molasse sediments deriving from the Indo-Burman Ranges are deposited above flysch sediments in their W foreland. Outer Island Arc with Indo-Burman Ranges and Andaman Islands is formed. Seafloor spreading begins in Andaman Sea E of subduction zone. – g, Recent: Subduction along N Indo-Burman Ranges ceases in favour of thrusting, it continues W of Arakan Coastal Area, Andaman and Nicobar islands. No further magmatic activity along Inner Volcanic Arc. Seafloor spreading continues in Andaman Sea, causing lateral movements along Shan Boundary Fault Zone.

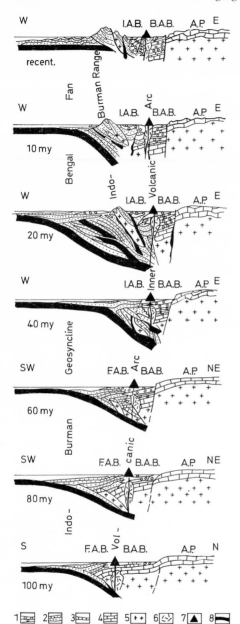

Fig. 67. Schematized cross-sections (not to scale) through Central and W Burma near 20° N, showing geological evolution since Upper Cretaceous (after BANNERT & HELMCKE 1981).

1: Molasse type sediments, 2: Flysch type sediments, 3: Mesozoic limestone, 4: Paleozoic shelf sediments, 5: Crystalline rocks of Asian Plate, 6: Granodioritic intrusions, 7: Volcanic rocks of Inner Volcanic Arc, 8: Oceanic crust. A.P.: Asian Plate, B.A.B.: Back-Arc Basin, F.A.B.: Front-Arc Basin, I.A.B.: Inter-Arc Basin (Interdeep).

Flysch sediments continued to accumulate throughout the Eocene in the Indo-Burman Geosyncline in the area of the present-day Indo-Burman Ranges and their W foreland (> 10,000 m).

In the near-shore shelf areas, local extrusions of andesitic lavas occurred (e.g. Wuntho Massif), which are intercalated as sills and flows in Upper Eocene and (?) Oligocene terrigenous sequences.

During the **Oligocene,** the supply of erosion products continued from the continental regions of E Burma into the shelf in the W. Sandstone and mudstone with faunas of the shallower (in the E) and deeper (in the W) shelf regions were laid down (Shwezetaw-Padaung and Okmintaung Formations).

Isopachs of Oligocene formations in the S Chindwin Basin (Figs. 68 a, b) indicate a further increase in thickness towards the W. This might be of local significance only and does not necessarily point to a regional westwards thickening Oligocene sequence. So far, no Oligocene sediments have been observed in the Indo-Burman Ranges. In the N part of the Inner-Burman Tertiary Basin (Chindwin Basin and further to the N), clastic and pelitic sequences were laid down in a dominantly continental depositional environment, locally with coal seams (Shwezetaw Formation). According to palynological studies they correlate with Oligocene sequences of identical facies in Assam.

Thick deep-water flysch sediments accumulated in the W foreland of the Indo-Burman Ranges. It is doubtful whether this means that the connection between the shelf and the Indo-Burman Geosyncline still remained fully open. Facies and isopachs of Upper Oligocene formations in the central and N Inner-Burman Tertiary Basin indicate uplift movements running transverse to the general N–S strike (at about 20°, 22° and 24° N). Orogenic movements, which resulted in the folding of the Indo-Burman Ranges in the course of the Miocene had commenced with local uplifts in the late Middle Eocene and continued during the Oligocene in the area of the present-day N and presumably also central Indo-Burman Ranges. As a result, the $\pm$N–S oriented Indo-Burman Geosyncline was shifted further to the W into the present-day Arakan Coastal Area.

The Oligocene/Miocene orogenesis in the area of the Lower Tertiary Indo-Burman Geosyncline also had an effect on the shelf area of the Inner-Burman Tertiary Basin. There fold belts originated which developed into mostly tight and steep, E-vergent or W-vergent synclines and anticlines which are frequently accompanied by overthrusts and steep normal faults and are broken up by numerous faults.

During the **Miocene,** too, the clastic sedimentation in the shelf W of the uplifted areas of E Burma continued. In the S section of the Inner-Burman Tertiary Basin limestone also formed during the Lower Miocene (Henzada District). The thickness of Middle Miocene formations decreases from the central and N parts of the Inner-Burman Tertiary Basin towards the Indo-Burman Ranges, because this geosynclinal area had in the meantime to a great extent been folded and uplifted. It is doubtful whether the northernmost parts of the Indo-Burman Ranges had already been completely uplifted, because terrestrial or terrigenous sequences that are correlatable with the Assam area were deposited, thus indicating certain links between the Inner-Burman Tertiary Basin and the Assam area.

During the Lower Miocene, the sedimentation of flyschoid sediments continued in the Arakan Coastal Area. Between the Middle and Upper Miocene the zone of major folding shifted from the Indo-Burman Ranges to the W to this region. Therefore, molasse sediments of the Upper Miocene formed here as erosion products of the Indo-Burman Ranges. They overlie with an angular unconformity structurally strongly deformed flysch deposits of earlier periods of the Miocene, Oligocene and Eocene.

With the further westward shift of the zone of orogenic movements, the Indo-Burman Geosyncline also shifted to the W from the Arakan Coastal Area into the Bay of Bengal. Flysch sedimentation continued here during the Miocene and post-Miocene. Towards the end

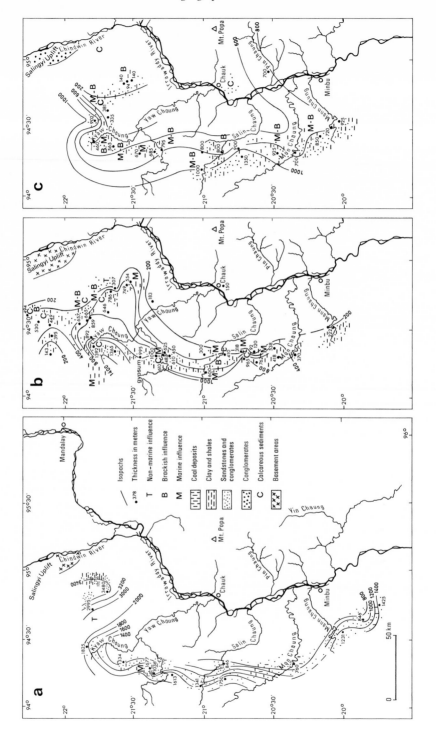

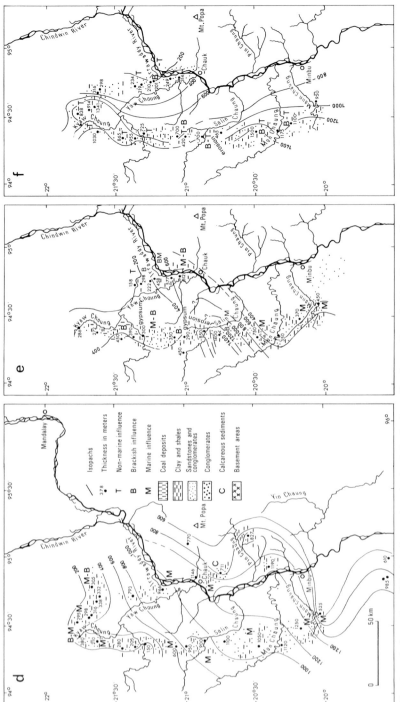

Fig. 68 a–f. Isopachs of Tertiary formations in the Salin Basin. a, Pondaung Fm. – b, Yaw Fm. – c, Shwezetaw Fm. – d, Padaung Fm. –
e, Okhmintaung FM. – f, Kyaukkok Fm.

of the Miocene, the subduction zone between the Indian Plate and the Southeast Asian Plate had shifted westwards to the area of its present-day location (Fig. 16).

In the latest Miocene or at the turn of the Miocene/Pliocene, the rock sequences were subjected to minor structural deformation in the Arakan Coastal Area and off the W coast. This led to the formation of $\pm$N–S-striking narrow and long anticlinal and synclinal trends which are frequently accompanied by parallel normal faults. This structural deformation affected also the Inner-Burman Tertiary Basin.

During the Upper Miocene, the volcanic activity continued in the Inner-Burman Tertiary Basin along the Inner Volcanic Arc. Acid, intermediate and basic magmas intruded; equivalent extrusive rocks are locally intercalated in the Upper Miocene-Plio/Pleistocene sequences of the Irrawaddy Group. Basaltic igneous rocks extruded in the Arakan Coastal Area.

Since the Upper Miocene, only terrestrial clastic rocks were deposited in large areas of the central and N Inner-Burman Tertiary Basin. Towards the end of the Miocene, the entire former shelf region had become a continental area of deposition and erosion.

Late Neogene and Quaternary

During the Late Neogene and Quaternary, thick, fluviatile erosion products from the Sino-Burman Ranges in the E, the E Himalayas in the N and the Indo-Burman Ranges in the W were transported into the Inner-Burman Tertiary Basin (Irrawaddy Group, > 2,000 m). Fluviatile and lacustrine sediments were deposited in the intramontane basins of the Sino-Burman Ranges. Locally lignite and peat, also bituminous rocks (oil shales) were formed.

Along the Inner Volcanic Arc the volcanic activity attained a peak towards the turn Pliocene/Pleistocene with the extrusion of mainly intermediate and basic lavas and tuffs. Locally, this volcanism brought with it a Cu/Au mineralization. It has not yet been clarified whether the Himalayan glaciations affected the sedimentation during the Quaternary. It might be expected that they were marked by terraces and loess deposits in the area of the Inner-Burman Tertiary Basin S of the E Himalayas.

From about 10.8 m.y. B.P. (Middle Miocene) onwards, a spreading centre developed under the Andaman Sea resulting in strike-slip movements during the latest Neogene and Quaternary along the Shan Boundary Fault System, which still continue. The structural deformation associated with the subduction of the Indian Plate also continued through to the present.

5. Energy, metallic and non-metallic raw materials, water and soil

The geological evolution of Burma has in many cases resulted in favourable conditions for the formation of natural resources.

Thick sediments in the shelf regions and along the borders of geosynclines, their facies and structure in the Inner-Burman Tertiary Basin and in the coastal area on the E side of the Bay of Bengal favoured the local formation and accumulation of **oil and natural gas.** This applies not only to the onshore but also to the offshore areas of Burma.

Oil shales formed in intramontane basins with lacustrine sediments in the area of the Sino-Burman Ranges.

During the Jurassic and Eocene, the preconditions existed for **coal** to form in paralic environments of deposition of the shelf. Lignite and **peat** were deposited in numerous intramontane basins with terrestrial sediments.

Locally, some of the conditions necessary for the formation of **nuclear fuels** are also satisfied: for uranium, these conditions may exist in pegmatites and alaskites and in deeply weathered, mineralized rhyolites, perhaps also in Tertiary sediments in suitable facies; for thorium, the conditions are met locally in heavy mineral placers.

Frequently, the geological evolution favoured the formation of **metallic raw materials.** Extensive intrusions of ultrabasic magmas and the ascent of ophiolitic and other igneous rocks on both sides of and in the interior of the Inner-Burman Tertiary Basin brought Cr, Ni, Pt, Cu and Mg mineralizations; granitic magmas in the S parts of the Sino-Burman Ranges were associated with Sn and W, the igneous rocks of the Inner Volcanic Arc with Cu and locally polymetallic mineralizations (Pb, Zn, Ag, Mo, Cu, Au, As). A number of different metallic mineralizations formed in pegmatites, in the dykes associated with the magmas, and in their contact-metamorphic aureoles. Strata-bound or structurally controlled, synsedimentary and postsedimentary Pb, Zn, Ag and Cu mineralizations formed in the Paleozoic, chiefly Ordovician, Carbonate Belt of the Sino-Burman Ranges, frequently in genetic association with the Paleozoic volcanism. Long-lasting weathering led to the formation of limonite-hematite deposits and of lateritic iron ore deposits; appropriate genetic conditions for the formation of magnetite and siderite deposits were fulfilled.

Favourable genetic preconditions existed for numerous **non-metallic raw materials** such as asbestos, steatite, baryte, feldspar, fluorite, graphite, gypsum, kyanite, magnesite, mica, quartz and rock salt, as well as for benthonite, building stones, cement raw materials, glass-sand, kaolin, laterite, pottery clay, tuff, sand and gravel.

Gemstone and semi-precious stone deposits formed in genetic association with aluminium-rich melts of acid magmas and carbonate metasediments. Rich jade deposits originated from the formation of jadeite and albitite under high pressure in the epizone as a result of regional metamorphism. Amber is found in Eocene shelf deposits.

The major deposits and occurrences of energy, metallic and non-metallic materials are shown in Fig. 72 (folder).

The physiographical and climatological evolution resulted in an essentially N–S orientation of the large **rivers,** namely the Irrawaddy, Chindwin, Sittang and Salween. Numerous **thermal springs** are related to the volcanism and the tectonic events in the Late Neogene-Quaternary.

Large **groundwater reservoirs** formed in suitable aquifers, mainly in Tertiary and Quaternary rocks, in low-precipitation areas (Dry Zone) as well as in areas where the need for water is particularly high (e.g. Bassein-Rangoon-Irrawaddy Delta Region) due to dense population and growing industrialization.

Extensive areas of the country are covered by agriculturally useful **soils,** i.e. soils of the wet tropical monsoon zone, of the dry tropical monsoon zone, of the subtropical monsoon zone, of the temperate high mountain zone, azonal soils and intrazonal soils. They can be utilized for agriculture in a very wide range of ways.

5.1 Energy raw materials

5.11 Hydrocarbons

In regional geological terms the following hydrocarbon-bearing or hydrocarbon-prone areas can be distinguished (Figs. 16, 27, folder):

– areas belonging to the Foredeep
  offshore Arakan and Arakan Coastal Area (Arakan Coastal Basin)
– areas belonging to the Interdeep (from S to N)
  W Gulf of Martaban
  W Irrawaddy Delta
  Prome Embayment
  Central Basin
  Chindwin Basin
  Hukawng Basin
– areas belonging to the Back Arc Basin (from S to N)
  E Gulf of Martaban
  E Irrawaddy Delta, Pegu Yoma, Sittang Basin
  Shwebo-Monywa Plain

5.111 Foredeep

The Arakan Coastal Basin is part of the "Assam-Andaman Trough"; it thus marks the N extension of the Indonesian-Andaman Foredeep. It runs W of and parallel to the Indo-Burman Ranges.

Offshore Arakan and Arakan Coastal Areas (Arakan Coastal Basin)

The Arakan Coastal Basin formed part of a marine depositional environment which was bounded in the E from the Upper Jurassic/Cretaceous onwards by the uplifted Sino-Burman Ranges. During the Upper Cretaceous and in the Early Tertiary, the margin of the Indo-Burman Geosyncline was located at the E edge of the present-day Indo-Burman Ranges.

To the W, flysch sediments formed. When the Indo-Burman Ranges were uplifted in the Oligocene and Miocene, the flysch trough as well as the subduction zone shifted westwards into the present-day Arakan Coastal Basin. From the Upper Eocene onwards, large quantities of flyschoid sediments, and from the Miocene onwards, molasse-like sediments accumulated in this rapidly subsiding trough.

As has been revealed by offshore drilling on elongated narrow anticlines in the Arakan Coastal Basin (7 exploration wells, dry), there was no euxinic depositional environment with potential source rocks developed in the flysch-type sediments because of the abnormally rapid rate of sediment accumulation (up to 80 cm/1,000 years in the Mio- and Pliocene). Low geothermal gradients and low C_{org} maturities, but very high overpressure, were observed. So far, only reservoir rocks with low permeability have been found in the Plio- and Miocene sequences. The reservoir rock qualities seem to decrease further towards the W.

The rock sequences drilled consist of mainly clay, mudstone, silty mudstone and siltstone, with occasionally some thin lignite, fine sandstone, limestone and dolomite intercalations. The upper portion of these sequences (a few metres to approximately 1,200 m) correlates in age with the Irrawaddy Group of the Inner-Burman Tertiary Basin (Quaternary – Late Neogene), the middle portion (>1,000 m) correlates with the Yenandaung Series of the Arakan Coastal Area (Late Neogene) and the lower portion (> 2,000 m) with the Kyauktan Formation and the Ngasanbaw Formation of the Arakan Coastal Area (Late, Middle and Early Miocene; Fig. 32, folder).

Close to the shore and in the Arakan Coastal Area, the Lower Miocene and Early Tertiary flysch series are uplifted and have undergone strong structural deformation (p. 105). Here, and in the deeply subsided Early Tertiary sequences of the Arakan Coastal Basin offshore in the W, the maturities are more favourable, as is shown by oil and gas seepages and mud volcanoes with natural gas on the islands of Boronga, Ramree and Cheduba and on the mainland in the area around and to the N of Akyab. Several tons/day of crude oil are produced from some sandstone layers within the Upper Miocene mudstone and from the flysch of the Lower Tertiary, e.g. at Yenandaung on the island of Ramree, and Hinzane on Cheduba Island. MALLET (1878, p. 207) has given a detailed description of how oil was extracted from pits, shafts and shallow boreholes in the last century. Also at the S plunge of the Boronga anticlinal trend some oil was produced from a number of wells drilled before World War II.

As in the Inner-Burman Tertiary Basin (p. 110), the mud volcanoes in this area are frequently located along fault lines. They are therefore often arranged in rows, as for example in the N part of the island of Ramree near Kyaukpyu (Fig. 23). Here 6 mud cones are located along an approximately 2.5 km long ridge. The craters at the top of the up to 9 m high mud cones intermittently or paroxysmically eject bentonitic argillaceous mud and angular rock fragments from the Neogene, together with natural gas (chiefly methane). MALLET (1878) mentioned also that small quantities of oil were produced from the mud volcanoes. On Cheduba, some mud volcanoes attain a diameter of more than 200 m. In the case of large eruptions, which also occur offshore, e.g. near Foul Island SE of Ramree, the natural gas occasionally ignited.

5.112 Interdeep

The Interdeep can be clearly defined between the Outer Island Arc and the Inner Volcanic Arc (Fig. 16).

Western Gulf of Martaban

The axis of the Interdeep runs SW–NE in the W part of the Gulf of Martaban. Above the "oil geological basement" (here probably Mesozoic rocks which correlate with those of the Arakan Yoma) follow marine series of the Eocene (> 500 m), of the (?) Oligocene and Miocene (> 3,600 m, in the lower portion argillaceous-calcareous-fine-sandy, in the upper portion mainly argillaceous with max. 30 m thick sandstone layers), and clastics of the Middle Miocene to Recent (up to > 400 m). Volcanoclastic rocks were observed below and interbedded in the Miocene sequence. Up to 1978, 4 wells had been drilled on seismically located, relatively broad and for the most part unfaulted anticlinal structures (one of these wells with subcommercial gas reserves, a second one with gas shows). Oil or condensate were not observed.

Western Irrawaddy Delta

The Western Irrawaddy Delta occupies the NNE extension of the Interdeep of the W Gulf of Martaban (Figs. 16, 27, folder). Like most of the Inner-Burman Tertiary Basin, this portion of it also was until the end of the Eocene the depositional environment of a shelf area of an ocean with its shores in the E along the Sino-Burman Ranges and open to the W, towards the Indian Ocean. Since the division into a Foredeep, Outer Island Arc (=Indo-Burman Ranges) and Interdeep did not commence before the Oligocene, the pre-Oligocene sequence is probably comparable with that of the S Indo-Burman Ranges and of the Arakan Coastal Basin (Fig. 32, folder).

None of the exploratory oil and gas wells drilled in the Western Irrawaddy Delta Basin (max. depth 3,900 m) have reached pre-Oligocene strata. The general subsidence of the Interdeep since the (?) Late Eocene lowered the pre-Oligocene rocks in the central parts of the Western Irrawaddy Delta Basin to their present depth of between 5,000 and 6,000 m. The overlying sediments of Oligocene (? m) and of Miocene to Quaternary (> 3,000 m) accumulated in a neritic depositional environment with normal oxygen content without euxinic facies having been developed. During the Miocene some limestone developed and the shale/sand ratio changed in favour of the sand probably as a result of the progressive uplifting of the erosional and source areas (Indo-Burman Ranges).

Since the Late Neogene, exclusively terrestrial clastic deposits have formed in what is now the onshore Western Irrawaddy Delta. In this area the Irrawaddy Group attains thicknesses of more than 1,200 m.

Gravimetric and seismic surveys revealed generally NNE and N-striking, elongated structural trends with relatively broad culminations. Of the ca. 15 exploratory wells drilled one struck gas in good reservoir sandstone of the Upper Miocene (Payagon 1). Subsequent boreholes struck the reservoir rock at points where it was waterlogged. *Miogypsina* limestone at the W side of the Western Irrawaddy Delta has recently been successfully tested for condensate and light oil.

Prome Embayment

From the Western Irrawaddy Delta Basin the axis of the Interdeep turns from NNE towards the N. Its E–W extent narrows between the Arakan Yoma and the Pegu Yoma to form the so-called Prome Embayment (Fig. 27, folder). Table 5 gives an overview of the stratigraphical sequence.

Table 5. Stratigraphical sequence Prome Embayment.

Age	Formation	Thickness (m)		
Quaternary-Miocene	Irrawaddy Group		0 –	> 1,200
Miocene	Obogon		450 –	900
	Kyaukkok		390 – approx.	1,250
	Pyawbwe		1050 – approx.	2,500
Oligocene	Okhmintaung		370 – approx.	600
	Tiyo	approx.	300 – approx.	400
	Alternating Clay/			
	Sand Member	approx.	900 – approx.	2,200
	Kyaukpon Sandstone		240 –	250
Upper Eocene			>	460

Hydrocarbon-producing and potential sandstone reservoir rocks are found in Late, Middle and Early Miocene and in Oligocene formations.

Structural and stratigraphic conditions are somewhat better known than further S because the pre-Irrawaddy Group sediments are partly exposed and relatively intense exploration has been going on here since oil was first discovered in the early sixties.

The Prome Embayment is characterized by very long (up to 80 km) and narrow, mostly asymmetrical folds with special culminations along their crests. Parallel and ±E–W transverse faults as well as NW and NE diagonal faults occur frequently. The folds are for the most part W-vergent with steep W flanks and flatter E flanks and considerable W-oriented overthrusts (with vertical displacements of up to 1,000 m) along N–S-striking faults. The fold trends are frequently offset by "en echelon" faults.

The following fields are producing hydrocarbons in the Prome Embayment:

– Prome oil field (recoverable reserves in 1975: about 1.6 million tons)
– Myanaung oil and gas field (roughly estimated recoverable reserves in 1975: 1.8 million tons of oil)
– Shwepyitha gas field (roughly estimated recoverable reserves of natural gas in 1975: 2.2×10^9 Nm^3).

Central Basin

An uplift area (p. 33) with locally exposed Oligocene rocks separates the Prome Embayment from the Central Basin further along the Interdeep to the NNW and N (Fig. 27, folder).

As in the Prome Embayment, the pre-Irrawaddy Group sequences are exposed over great distances along the structurally uplifted areas. The anticlinal crests have locally been eroded down to the Eocene (e.g. Chauk and Yenangyat). The geology of the Central Basin is relatively well known because oil exploration commenced here very early (Yenangyaung: the first mechanized well was drilled in 1887; Chauk: an exploratory well was sunk in 1891). Table 6 provides an overview of the stratigraphical sequence.

Table 6. Stratigraphical sequence Central Basin.

Age	Formation	Lithology	Maximal thickness (m)
Quaternary-Miocene	Irrawaddy Group	sandstone, med.- and coarse, congl., thick-bedded, massive	approx. 3,000
Upper Miocene	Obogon	shale, argill. sandstone	approx. 1,100
Middle Miocene	Kyaukkok	sandstone, fine- and med. grained, subord. shale	approx. 1,250
Lower Miocene	Pyawbwe	shale, subord. sandstone inter-cal.	approx. 1,300
Upper Oligocene	Okhmintaung	sandstone, fine- and med. grained, some shale	approx. 2.000
Middle Oligocene	Padaung	shale, fine- and med. grained sandstone intercal.	approx. 1,200
Lower Oligocene	Shwezetaw	sandstone, fine-grained, fine-sandy shale	approx. 1,100
Eocene	Yaw	shale, some sandstone, and lignite intercal.	approx. 900
	Pondaung	sandstone, fine-med., grained shale and marl	approx. 2,300
	Tabyin	shale with yellow-brownish sandstone layers	approx. 3,200
	Tilin	sandstone, blue-grey, fine-medium grained	approx. 3,000
	Laungshe	shale	approx. 4,500

Reservoir rocks (sandstones) are found in the Middle and Lower Miocene, Upper Oligocene and in the Eocene.

The generally $\pm$NNW–SSE-striking structures are elongated and narrow and, as in the Prome Embayment, cut by a great number of faults. Almost all the hydrocarbon-bearing structures are located at the E edge of the Salin Basin (Figs. 27, folder, 49), namely (from S to N): Mindegyi, Tagaing, Minbu, Mann, Yenangyaung, Chauk-Yenangyat, Ayadaw, Sabe, Letpanto, Kyaukwet and Sawin. The outcropping Miocene, Oligocene and Eocene strata ("Western Outcrops") at the W side of the Salin Basin contain many surface oil shows. Large numbers of oil seepages along anticlinal crests at the E side of the Salin Basin resulted in oil mining already in Medieval times (Yenangyaung). Mud volcanoes occur along a fault in the Minbu structure (Figs. 69, 70).

Chauk-Yenangyat is a typical example of a hydrocarbon-bearing structure of this area. It can be traced over a distance of 85 km and it attains closures of up to 4 km×0.5 km. It is an asymmetrical fold with an E-overthrust along its E flank. The vertical displacement at this overthrust is up to 1,000 m. The tangential stress which led to this asymmetry was here directed from W to E. Countless transverse and diagonal faults with throws between 30 and 150 m break up the anticline into individual sections. While oil is produced exclusively from sandy reservoir rocks above the overthrust, gas is produced from stratigraphically identical horizons below it.

Fig. 69. Mud volcano near Minbu. Photo: F. BENDER.

The probable and potential oil reserves in the Central Basin were estimated in 1975 to be about 40 million tons, and the natural gas reserves were put at about 11 billion Nm^3. The hydrocarbon potential of the Central Basin, but also of the Prome Embayment, is certainly much higher than that, particularly if one includes the potential reserves of the Miocene, Oligocene and Eocene sequences in structures below 2,500 m. Testing of these structures did not commence until 1979.

Chindwin Basin

As in the S, the Central Basin ends also in the N at a structural uplift, the "22° N uplift area" (Fig. 27, folder).

Apparently, the Interdeep undulates along its general S–N strike with structural lows, with from the Oligocene and Miocene onwards also thicker sediments (W Gulf of Martaban, Western Irrawaddy Delta Basin, Prome Embayment, Central Basin, Chindwin and Hukawng Basins), and interposed uplift areas with thinner Late Tertiary sequences.

Fig. 70. Crater of mud volcano with natural gas bubbles (diameter approx. 1 m each) immediately before eruption; near Minbu. Photo: F. BENDER.

To the N of the Central Basin the axis of the Interdeep turns towards the NNE to NE and marks the course of the Chindwin Basin between approximately 22° N and 26° N. With a length of almost 500 km and a width of up to 160 km, it is the largest sub-basin in the Inner-Burman Tertiary Basin and it probably also contains the thickest deposits of Late Tertiary-Quaternary rocks. The sequence is only known to some extent from the "Western Outcrops" and "Eastern Outcrops" (outcropping strata at the W and E flanks of the S part of the Chindwin Basin). Wells sunk in the only structures that have been tested so far (Indaw, Yenan) in the central part of the S part of the basin did not reach Early Tertiary strata. According to an aeromagnetic survey, the central parts of the basin probably contain sediments up to 9,000 m thick above the magnetic basement. The Tertiary and Quaternary sequence that is exposed in the "Western Outcrops" (Kalemyo-Kalewa-Thetkegyin) at the SW side of the Chindwin Basin is estimated to be about 14,000 to 16,000 m thick. Comparisons show that the Eocene to (?) Oligocene sequences increase in thickness towards the W (Indo-Burman Ranges). Isopachs of Upper Miocene sediments indicate the beginning of a basin configuration corresponding to that existing nowadays.

As far as is known, definite marine (shelf) sedimentation occurred only in the lower part of the Lower Eocene Tabyin-Laungshe Formation and in the Miocene Natma Formation. Otherwise the Tertiary sequence formed in a near-shore marine-brackish shallow-water environment (with paralic coal in the Middle and Upper Eocene (p. 94)) and presumably also in onshore areas near the coast under fluviatile conditions.

Potential reservoir sandstones alternating suitably with thick sealing claystone and shale are found throughout the pre-Upper Miocene sequence.

Geological reconnaissance mapping, aerial photograph and satellite image interpretations and some seismic reconnaissance in the S section of the basin have led to the discovery of a number of interesting structures in terms of petroleum geology, e.g. at Indaw, Mahudaung, Makkadaw, Yenan and Patolon.

Before the Second World War, some oil was produced from near-surface sandstones of the Natma Formation in a limited portion of the crest area of the Indaw Anticline which is more than 75 km long. Small amounts of oil were also extracted from shallow wells drilled in Eocene sandstones in the Yenan Structure in the area of the "Western Outcrops". Numerous surface oil and gas shows mark the "Western Outcrops" in the S section of the basin, as well as the "Eastern Outcrops" along the Mettaung Thrust and the associated structures E of Indaw (Fig. 27, folder). Mud volcanoes are found about 25 km NNW of Kalewa ("Western Outcrops") and on the Indaw Anticline. It is remarkable that very few surface oil or gas shows are known to occur N of 24° N.

Together with the N Central Basin, the S part of the Chindwin Basin is probably one of the most promising areas in Burma for the discovery of further accumulations of hydrocarbons.

The N areas of the basin are even less explored than its S half. According to an aeromagnetic survey, the N section probably contains sequences up to 6,000 m thick above the magnetic basement, most of them of post-Oligocene age.

Table 7 provides an overview of the stratigraphical sequence of the Chindwin Basin.

Table 7. Stratigraphical sequence Chindwin Basin.

Age	Formation and thickness (m)		
	Southwest	Northwest of 24° N	East
Quaternary-Miocene	Irrawaddy Group and	? Irrawaddy Group	? Irrawaddy Group (approx. 2,600)
L. Plioc./U. Mioc.	Mingin (approx. 2,800)	Kaungton	Kaungton (approx. 1,500)
U. Mioc./M. Mioc.	Shwethamin (approx. 2,100)	Upp. Thaungdut	Shauknan (approx. 2,700)
M. Mioc./L. Mioc.	Natma (approx. 1,300)	Mid. Thaungdut (> 6,000)	Nandawbee (approx. 700)
L. Mioc./U. Oligoc.	Letkat	Low. Thaungdut	Inga (approx. 200)
Oligocene	(approx. 2,400)	Tonhe (approx. 2,500)	Upp. Yeyein
U. Eocene	Yaw (approx. 1,200)	Nankyila (> 1,800)	Upp. Yeyein (approx. 1,500)
U. Eocene	Pondaung (approx. 2,700)	Pondaung (?)	Mid. Yeyein and
M. Eoc./L. Eoc. Paleocene	Tabyin-Laungshe (?) Paunggyi Congl. (?)		Low. Yeyein Maingwun (?) Naunggauk (> 3,600) Nanhizin (?) Nantat (?)
U. Cret./? Paleocene	Kabaw (> 500)		

Hukawng Basin

At about 26° N, the Chindwin Basin borders on plutonic and metamorphic rocks of the N and NNW-striking uplifted Central Ridge (Fig. 27, folder, p. 33). The approximately 90 km wide Central Ridge running transverse to the strike of the Interdeep separates the Chindwin from the Hukawng Basin which is located along the assumed NE extension of the Interdeep (Fig. 16). It is bounded in the N by the arcuate Indo-Burman Ranges running NE and ENE. Its subsurface geology is almost totally unknown. According to an aeromagnetic survey it is likely that sediments more than 5,000 m thick exist in its central part above the magnetic basement. Nothing is known of the presence of hydrocarbons in this area and so far no drilling for hydrocarbons has been carried out.

5.113 Back Arc Basin

The Inner Volcanic Arc separates the Interdeep from the Back Arc Basin adjoining to the E (Fig. 16). A number of large sub-basins and uplift areas occur also along the course of the Back Arc Basin. Towards the E, the Back Arc Basin is bordered by the Sino-Burman Ranges.

Eastern Gulf of Martaban

The area in the E Gulf of Martaban is characterized by sediments that increase in thickness rapidly from the Inner Volcanic Arc in the W and from the Tenasserim region in the E towards the centre (Fig. 28). The axis of the Back Arc Basin runs S−N in the area of this "Central Trough". None of the offshore wells drilled so far in this area have entered the "oil geologic basement".

Fine clastic marine sediments of the Quaternary were encountered in thicknesses of 250 m to 300 m. They overlie mainly terrestrial Quaternary to (?) Miocene and marine alternating mudstone and sandstone of the Miocene (totalling > 4,000 m thick). Below follow 500 m to 1,500 m thick shale and subordinate sandstone of a higher degree of diagenesis (Miocene?) or, in the vicinity of the Inner Volcanic Arc in the W of the "Central Trough", an alternating sequence of volcanoclastics and shale. The oldest strata so far drilled consist of thick, hard, argillaceous fine sandstone and shale which can be assigned to the Lower Cretaceous/Upper Jurassic. The change to a much higher degree of diagenesis, which has also been verified by a sudden change of the vitrinite reflectance (Fig. 71), takes place independently of the stratigraphic level at depths between about 2,500 m and 3,600 m. Above this change the reservoir rock properties of the Late Tertiary sediments are good to very good, below the change they are almost totally lacking.

Transverse structural elements ("uplift areas", ±E−W) occur between 15° N and 14° N and between 11° N and 10° N. From the latter area (W of the S Tenasserim coast) the axis of the hydrocarbon-bearing North Sumatra Tertiary Basin plunges towards the S.

The E Gulf of Martaban area is structurally complicated and broken up by a network of faults into numerous antithetic and synthetic block steps (Fig. 28). The structure is further complicated by ±N−S running, less than 10 km wide and up to about 250 km long diapiric

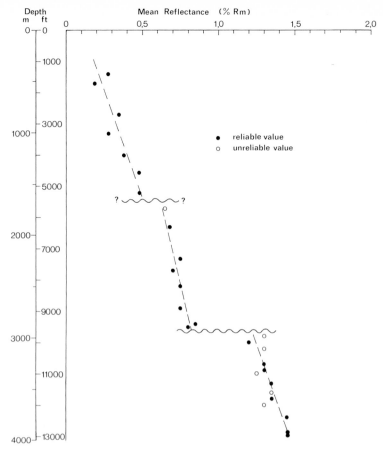

Fig. 71. Mean vitrinite reflectance in sediments of the Gulf of Martaban.

mud intrusions in the E part of this area. The exploratory oil wells drilled in uplifted blocks and in crests of arched blocks encountered 5 occurrences of natural gas, whose economic potential has not yet been proved.

Eastern Irrawaddy Delta, Pegu Yoma, Sittang Basin

The axis of the Back Arc Basin runs here N–S. It has not been established whether the offshore E Gulf of Martaban Sub-basin in the S is separated by an uplift area or swell from the onshore continuation of the Back Arc Basin comprising the areas of the Eastern Irrawaddy Delta, Pegu Yoma and Sittang Basin. The Back Arc Basin narrows to less than half its original width from the E Gulf of Martaban Sub-basin through to the area N of Rangoon because of the tectonic frame: the Inner Volcanic Arc runs NNE and the Shan Boundary Fault Zone strikes NNW.

Table 8 gives an overview of the section of the stratigraphical sequence in which drilling has been carried out.

Table 8. Stratigraphical sequence Eastern Irrawaddy Delta, Pegu Yoma, Sittang Basin.

Age	Formation	Lithology	Maximal thickness (m)
Quaternary-Miocene	Irrawaddy Group	sandstone, med.-coarse, gravel, congl., thick-bedded, cross-b., massive	approx. 1,200
		angular unconformity	
Upper	Obogon	altern. fine med. sandstone, argill. siltst., sandy shale	approx. 1,000
Middle Miocene	Kyaukkok	fine-med. grained sandstone, little shale	approx. 900
Lower	Pyawbwe	sandy shale, shale, some fine-grained sandstone intercalations	approx. 850

Exploratory oil wells have mostly been sunk in the crests of the Pegu anticlines (p. 110) which plunge to the SSE from the Pegu Yoma below the clastics of the Irrawaddy Group. Some of these wells yielded good indications of natural gas from sandy reservoir rocks directly below the Irrawaddy Group and in Lower Miocene rocks, although so far gas has not been produced on an economic scale in this area.

The Sittang Basin, which is located between the Pegu Yoma in the W and the Shan Boundary Fault Zone ("Sittang Fault Zone") in the E, has so far not been examined for its oil possibilities. No hydrocarbon exploration at all has been so far carried out in the area of the block-folded and faulted Miocene sediments of the Pegu Yoma.

Shwebo-Monywa Plain

In the area of the sharp bend of the Irrawaddy near Mandalay, the Pegu Yoma Anticlinorium, which consists chiefly of Miocene sequences, plunges gradually to the N below the Irrawaddy Group. There the Shwebo-Monywa Plain commences; in the W it is bounded by the Cenozoic volcanic rocks of the Inner Volcanic Arc, in the N by the Wuntho Massif and in the E by the Sino-Burman Ranges. An aeromagnetic survey revealed the NNE continuation of the Inner Volcanic Arc N of Mt. Popa-Monywa through to the Wuntho Massif. It also clarified the N continuation of the Back Arc Basin with sediments up to 8,000 m thick above the magnetic basement.

No oil exploratory drilling has been carried out in this area. The Miocene and Oligocene sediments exposed in the cores of anticlines of the N–S running structures S and N of Shwebo reveal mainly terrestrial or shallow marine, sandy facies. There are no seepages of oil or gas, apart from a number of mud volcanoes NNW and SSE of Shwebo which show signs of methane.

The question has not been settled whether the Back Arc Basin N of the Wuntho Massif continues into the E section of the Hukawng Basin (Figs. 17 a, 27, folders).

5.12 Bituminous rocks

HELFER (1838b) reported on a "slaty coal, 5–6 feet thick" from a tributary valley (Thein Kun) of the Little Tenasserim River near Mergui. OLDHAM (1856) recognized that it was bituminous shale in a shale-sandstone sequence in which coal seams up to about 10 cm thick also occurred. RAO (1922) visited the site (Naungbwa, 11° 51′ 40″ N/99° 29′ 10″) and assigned a Tertiary age to the oil shale-bearing sequence in the valley of the Thein Kun River.

GREGORY (1923) and COTTER (1924) mentioned this and further occurrences near the locality of Bonkum in the Lenya Valley (Mergui District) and in the Great Tenasserim Valley of the adjoining Tavoy District. Extensive exposures of oil shales were described by these authors from the intramontane Late Tertiary Basin of Kaw Kaweik (Phalu Tertiary Basin) at the border with Thailand (No. 60 in Fig. 72, folder) which extends to the N into Thailand (Mesauk-Methalaun-Melamat Basin). From the Mepale Valley they described the Htichara (16° 46′ N, 98° 28′ E) Basin. This basin is about 21 km long and 14 km wide and contains fresh-water sediments of Late Tertiary age (probably Pliocene; p. 103) with several oil shale beds up to 2,00 m thick. CHHIBBER (1934b, p. 241) gave the following analysis of an oil shale sample from the Htichara Basin:

water	at 50° to 170 °C	42.00%
oil	at 50° to 170 °C	5.00%
oil	at 170° to 230 °C	0.17%
oil	at 230° to 270 °C	1.50%
oil	above 270 °C	41.00%
residue	non-volatile at 400 °C	10.33%

Oil shales from the same locality, which were analysed by the Geological Survey of India, yielded between 9.78 and 16.20 gallons of crude oil per ton. They had the following composition (J. C. BROWN & DEY 1975, p. 457):

water	13%
light naphtha	4%
heavy naphtha	3%
kerosene	23%
lubricating oil	40%
residue	17%

The Htichara oil shale could be open-cast mined over wide areas because of little overburden, gentle dipping and only minor faulting of the deposit. Part of the deposit was drilled in the early thirties, but the Second World War prevented any major exploitation. Exploration has still not been recommenced.

5.13 Coal, peat

Occurrences of **coal** are located in the Arakan Coastal Area, in the Inner-Burman Tertiary Basin and in the Sino-Burman Ranges (Fig. 72, folder).

In most cases only indications of coal were observed, and so far the few deposits that have been discovered are of little economic significance (J. C. BROWN & DEY 1975).

In the **Arakan Coastal Area** and on the islands of Ramree and Cheduba numerous cm – to decimetre-thick carbonaceous clay layers, coal seams and streaks have been observed in the Eocene sequence (p. 91). They are, however, of no economic value. The exploratory oil wells drilled in this area have also failed to provide any evidence of thicker coal seams.

Very frequent coal indications and occurrences are observable in the **Inner-Burman Tertiary Basin.** Some of the deposits are of local economic importance.

Coal of Eocene age is exposed in seams usually less than 0.60 m thick in the Upper Chindwin District in the valleys of the Nantahin, Peluswa, Maku and Telong rivers N of the Kale River in an area measuring about 100 km from N to S and about 20 km from E to W. Occasionally seams up to 3.60 m thick (Nantahin-Peluswa area) are observed. Near Kalewa some of the seams (maximum thickness 2.70 m) are mined. The production is around 10,000 tons/a of coal having a calorific value of approximately 6,200 kcal/kg. The reserves are estimated to be about 123 million tons.

A likewise Eocene, structurally highly deformed and fractured coal deposit with a seam thickness of 2.40 m occurs near Kywezin in the Henzada District. Attempts to mine this coal have been abandoned. Other, but thinner seams have been found in this region near Hlemauk and Posugyi. According to J. C. BROWN & DEY (1975) the coal from this area (Kyewezin) has the following characteristics: moisture: 1.6%; ash: 5.6–6.8%; volatiles: 13.0–26.6%; C_{fix} : 65.1–79.9%; S: 0.91%.

At the end of the last century, Tertiary lignite of the following quality was mined in the area of the Shwebo-Monywa Plain between the bend of the Irrawaddy at Kabwet and the Man Chaung: moisture: 12.3%; volatiles: 37.4%; C_{fix} : 39.0%; ash: 11,3%. Geological details are lacking. Mining was abandoned.

Further Tertiary lignite occurs in the districts of Bhamo (near Mithwe), Katha (near Pinlebu), Minbu (Kyaukwet) and Pakokku (Letpanhla and Tazu) (Fig. 72, folder). Near Gyobin (S of Pinlebu), lignite of Upper Cretaceous to Eocene age occurs in several seams. In these districts, seams between 1.20 m and 2.10 m thick have been observed and some of them have been mined on a small scale near Kyaukset in the Minbu District. They possess the following characteristics: moisture: 4.8–26.4%; ash: 4.1–29.8%; volatiles: 24.1–48.7%; C_{fix} : 12.9–52.8%; S: 0.53–2.55%; calorific value: 4,960–5,750 kcal/kg; near Tazu and Letpanhla in the Pakokku District: moisture: 18.7%; ash: 11.3%; volatiles: 34.1%; C_{fix} : 35.9%; S: 2.28–5.63%; calorific value: 4,500–4,835 kcal/kg.

In the **Sino-Burman Ranges,** Silurian anthracitic coal, Jurassic coal and Late Tertiary/Pleistocene lignite occur, the latter in intramontane basins.

In the Wabya Formation of the Llandoverian (p. 63) approximately 10 km N of Heho, and up to 80 cm thick, mylonitized seam of anthracitic-meta-anthracitic coal is exposed. The coal is a mixture of finely fragmented detritus and about 30 vol.% iron hydroxide-rich clay. The organic detritus consists of isotropic particles in which fragments of cell walls are recognizable, and of anisotropic, dense and structureless fragments (Figs. 73 a, b). The cell wall fragments were interpreted as fusinite (inertodetrinite). The anisotropic structureless particles are probably vitrinite, which is possibly sapropelitic in origin. Since, however, fusinized remains of plant tissue are very rare in sapropel coal, the particles may be a humus coal substance. The maximum reflectance of the probably vitrinitic substance is 4.1%, which corresponds to the transition between anthracite and meta-anthracite (unpubl. rep. BGR. W. HILTMANN 1979).

Fig. 73 a, b. Fragments of fusitinized cellular tissue of Silurian coal (400×); near Heho. Photo: H. JACOB.

Jurassic coal is found in the Loi-An Series (p. 84) in the area of Loi-An between Myinka and Konhla and in the Pan Laung Valley N of Loi-An (Southern Shan State; Fig. 72, folder). The coal occurs mostly as streaks or irregularly formed seams up to 0.45 m thick which have undergone considerable faulting and folding. The coal from Loi-An contains 0.6–6.5% moisture; it has an ash content of between 9.5 and 73.6%; samples with <10% ash contain between 24.0 and 24.3% volatiles; the C_{fix} content is 59.7%; the sulphur content is between 0.33 and 0.65%; the calorific value is 6,340 – 6,420 kcal/kg. This coal is cokable.

Late Tertiary (Miocene, Plio-Pleistocene) lignite is bound to fluvio-limnic sediments of intramontane basins in the Sino-Burman Ranges. For example:

– the Lashio Basin (Northern Shan State) covering approximately 130 km², with exposed soft lignite in thicknesses between 0.9 m and 10.5 m. It has the following characteristics: moisture: 17.8–19.8%; ash: 9.2–13.6%; volatiles: 34.8–35.6% C_{fix}: 30.4–37.4%; S: 1.45%
– the Namma Basin, which is about 20 km S of the Lashio Basin, and also about 130 km² in extent, contains some seams of black lignite with thicknesses of 2.10 m to 6.30 m. This coal, which is mined on a small scale, contains 16.6% moisture, 7.7% ash, 36.9% volatiles and 38.8% fixed carbon (mean of 5 samples)
– the Mansang (approximately 30 km S of Namma) and the Mansele (approximately 45 km E of Namma) basins with numerous exposures of lignite with seams up to 1.35 m thick
– the Theindaw-Kawmapyin Basin (Mergui District, Tenasserim), with a number of lignite seams between 1.2 m and 4.5 m thick. Average samples gave the following values: moisture: 13.9%; ash: 6.6%; volatiles: 35.7%; C_{fix}: 43.8%.

These basins are highly prospective for lignite and detailed exploration would be justified. There are a large number of other intramontane basins in the Sino-Burman Ranges sometimes covering considerable areas (e.g. the Heho-Taunggyi, Inle and Namsang basins). They can also be considered prospective areas for lignite.

So far, **peat** deposits have not been studied or prospected anywhere in Burma. CHHIBBER (1934a, p. 285) mentioned surface peaty deposits in the vicinity of Nawnghkio on the S side of the Gokteik Gorge. It is very likely that further surface and also underground peat deposits will be found in the Late Neogene-Quaternary lacustrine sediments of the intramontane basins of the Sino-Burman Ranges.

5.14 Nuclear fuels

Uranium minerals have long been known from the area of Mogok (Fig. 72, folder). Uraninite occurs in very small quantities disseminated in alaskite (p. 123), in pegmatites and in gemstone placers and tailings (p. 207) where the following mineral assemblages were observed in the gemstone-bearing valley fills of Mogok-Kyatpyin:

ilmenite, magnetite, topaz, spinel (very frequent)
uraninite, zirconium (frequent)
corundum, tourmaline, alunite, sphene, beryl, columbite, pitchblende (rare).

Analyses of selected samples of alaskite, syenite and "hornblende rock" revealed some uranium content. The radioactivity comes for the most part from thorium, and the uranium content in the alaskite samples is usually well below 100 ppm. The thorium content is in the range from 100 ppm to several 100 ppm. The radioactivity is mostly bound to coarse-grained portions of the alaskite in which thorite ($ThSiO_4$) and zirconium are relatively frequently encountered. Pegmatites were also found to contain thorite and occasionally thorianite (ThO_2) and brannerite (UTi_2O_6).

When the alaskites and pegmatites were weathered, some of the radioactive minerals found their way into the unconsolidated Quaternary sediments which locally bear gemstones in the Mogok area. Gravity separation, which is the common procedure used in mining gemstones, resulted in a certain concentration of the radioactive minerals in the tailings. So far, no concentrations of economic interest have been found.

Slightly increased radioactivity was observed while carrying out iron ore exploration near Pang Pet (NEUBAUER 1965, GOOSSENS 1978a) about 10 km E of Taunggyi. The iron ore is mylonized hematite, large parts of which have been converted into limonite. The very irregularly distributed uranium content is for the most part bound to the limonite and varies from a few ppm U up to 0.2%U. Thorium and lead are always well below 100 ppm. Yttrium was present in concentrations of up to 33 ppm. The average copper content can attain 0.2% over a thickness of 10 m to 12 m.

Remnants of rhyolitic rocks, deeply weathered and altered, have been identified in the iron ore. Therefore the view expressed by NEUBAUER (1965) regarding the genesis of the ore deposit may be correct. He believed that a rhyolite body containing primary hematite and sulphidic mineralizations was largely altered and converted by weathering processes. As a result the hematite was mostly converted into limonite, and the uranium was absorbed on goethite. The sulphides intensified the kaolinization and almost complete chemical conversion of the rhyolite. In the process, the sulphidic minerals were transformed into oxides.

HERON (1917, p. 179) observed **monazite** in the sediments of the Shwe Du and the Lamawpyin Chaung, which drain the Anatholin Range near Mergui to the SW. Apart from ilmenite, he found a little magnetite, traces of gold and tin and up to 0.18% ThO_2 in the investigated heavy mineral placers. JONES (1920, p. 156) mentioned monazite from Mong Klang near Wan Hapaum (21° 32′ N, 97° 29′ E) which he classified as of no economic importance. J. C. BROWN & HERON (1919) reported on monazite in the cassiterite concentrates and in sands of the rivers which drain the granites in the Tavoy District. HERON (1917, p. 180) found monazite also in black heavy mineral sands which occasionally occur along the coast in the S Tavoy District (e.g. NW of the mouth of the Kyanchaung). Heavy mineral beach sands have also formed as patches over a distance of more than 20 km along the coastline to the W, SW (mouth of the Salween) and S of Moulmein. They probably also extend further to the S. Near Amherst, samples taken from a 1.6 m thick black sand contained about 2% monazite, 5% zirconium and 1% rutile in addition to 65% magnetite and ilmenite. GOOSSENS (1978a, p. 487) mentioned also the area of the Yadanabon Mine (11° 17′ 05″ N/99° 17′ E) and the area of Sakangyi (22° 55′ 58″ N/96° 13′ E) with monazite in river sands and granites (from unpubl. reports).

5.2 Metallic raw materials

The distribution of the metallic raw materials (Fig. 72, folder) shows very clearly that certain ores are bound preferentially to certain geological units, as was pointed out already by J. C. BROWN (1924) and CHHIBBER (1934b). SOMMERLATTE (1948) differentiated between the "Shan-Yunnan sulphide ore province", the "Tenasserim tungsten/tin ore province", the "Mingin gold ore province" and the "Arakan-Naga serpentine area" with chromite and magnesite occurrences. BA THAN HAQ (1972) defined 7 metallogenetic provinces, namely

- the Cr, Ni and Pt province (Arakan-Chin Ranges, Naga Mountains and the area N of Myitkyina)
- the Sn-W provinces (Tenasserim, Karen State, W Shan State)

- the Fe-Mn provinces (near Myitkyina, parts of the S Shan State, near Moulmein, Tavoy and Mergui, along the Tenasserim Archipelago)
- the Cu provinces (E of Mandalay-Kyaukse, E of Yamethin, SW of Moulmein and E of Mergui)
- the Cu provinces bound to Tertiary volcanic rocks (near Monywa, S of Myitkyina)
- the Pb-Zn-Ag province (from northernmost Burma to the S along the Chinese border via the area of Lashio-E of Mandalay-Heho-Bawsaing to the border with Thailand)
- the Sb province (central and E Shan State, the area from Lebyin via Thaton and Moulmein to the border with Thailand).

Like J. C. BROWN (1924) and CHHIBBER (1934 b), SOMMERLATTE (1948) considered more the genetical aspect of the metallic ores and in assigning the occurrences to geological units he differentiated between

- the Arakan-Chin metallogenetic province with Cr, Ni, Pt, Cu, Fe
- the NE Burma province with Cr, Cu, Fe, Pb, Mn, Zn, Ag, Mo
- the Shan-Tenasserim province with Sn, W, Cu, Sb, Pb, Zn, Fe, Mn, Ba, F.

Since ultrabasic rocks occur at the E edge of the Arakan Yoma, in the Chin-Naga Ranges and N of Myitkyina, over a total N–S distance of more than 1,000 km, a metal assemblage containing mainly Cr, Ni, Cu, Mg, and Pt is just as likely to occur as in NE Burma where, apart from ultramafites (Cr, Ni, Cu, Fe, Mn, Mg), acid intrusives and extrusives (Pb, Zn, Ag, Mo, Cu, Au) and Precambrian rocks with polymetallic mineralizations are known.

The Ordovician carbonate belt with strata-bound or structurally controlled Pb-Zn-Ag mineralizations can be traced from Kanchanaburi (Thailand) through the Shan States (Bawsaing-Heho-Lashio-Bawdwin) and NE Burma to Yunnan (China) over a N–S distance of approximately 2,000 km and an E–W width of up to around 300 km (including parts of W Thailand). The sedimentation of the Lower Paleozoic carbonate series is locally associated with contemporary volcanism (e.g. Bawdwin; p. 178). This is the Pb-Zn-Ag (plus Cu, Co, Ni in the case of Paleozoic volcanism) province.

The fourth metallogenetic province with chiefly W, Sn and polymetallic mineralizations is associated with intrusions of granite of varying age into Paleozoic sedimentary sequences (in S Burma, mostly into the Mergui Series (p. 70)). It extends from the southernmost Tenasserim Archipelago (Davis Island) through Tenasserim to the Southern Shan State.

The Tertiary to Pleistocene (?) Cu mineralizations (traces of Au, Ag) that are bound to intermediate to acid volcanic rocks of andesitic-dacitic composition, can be regarded as a fifth metallogenetic province. They belong to an approximately 40 km wide, N–S-striking zone of unknown length S and N of Monywa. GOOSSENS (1978 a, p. 475) described the Monywa Cu occurrence as a "unique deposit situated in a significant tectonic location: in addition to being along a calc-alkaline volcanic chain, it is also located at the intersection of this line (fault zone?) and the westward extension of the Lashio NNE Fault." However, since such volcanic rocks with Cu indications also occur further N of this intersection, a connection with a hypothetical WSW continuation of the Lashio Fault to a point of intersection with the N–S-running chain of volcanic rocks is improbable (Fig. 17 a, folder).

The distribution of the post-Jurassic metal ore occurrences can, to a certain extent, be explained by the collision of the Indian Plate with the Eurasian (China) Plate (p. 16). MITCHELL (1973, 1977) and GOOSSENS (1978 a) assumed that the ore-supplying extrusions and intrusions of igneous rocks in the Late Mesozoic and Cenozoic came from a depth of about 150 km (Monywa) and of approximately 300 km (Shan Scarp Area) from the subducted oceanic plate that is dipping from the Arakan Yoma to the E.

5.21 Noble metals, iridium, osmium, selenium, tellurium, niobium, tantalum, rare earths

So far, only traces of **platinum** have been observed in Burma, and the metal has not been mined. Such traces have been found in the fluviatile sediments of the waterways that drain the N section of the chain of olivine-pyroxenite, picrite, dunite, norite, gabbro, diorite and serpentinite, which accompanies the E edge of the Arakan Yoma and Naga Hills and also occurs in the Hukawng area. SOMMERLATTE (1948) mentioned placers along the upper Irrawaddy and its tributaries, particularly in the Myitkyina area, in the Uyu Valley and near Katha on the Meza River. In the sediments of the Lower Chindwin Valley washed placer gold contained 2.53% platinum and 1.04% **iridium** and **osmium.**

GOOSSENS (1978a, p. 487) mentioned a sample of gold with 2.35% Pt and 7.05% iridium plus osmium from the Meza River ($25° 08'$ N, $96° 04'$ E) and from Kani, associated with gold dust, grains with 20% Pt and 40% iridium plus osmium ($22° 27'$ N, $94° 53'$ E) (from unpubl. reports).

Primary **gold** and placer gold occurrences are widespread. Their exploitation has so far had no economic importance, despite the fact that in many localities panning for gold has gone on for centuries.

The **Wuntho Massif-Myitkyina-Hukawng area** is of interest for Au-prospecting and exploration. The approximately 150 km long and about 60 km wide, NNE–SSW running Wuntho Ranges (southernmost point at $24° 10'$ N) are composed of quartz diorite and andesitic volcanites which contain quartz veins with auriferous pyrite, magnetite, chalcopyrite, galenite and altaite (NOETLING 1894). According to SOMMERLATTE (1948), a vein of this type was mined to a depth of 130 m near Kyaukpasat, 45 km N of Wuntho. The ore contained on average 12 g Au/t over a period of 7 years of mining. J. C. BROWN (1935) correlated the andesitic volcanic rocks of the Wuntho Massif with the Tertiary andesites (dacites) that occur further to the S in the Monywa area in which copper ore impregnations containing small amounts of gold were reported (p. 134). CHHIBBER (1934b) mentioned also gold quartz veins in crystalline rocks from the Myitkyina area near Jamaw ($25° 25'$ N, $96° 18'$ E), near Hkamti Long, on the upper Chindwin and Shweli rivers in the Northern Shan State (Fig. 72, folder). No details are available from this geologically almost unknown area.

NOETLING (1934) described gold occurrences about 10 km S of Kalaw. They are bound to contact zones between diorite and biotite granite with aplites and Jurassic limestone and slate. The sedimentary rocks are highly silicified, and wollastonite occurs frequently. The diorite is pyritiferous. Small quantities of gold are also known from quartz veins that occur in the gneiss in the Kyaukse area about 17 km NE of Myogyi.

Placer gold was and still is locally mined in small quantities from Pleistocene and Recent river terraces on the **upper Chindwin** and its tributaries, in particular the **Uyu River.** The Chaunggyi, which, like the Uyu River, drains part of the Wuntho Massif, flows into the Uyu near Gyobin ($24° 53'$ N, $95° 20'$ E), where river terraces up to 15 m thick with about 0.17 g

Au/m³ are found (SOMMERLATTE 1948). To the N, near Mamon, about 3,000 g gold also containing platinum was produced in a small pilot test conducted in 1914, which was abandoned soon afterwards. The Uyu Boulder Conglomerate, a Pleistocene terrace deposit (p. 103), also contains gold and platinum. CHHIBBER (1934b) mentioned further placer gold occurrences in the fluviatile sediments of the streams draining the Kumon Ranges to the Hukawng Valley.

Small amounts or traces of placer gold are also observed at many locations along the Irrawaddy in Recent and Pleistocene river terraces, e.g. N of Myitkyina, near Nalon (60 km NNE of Bhamo), near Myothit (35 km NE of Bhamo) and at Shwegu, Prome and Shwedaung (SOMMERLATTE 1948, p. 53).

SOMMERLATTE mentioned further placer gold from the Namma River, a W tributary of the Salween (average contents 0.35 g/m³) and in the drainage area of the Paong River, which flows into the Salween at $21° 56' N/97° 43' E$; furthermore from the area of the Shweli River (flowing into the Irrawaddy downstream of the village of Katha), from the Sittang Valley near Shwegyin, from the area of the Tavoy and Tenasserim rivers and from the Heinze Basin in the Tavoy District. Here, as in other tin placer areas, insignificant quantities of fine gold also accumulate when the tin ore is dressed.

GOOSSENS (1978, p. 481) listed numerous other locations in which, according to unpublished reports, indications of placer gold deposits were said to have been found.

Silver is associated with the strata-bound and structurally controlled lead-zinc ore deposits in carbonates, e.g. Bawsaing (p. 179), the sedimentary-volcanogenic lead-zinc ore deposits, e.g. Bawdwin (p. 177) and lead-zinc ore veins, e.g. Yadanatheingyi (p. 182).

NE of Putao, in the area of the upper reaches of the Nam Tamai, numerous argentiferous lead-zinc ore veins occur in partially marmorized limestone and dolomite. The carbonate rocks occur mainly together with quartzite in the contact zone with granitic rocks. According to JURKOVIC & ZALOKAR (1961), the area with Ag-Pb-Zn mineralization in which mining is said to have been carried out in the 19th century is located between $97° 25' - 97° 55' E$ and $27° 15' - 27° 51' N$. The following parageneses were observed:

– Pangnamdin (Fig. 72, folder): quartz, pyrite, sphalerite, chalcopyrite, pyrrhotite, chalco-pyrrhotite, galena, cerussite, anglesite, smithsonite, hemimorphite, goethite, lepidocrocite
– Hsamra Razi (Fig. 72, folder): quartz, pyrite, galena, freibergite, calcite, barite, chalce-dony, anglesite, cerussite.

Silver is produced only in Bawdwin (Figs. 75, 76). The mineralization took place in two phases (CARPENTER 1964, SOE WIN 1968). The first phase comprised the following paragenesis: quartz, pyrite, sphalerite, sericite, loellingite, tetrahedrite, bismuthinite, boulangerite, galena and pyrargyrite.

The paragenesis of the second phase comprised: gersdorffite, cobaltite, chalcopyrite, sphalerite, galena, bournonite, calcite, pyrargyrite.

The silver/lead ratio in the Bawdwin ores is on average 1 oz. Ag per 1% Pb per ton of ore, i.e. 31.1 g Ag/t ore. The higher the lead content of the ore, the higher the silver content. The upper zones of the Chinaman Lode, of the Shan and Meingtha veins, contained 400–600 g silver per ton of Pb, while the Ag contents declined considerably with depth.

Covellite, cerussite, anglesite, smithsonite, erythrite (cobalt bloom) and annabergite (nickel bloom) occur in the oxidation zone and, in addition to native copper, native silver also occurs in the cementation zone. Mining was carried out for hundreds of years by Chinese in these silver-rich portions of the ores in the Bawdwin area.

BROWN & DEY (1955, p. 301) mentioned **selenium** in the Pb-Zn-Ag-Cu ores from Bawdwin. The same authors mentioned **tellurium** (altaite) together with gold from the area of Kyaukpasat (24° 06′ N, 95° 05′ E) and from Banmauk (24° 23′ N, 95° 54′ E).

Traces of tellurium and **germanium** together with As, Mn, Fe, Cu, Cd, Sn, P, V and Mo were described by KUTINA (1969) in a sample of a colloform sphalerite from the Kawlin-Wuntho area (Katha District).

Niobium/tantalum-bearing minerals occur as samarskite and euxenite (columbate-tantalates of rare earths) as well as in the form of unidentified niobium-bearing minerals in pegmatites and gemstone-bearing alluvia (p. 208) in the Mogok area. Columbite was found in river sands near Mong Kung (21° 37′ N, 97° 31′ 30″ E) together with monazite, cassiterite, tantalite and magnetite, as well as in pegmatite N of Tavoy (GOOSSENS 1978, p. 480, from unpubl. reports).

Cassiterite from Heinda (p. 188) contained:

SiO_2	TiO_2	Fe_2O_3	MnO	SnO_2	WO_3	tantalum
0.35%	0.45%	0.43%	0.00%	98.00%	0.54%	25 ppm

Samples of slags from abandoned tin smelters in the area around Tavoy contained:

SiO_2	TiO_2	Fe_2O_3	MnO	SnO_2	WO_3	tantalum	niobium
35.72%	4.44%	7.79%	2.10%	16.73%	6.46%	3,000 ppm	3,500 ppm

(Bundesanstalt für Geowissenschaften und Rohstoffe 1980; Analysis No. 25812/13).

So far, monazite is the only **rare earth** to have been found in Burma (p. 173). Like selenium, tellurium, niobium and tantalum, it is not mined anywhere in the country.

5.22 Non-ferrous metals, tin, tungsten

5.221 Lead, zinc

Lead-zinc-silver ores are produced in Bawdwin/Northern Shan State (Fig. 72, folder). The mines at Bawsaing/Southern Shan State and Yadanatheingyi are of minor importance. Apart from these deposits, about 50 more Pb-Zn occurrences are known in the Sino-Burman Ranges from the Myitkyina District through to the Tenasserim Division. So far these occurrences have not been sufficiently studied as to their economic significance.

The **Bawdwin ore deposit** (approximately 50 km W of Lashio; 97° 18′ E, 23° 07′ N) consists of three sulphidic high-grade lead-zinc ore bodies and a small pyrite-chalcopyrite ore body (Chinaman Lode). The mineralization is bound to an approximately 4 km long and about 100 m wide NW–SE oriented fault zone ("Bawdwin Fault Zone") (Figs. 74–76) in which the three ore bodies, from N to S, are arranged as follows:

	maximum length	maximum vertical dimension	average width
Shan Lode	380 m	350 m	6 m
Chinaman Lode	400 m	350 m	42 m
Meingtha Lode	550 m	450 m	6 m

Around each of these high-grade ore bodies impregnated lead ore zones have developed. The broadest of these low-grade ore zones (approx. 200 m) occurs around the Chinaman Lode. Numerous, up to m-thick, barite dykes are bound to the ore zone. The main ore minerals are as follows (DUNN 1937): Argentiferous galena (PbS), sphalerite (ZnS), chalcopyrite ($CuFeS_2$), pyrite (FeS_2), loellingite ($FeAs_2$), tetrahedrite ($Cu_3SbS_{3.25}$), bismuthinite (Bi_2S_3), boulangerite (Pb_5Sb_4S), pyrargyrite (Ag_3SbS_3), gersdorffite (NiAsS), cobaltite (CoAsS), bournonite ($PbCuSbS_3$), cubanite ($CuFe_2S_3$). In addition, the following minerals were mentioned from the oxidation zone (SOE WIN et al. 1967): covellite (CuAsS), cerussite ($PbCO_3$), anglesite ($PbSO_4$), smithsonite ($ZnCO_3$), erythrite and annabergite ($Co_3(AsO_4)_2 \cdot 8\,H_2O$, $Ni_3(AsO_4)_2 \cdot 8\,H_2O$).

Mining in Bawdwin dates back to the beginning of the 15th century and was carried on by Chinese miners for the purpose of obtaining silver. Up until 1919, the slags produced by this mining activity – approximately 180,000 tons containing on average 60% Pb+Zn and 2 oz. of silver/ton – were smelted first in Mandalay and later in Namtu (approximately 10 km E of Bawdwin). The period of modern mining commenced in 1918 when the Burma Corporation Ltd. completed the Marmion Shaft and the 2.4 km long Tiger Tunnel which served both as a haulage and drainage gallery. Maximum output was achieved in the years 1928–1938 as the following table shows:

	1938	1960	1974/75
Refined lead (t)	77,700	16,518	4,843
Refined silver (kg)	175,000	24,000	10,000
Zinc concentrates (t)	65,982	17,473	5,575
Copper matte (t)	6,300	358	78
Nickelspeiss (t)	3,345	327	75
Antimonial lead (t)	1,200	530	104

The metal content of a representative sample of run of mine (ROM)-ore obtained during the heyday of the mining operations had the following composition: Pb: 21%, Zn: 15%, Cu: 0.3%, Ni: 0.23%, Co: 0.08%, Ag: 0.055%, Fe: 4.4% Sb: 1.2%, As: 0.62%, Bi: 0.052%, S: 13.75% (DUNN 1973). Since the fifties, production at the mine has declined sharply because of the decrease in metal content of the ROM-ores. The reserves of the three high-grade ore bodies are now almost depleted. Therefore, activity is concentrated on producing low-grade ores from the impregnation halos, chiefly by open-cast mining.

Since the geology of the Bawdwin area was described by J. C. BROWN (1918a) it has been assumed that the ore bodies are of epigenetic origin and thus are true replacements. In contrast, HANNAK (1972) attempted to interpret the deposit as syngenetic-synsedimentary. On the basis of detailed geological mapping BRINCKMANN & HINZE (1981) were able to show that the regional setting of the mineralization is closely associated with the Lower Paleozoic rhyolitic volcanism (p. 124). The volcanism occurred during an early phase characterized by the deposition of volcanoclastics, a subsequent phase with intrusions of large

Fig. 74. Bawdwin Valley as seen from NW; in background open cast Pb-Zn-Cu mine Chinaman Lode; surrounding area mainly occupied by Pangyun Formation, Cambro-Ordovician. Photo: F. BENDER.

rhyolitic masses and a late phase characterized by the ascent of pneumatolytic-hydrothermal solutions. Because of the better permeability, these solutions precipitated their metal content chiefly in fissures and unconsolidated zones in the volcanoclastics and to a lesser extent in the rhyolites and in the sediments of the Pangyun Formation (p. 53). The three high-grade ore bodies were formed more or less within the system of extensive fissures of the Bawdwin Volcanic Rocks. It was probably associated with the synsedimentary-syngenetic Bawdwin Fault System and continued to be effective even after the ore had formed (Yunan and Hsenwi Fault; Figs. 75, 76).

The genesis of the Bawdwin ores is thus comparable to the massive Kuroko Type A ore (TATSUMI & WATANABE 1971, Fig. 1; LAMBERT & SATO 1974) within volcanic fissures. Their age is identical to or barely younger than the Cambro-Ordovician rhyolite volcanism. LENZ & MÜLLER (1981) arrived at a Triassic age of 220 ± 20 m.y. (Rb/Sr total rock) for rhyolites and rhyolite components from the Bawdwin volcanoclastics. This age, however, is in all likelihood due to thermal events which also affected the Sino-Burman Ranges elsewhere.

The argentiferous lead ore deposit at **Mohochaung** (approximately 30 km N of Namtu) was also mined. This is a stockwork mineralization of galena in calcite gangue. The country rock is formed by sandstone of the Chaung Magyi Series and of the Pangyun Formation (p. 51). Nothing is known about the presence of the rhyolitic volcanoclastics and intrusives that are typical of the Bawdwin deposit.

The sulphidic and carbonaceous Pb-Ag-Zn ores of the **Bawsaing** area/Southern Shan State (approximately 25 km N of Heho, $96° 47' 30'' $ E, $20° 57' $ N) occur in the Ordovician limestone of the Wunbye Formation (p. 59). The sulphidic ores are found in numerous small occurrences in a narrow NNW–SSE-striking zone approximately 6 km long, of which

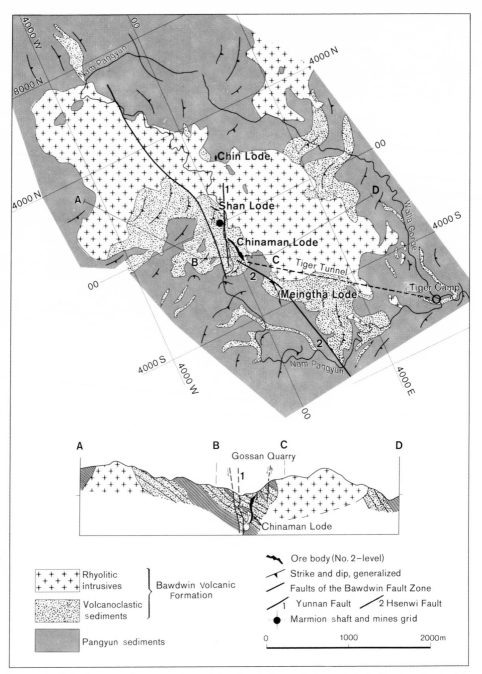

Fig. 75. Geological map and cross-section of the Bawdwin ore lodes and their Cambro-Orodovician host rocks (after BRINCKMANN & HINZE 1976).

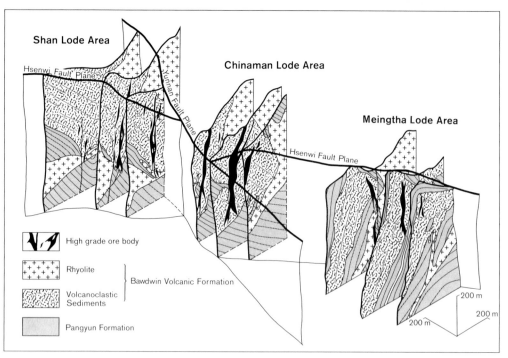

Fig. 76. Isometric projection of the Bawdwin ore bodies (modified after CARPENTER 1969).

the Theingon deposit is currently being mined. Cerussite occurs in cm-thick crack fillings in limestones of the "Carbonate Hill" to the W of Bawsaing and in up to 1 m thick NNW–SSE-striking fault zones. It originated from galena; its country rock is deeply weathered, loamy and limonitized, but primary textures which resemble those of the sulphidic ores in the E have been preserved.

The unweathered sulphidic ores were described as fissure fillings by J. C. BROWN (1932, pp. 420, 422). They consist of mineralized clay which forms a matrix surrounding fragments of limestone in crack and cavity fillings, or they form ore bands in thin-bedded carbonaceous shale. GOOSSENS (1978a) differentiated between 8 types, among others:

– galena (up to 1 cm in size) disseminated in oolitic dolomitic limestone
– crystals or decimetre-sized aggregates of pure galena with calcite and barite in red clay as filling of channels and cavities in dolomitized limestone
– stratiform beds of galena and barite, often rhythmically banded in cavities of karstic limestone
– accessory galena in the massive barite of Bawetaung in the Upper Wunbye Formation (p. 59).

GOOSSENS (1978a, p. 469) assumed that the ores from the Bawsaing area were formed by enrichment of a presumably low-grade primary mineralization in the carbonate rocks. In an intraformational karst topography followed by the formation of collapse breccias, the mineralization was dissolved and precipitated again as sulphidic ores.

Numerous occurrences, such as that at the Theingon mine, are presumably due to tectonic remobilization of galena and barite. They attain their maximum thickness in the culminations of overturned anticlines. The oxidation of the sulphidic ore ("Carbonate Hill") took place mainly in upthrown blocks where there was vigorous circulation of groundwater, while thick residual clay to a large extent prevented oxidation in the downthrown blocks.

The following ores were produced in the Bawsaing area:

	1972/73	1974/75
PbS concentrate (t), 50–54% Pb with approx. 10 oz. Ag/t	498	208
PbCO$_3$ concentrate (t)	395	520
lead slag from early mining	717	875

Some further galena-barite ores occur in Ordovician carbonate rocks outside the Bawsaing area, e.g. at Ye-U (20° 22′ N, 96° 16′ E). Galena mineralization in cm-thick layers in marbles is known from Phaungdaw (20° N, 96° 21′ E).

The lead ore of **Yadanatheingyi** is located 60 km NNE of Maymyo (96° 31′ E, 22° 32′ N). It contains accessory pyrite and arsenopyrite, and quartz, barite and calcite as gangue. It is mined from a shear zone about 10 m thick which cuts across the sediments of the Chaung Magyi Series (p. 51) in an NW–SE direction. The ore is found in vein fissures and stockworks, as well as in interstices in the gangue materials, and frequently peters out. The Pb contents of the run-of-mine ore are correspondingly low at about 5–10% Pb+Zn. In 1972/73 the mine produced 1,532 t and in 1974/75 1,634 t of lead concentrate containing 50% Pb and 1% Zn.

About 75 km ESE of Taunggyi and 35 km S of the village of Mong Pawn is located the apparently important zinc carbonate deposit of Loungh Hken (Lough Ken; 20° 33′ N/ 97° 29′ E). According to investigations carried out by the Burmese Geological Survey in the early fifties, and also according to GRIFFITH (1956, p. 14), the massive zinc ore bodies (smithsonite) are found in Ordovician limestone or at the contact between limestone and shale. The T-shaped ore body exposed in "Hill 1" is about 500 m long and approximately 30 m wide dipping almost vertical. The ore body exposed in "Hill 2" strikes 260° and is approximately 150 m long and 18 m to 20 m wide. GRIFFITH (1956, p. 15) estimated the exposed ore reserves alone to be 500,000 t of zinc ore with average Zn contents ranging between 35% and 40%. In 1951/52 approximately 2,250 t of zinc ore were mined in a small quarry operation; in 1956 mining was abandoned.

5.222 Copper

Although there are more than 50 known occurrences of copper mineralization in Burma, copper by itself has not yet been mined except at Monywa (p. 134). Most of the occurrences are located in the Sino-Burman Ranges and are associated with other non-ferrous metal sulphides. They occur in association with the Cambro-Ordovician volcanism (e.g. Bawdwin, p. 124), and frequently strata-bound in the belt of Paleozoic platform carbonates.

The copper minerals chalcopyrite, tetrahedrite, bournonite, cubanite and covellite are known as accessory minerals in the Bawdwin Pb-Ag-Zn deposit (p. 177). About 300 tons of copper matte containing approximately 51% Cu are obtained per annum in the Namtu Smelter.

Copper mineralization is known from Nyaung Chaung (22° 20′ 20″ N, 96° 21′ E), approximately 2.5 km E of Peinnebin on the Chaung Magyi. Here, malachite is observable in shales of the Chaung Magyi Series (p. 51). Copper mineralization has also been found in the Shan Dolomite Group (p. 72) at Hkwe-aik on the W side of the Heho Range (21° 44′ N, 96° 51′ E) with a quartz vein which contains covellite, chalcopyrite, tetrahedrite, malachite und azurite; furthermore, at Myededwin (20° 51′ N, 96° 29′ E) with an up to 3 m thick, strata-parallel, highly ferruginous quartz vein running in limestone and containing a complex oxidic copper mineralization. Near Myindwin (20° 57′ N, 96° 36′ E) stringers of antimonial tetrahedrite and copper carbonates are contained in intercalations of slates in the limestone sequence.

The volcanogenic copper ore occurrences in the Inner-Burman Tertiary Basin are of some economic significance. The copper ore deposit at Monywa (22° 15′ N, 95° 05′ E) on the lower Chindwin is associated with volcanic rocks which are located within the Inner Volcanic Arc (Fig. 77; KRISL 1975). The volcanic rocks (p. 134) consist of pillow basalt, porphyritic felsite and rhyolitic tuff (> 1,000 m). They are interbedded with marine sandstone and mudstone of the Miocene and brackish clastics towards the top. The copper mineralization occurs near the feeding vent of a biotite-porphyry-lava dome that is transected by rhyolite, quartz-biotite-porphyry and hornblende-biotite-prophyry dykes. The latter altered the tuff breccia with pyritic ore and thus probably form part of the most recent volcanic activities of Monywa (5.8 m.y. B.P.; p. 134). The mineralized areas belong to a complex of submarine lahars, volcanogenic turbidites and pyroclastic rocks at the edge of a caldera (Fig. 78). The

Fig. 77. Copper ore-bearing acid and intermediate volcanic rocks of Upper Miocene and Pleistocene age (Kyising Taung and Letpadaung Taung, ca. 15 km W of Monywa). Photo: F. BENDER.

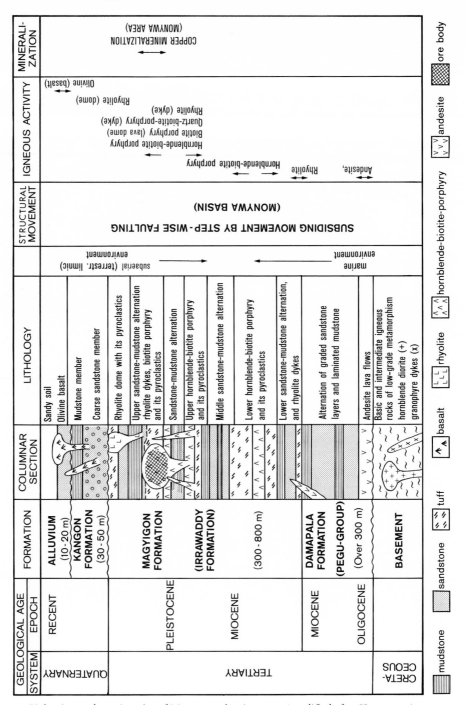

Fig. 78. Volcanism and stratigraphy of Monywa volcanic center (modified after KRISL 1975).

ore consists chiefly of chalcocite, pyrite and bornite. The greatly altered supergene enrichment zone is more than 300 m thick. There is a well-developed alunite zone. Because of the local development of massive ore mainly in the upper stockwork of the volcanic rocks, the Monywa copper ore belongs to a transitional stage between the Kuroko and the disseminated ore type in porphyries (KRISL 1975).

The ore occurs in the areas of Kyising Taung, Sabe Taung and Letpadaung Taung; in recent years it has been extensively drilled. Reserves of about 200 million tons of ore containing on average just below 1% Cu have been proven.

To the N of Monywa, near Shangalon (24° 30′ N, 95° 31′ E), further copper occurrences are found in the volcanic rocks of the Inner Volcanic Arc. They are located in a granodioritic intrusive body in the S Wuntho Massif. According to JURKOVIC & ZALOKAR (1959 b, p. 259) the 18 occurrences so far discovered are bound to hydrothermal alteration zones in the vicinity of recent volcanic protrusions of diabase, dacite, andesite and aplitic granite. The vents from which these volcanic rocks and, later, the ore-bearing thermal solutions issued are controlled by an NW–SE-striking fault system. Most of the occurrences exhibit a typical paragenesis of plutonic-hydrothermal ore deposits, namely quartz-tourmaline, molybdenite, pyrite, arsenopyrite, loellingite and chalcopyrite. All the occurrences are intensively brecciated. JURKOVIC & ZALOKAR (1959 b) believe that the emplacement of a younger mineralization phase with a complex polymetallic sulphidic paragenesis is linked with the formation of the breccia. All these copper ore occurrences are small and developed as cm-thick vein and fissure fillings along the NW–SE oriented fault zone.

Apart from some cupriferous galena veins in the Putao region NNE of Myitkyina, numerous copper mineralizations have also been reported from serpentinized ultrabasites of N Burma. One of these is the occurrence in the Taungbalaung Forest Reserve in the Nankesan-Sinbo region (24° 40′ N, 97° 03′ E) from which approximately 500 tons of copper ore were mined in 1910/1911. The mineralization (chalcopyrite, pyrrhotite) occurs in stringers together with quartz and pyrite. It is bound to a N–S and a NE–SW-striking fissure system which cuts gabbroid and serpentinized peridotitic rocks. Copper ore impregnations in serpentinite have also been discovered at Longyi in the Pakokku Hill Tract (20° 46′ N, 94° 05′ E) and in gabbro and peridotite along the E side of the Chin Hills. J. C. BROWN (1924 a) described native copper and chalcocite associated with chromite at these locations.

Within the Kanpetlet Schist (p. 77), epimetamorphic deep-sea sediments overlie thick, massive greenschist with amphibolite and relicts of gabbro which in turn are in contact with serpentinized ultrabasites. In the area W of Mindat, about 15 km NNW of Mt. Victoria in the Central Chin Hills, several zones of greenschist are associated with black slate, phyllite and volcanoclastics with disseminated Fe- and Cu-sulphides. These zones with sulphidic mineralization were observed over a distance of more than 8 km (BGR, unpubl. rep. 1981).

5.223 Tin and tungsten

The Tenasserim region of S Burma and parts of E Burma belong to the SE Asian tin/tungsten belt which runs from the tin islands of Banka and Billiton in the Indonesian Archipelago via Malaysia, Thailand and Burma as far as China; it supplies 65% of the world's current tin production and a large percentage of the world's tungsten. Most of the tin ores are mined in placers while the tungsten is extracted from veins.

The tin/tungsten ores are bound to pneumatolytic-hydrothermal dyke systems of granites. These tin- and tungsten-bearing granites of Burma belong to the westernmost of the three granite belts of the SE Asian Peninsula. For the most part they belong to the Tertiary intrusion generation of this region. The country rock of these intrusive masses consists of the clastic Mergui Series, Taungnyo Group, Mawchi Series and Lebyin Group (p. 168).

Primary cassiterite, scheelite and wolframite ores occur

- in granite (cassiterite, finely dispersed with concentrations up to 0.15% Sn; BLEECK 1913, p. 59, and wolframite; CAMPBELL 1919b, p. 3)
- in aplite granite dykes (J. C. BROWN & HERON 1923a, p. 227)
- in pegmatites (e.g. in the Pagaye mine: wolframite, scheelite and cassiterite in a microcline-quartz-muscovite-pegmatite; J. C. BROWN & HERON 1923a, p. 228)
- in greisen
- in quartz veins and fissure fillings; the Sn-W mineralization is often discontinuous; it frequently occurs in parallel or intersecting sets of veins, in lenses and stringers of rapidly varying thickness; the contact with country rock is either distinct, often characterized by a muscovite-biotite selvage or it is formed by a greisen zone.

The veins frequently strike parallel to the granite bodies arranged in the general NNW–SSE strike of the mountain range. The E–W-striking vein systems of the Kanbauk deposit in the Tavoy District are an exception. The veins attain lengths up to 400 m. The longest NNW–SSE-striking "Kadando Vein" near Yewaing was traced over several kilometres (CAMPBELL 1919b, p. 5). Ore-bearing quartz veins occur in granites in the contact zone to their country rocks, in particular in roof regions of intrusions and in the surrounding country rock. According to J. C. BROWN & HERON (1923a, p. 231), high concentrations of cassiterite and wolframite are found in particular in cm-thick stringers and lenses in the contact zones.

Apart from cassiterite, scheelite and wolframite, the paragenesis of the ore-bearing quartz veins consists of traces of gold (so far only observed in placers), molybdenite, pyrite, chalcopyrite, covellite, galena, sphalerite, stibnite, bismuthinite, pyrrhotite, arsenopyrite, siderite, calcite, cerussite, fluorite, topaz (frequently observed in placers) and beryllium (J. C. BROWN & HERON 1923a, pp. 219 et seqq.).

Tin placers are found in eluvial, colluvial, fluviatile and lacustrine sediments. They are generally comparable to those of the Kaksa and Mintjan placers of Indonesia (OSSBERGER 1968). Also near-shore marine tin placers were found offshore of the coast of the Tavoy and Mergui districts.

Altogether, more than 120 tin/tungsten occurrences are known. Many of these are located in the Tavoy District where they are associated with the three NNW–SSE-oriented granite belts of the Coastal Range, the Central Range and the Frontier Range (Fig. 79; J. C. BROWN & HERON 1923a). Of the three important producing mines in this district, the primary ore deposit at Hermyingyi and the tin placer deposit at Heinda are located in the Central Range, and the primary and placer deposits at Kanbauk are located in the Coastal Range Granite.

The **tin/tungsten deposit at Hermyingyi** (Fig. 80), where mining commenced in 1910, is located about 40 km NE of Tavoy (14° 15′ N, 98° 21′ E). The mineralization is bound to an

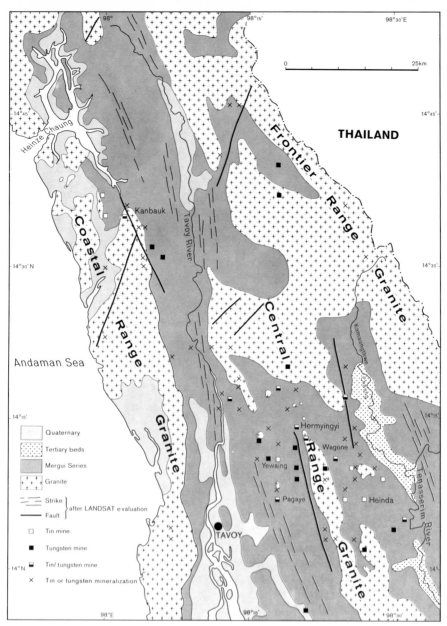

Fig. 79. Tin/tungsten occurrences in the Area of Tavoy (modified after CLEGG 1944).

N–S-striking fissure system that cuts an NNW–SSE-oriented, approximately 1,400 m long and max. 550 m wide granite body and its country rock in the contact zone. The quartz veins are up to 2 m thick and contain the following ore paragenesis (DUNN 1938b): wolframite, cassiterite, molybdenite, pyrite, sphalerite, chalcopyrite, galena, bismuthinite and bismuth.

The most frequent constituent of the highly discontinous ore veins is wolframite. Most of the cassiterite occurs in the form of small crystals in greisen, less frequently of up to decimetre-sized zonally structured cassiterite crystals. The metal sulphides occur merely as accessory minerals. Of the more than 200 ore veins covered in the mine workings over a length of 300 m and a height of 200 m, about 40 veins have at one time or another been mined. WNENDT (1979) observed a decrease in the ore content of the veins with increasing depth, as is shown in Table 9.

Table 9. Decrease of W + Sn content with increasing depth in Hermyingyi mine.

	$\%WO_3$	$\%Sn$	$\%WO_3 + Sn$	$\%WO_3 : Sn$	Average thickness of the veins (cm)
(uppermost) 2nd level, total average	0.276	0.061	0.337	4.52 : 1	119.84
vertical distance 29 m					
1st level, total average	0.311	0.049	0.360	6.35 : 1	103.37
vertical distance 47 m					
zero level, total average	0.171	0.018	0.189	9.5 : 1	122.75

From 1915 to 1918 the mine produced between 640 and 1,040 tons of mixed tin/tungsten concentrate per annum (J. C. BROWN & HERON 1923a, p. 288), and in 1939 the output was 126 tons of tin and 345 tons of tungsten concentrate (CLEGG 1944a, pp. 126, 127); from 1967 to 1975 it produced 102 to 109 tons/a of a mixed tin/tungsten concentrate.

The **tin placer deposit at Heinda** (98° 26′ E, 14° 7′ N) is located approximately 40 km to the E of Tavoy (Hpolontaung Hill) in the drainage area of the W contact zone of the Central Range granite with the Mergui Series (p. 69). Because a number of cassiterite-wolframite-bearing granites and greisen are known from this area, it is regarded as the source region for the cassiterite-bearing sediments of the Heinda deposit, particularly since the orientation of the detrital material indicates that it was transported from the W.

The tin-bearing clastics of the Hpolontaung Hill at Heinda (CLEGG 1944a, p. 89) can be subdivided into 12–13 sedimentation cycles, each of them is about 8 m thick and grades from conglomerate at the base to sand and clay at the top. The components of the conglomerate attain diameters in the range of 1 m and above. The entire sequence is characterized by current bedding, fluviatile resedimentation and by bimodal distribution of grain sizes in the conglomerate. These features identify the tin placer as a fluviatile formation and not a lacustrine one filling a lake basin and causing it to subside as CLEGG (1944a, p. 90) had assumed.

The age of the tin-bearing clastic rocks has not been biostratigraphically clarified. They differ from Quaternary unconsolidated clastics because of their lateritic consolidation, their structural displacement by 5°–10°, and also because of normal faults with throws of up to 3 m and because of their morphological position. It is therefore possible that their age is Late Tertiary.

The mineral contents per ton of concentrate are as follows: 973.4 g cassiterite, 1.3 g topaz, 1.0 g zirconium, 22 g hematite. Magnetite, tourmaline, wolframite, andradite and quartz appear as accessory minerals (WERNICKE 1961). The amount of cassiterite is very irregular and varies throughout the deposit in the vertical and horizontal direction. DRESCHER &

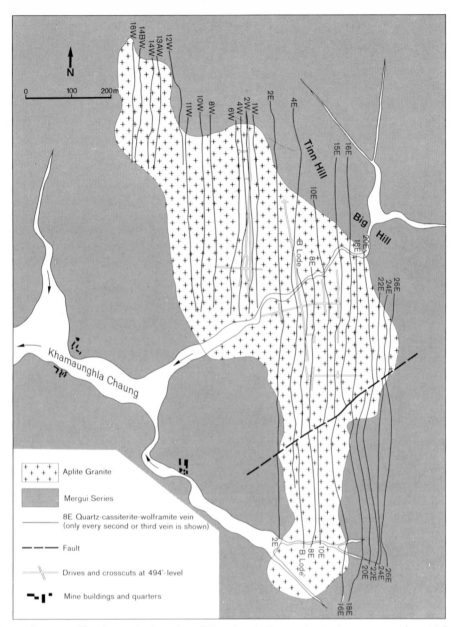

Fig. 80. Quartz-wolframite-cassiterite veins of Hermingyi, Tenasserim (only every second or third vein is shown).

GOCHT (1978) gave values of $0.06-3.9 \text{ kg } SnO_2/m^3$. The free cassiterite is enriched in the silty matrix of the placer, subordinately also in the conglomerate. The silt layers and lenses were formed by secondary enrichment caused by groundwater circulation. The tin content decreases distinctly in the upper part of the placer. The decline in the tin content

Fig. 81. Open cast tin/tungsten placer mining near Kanbauk, Tenasserim. Photo: F. BENDER.

from the bottom upwards coincides with the continuous alternation in the composition of the conglomerate from Mergui sediment detritus to granite detritus. Thus, the tin placer documents the sequence of erosion from a source area characterized initially by an ore-free cover of Mergui sediment, then by a mineralized contact zone and finally by deeper sections of the granite.

In the period from 1959 to 1975, the mine's output was about 230 tons per annum. According to DRESCHER & GOCHT (1978), the total amount of ore available as of 1977 is about 19 million m³ with an average cassiterite content of 0.7 kg/m³ equivalent to 0.282 kg/t. On the basis of a recovery rate of 80–90%, a minimum of 10,000 tons of cassiterite concentrate (72%) can be produced from the Heinda deposit. Because of the favourable reserves situation, the mine is being modernized.

The **tin/tungsten deposit of Kanbauk** (14° 36′ N, 98° 03′ E) is located approximately 90 km N of Tavoy (Fig. 72, folder). The primary ore, which is found in quartz veins ranging in thickness from several centimetres to 0.5 m, contains wolframite and cassiterite and, as accessory minerals, pyrite, chalcopyrite, galena, sphalerite and native bismuth. The veins strike in general W–E and dip steeply to the S. Their country rock consists of slate and sandstone of the Mergui Series (p. 69; J. C. BROWN & HERON 1923 a, p. 265). The area being mined is approximately 500 m long, 250 m wide and contains at least 20 ore veins.

Most of the tin ore, and to a lesser extent also the wolframite, is obtained from eluvial and alluvial placers up to about 30 m thick (Fig. 81). The more productive areas run along the Heinze Chaung on the NE side of Home Hill. The placer tin originated in the erosional

debris of Home Hill. In the already mined-out eluvial slope area close to the source, wolframite had constituted up to 90% of the concentrate. The ore-bearing unconsolidated sediments that are currently being mined (Fig. 81) are debris flows and deposits that have been resedimented by slope run-off and torrents, and as a result they display a broad grain-size spectrum ranging from cobbles via gravel and sand to clay. In keeping with the inhomogeneous nature of these deposits, the tin content also varied from several kg/m^3 to a few g/m^3. Particularly ore-rich unconsolidated sediments of this "unique" placer deposit (J. C. BROWN & HERON 1923a, p. 269) were encountered above the eroded limestone beds of the Mergui Series which acted as "traps".

Together with other small mining sites located in the valley of the Heinze Chaung, the Kanbauk mine produced 829 tons of tin concentrate and 187 tons of tungsten concentrate in 1936 (CLEGG 1944a, p. 118). In the last 10 years, the annual output varied between 45–160 tons of cassiterite concentrate, 31–252 tons of wolframite concentrate and 5–27 tons of cassiterite/wolframite concentrate.

The Tavoy District contains numerous other veins and placer deposits which were mined in particular in the years preceding the last war (CLEGG 1944a, pp. 118–129). Apart from the "Widness mine" (14° 1′ N, 98° 24′ E), other important mines were the "Paungdaw mine" (14° 05′ N, 98° 31′ E), "Wagone mine" (14° 11′ N, 98° 24′ E) and the "Pagaye mine" (14° 35′ N, 97° 59′ E). All these vein deposits are bound to the contact zone between the Mergui Series and granite intrusions. The Kyaukmedaung deposit (14° 10′ N, 98° 26′ E), approximately 5 km to the E of the Central Range granite, is at present being mined. The deposit consists of NNW–SSE-striking quartz veins which cut through a series of folded argillite and quartzite of the Mergui Series. Thirty-one tons of tin concentrate, 35 tons of tungsten concentrate and 37 tons of tin/tungsten concentrate were produced in 1974/75.

TREMENHEERE (1841, 1843) described **tin deposits in the Mergui District.** BROWN & HERON (1923a, pp. 259, 260) show that up until 1917 tin production in the Mergui District was higher than in the Tavoy District, while production of tungsten was always higher in the Tavoy District. In 1937/38 the Thabawleik placer deposit, which is the largest in the Mergui District (12° 1′ N, 99° 12′ E; CLEGG 1944a, pp. 142, 143), attained an output of 400–500 tons of tin concentrate.

Of the numerous deposits in the Mergui District, the vein deposit at Yadanabon (11° 17′ N, 99° 17′ E) is currently being worked. The quartz veins at the NW flank of a granite intrusion strike NNE–SSW and dip at a shallow angle to the E. Of the four locations being mined, two are located in granitic country rock, one in the contact zone and one in the Mergui sedimentary cover. The quartz stringers and veins contain clusters of wolframite, while cassiterite is disseminated. Molybdenite, pyrite, chalcopyrite, tourmaline, fluorite and lepidotite occur as accessory minerals. The veins in the granitic country rock are often accompanied by greisen up to 2 m thick which contain bismuth in addition to tin. Between 1970 and 1975 the mine produced approximately 100 tons of tungsten concentrate and 5 tons of tin concentrate per annum.

Also to the N of the Tenasserim Division the granite intrusions are characterized by a pneumatolytic-hydrothermal system of quartz-cassiterite-wolframite veins. They include the **Mawchi deposit in the Karen State** (Fig. 82). The tin/tungsten mine at Mawchi (18° 48′ N, 97° 10′ E) is the most important tungsten deposit in Burma in terms of ore reserves. The steeply dipping, N–S-striking ore veins (DENYER & HEATH 1940) occur in the contact zone of an isolated granite intrusion. The veins cut the granite and the contact-metamorphic sediments around it. The latter are composed of slate, fine-grained sandstone, calcareous shale and limestone of the Mawchi Series (p. 70). In addition to quartz, the veins, which in the N section of the deposit are interrupted by limestone beds and attain thicknesses of up to 1.20 m, also contain cassiterite, wolframite, scheelite and, as accessory minerals, arsenopyrite, pyrite, chalcopyrite, galena and tourmaline. A maximum of 64 veins were being mined. In 1938 a total of 143,000 tons of ore were produced (4,774 tons of mixed cassiterite/wolframite concentrate containing 32% Sn and 34% WO_3; SOMMERLATTE 1948). According to CLEGG (1944a, p. 67), the final concentrate contained 38% cassiterite and 32.5% wolframite, 0.6% pyrite and 0.6% arsenopyrite.

In 1974/75 approximately 400 tons of tin/tungsten concentrate were produced. The average content of the ore is about 1.54% Sn and 0.75% WO_3.

The **deposits to the E of Pyinmana** were mechanically mined in the thirties but at present they are being worked manually by tributors only in the area near the village of Padatgyaung (18° 42′ N, 96° 34′ E). To the SW in Peinnedaik, where 197 tons of wolframite concentrate were mined in 1939 (CLEGG 1944a, pp. 112, 113), tabular wolframite in the cm-size range occurs in NNW–SSE-striking quartz veins which are up to 2 m thick. The country rock consists of highly weathered argillaceous-arenaceous metasediments (BATESON et al. 1972). Ore-bearing veins from other mining districts, e.g. Padatgyaung I and II and Thavagon, cut through granite intrusions or are located in the contact zone with metasediments. In 1975 approximately 80 tons of wolframite concentrate were mined manually in the Padatgyaung District.

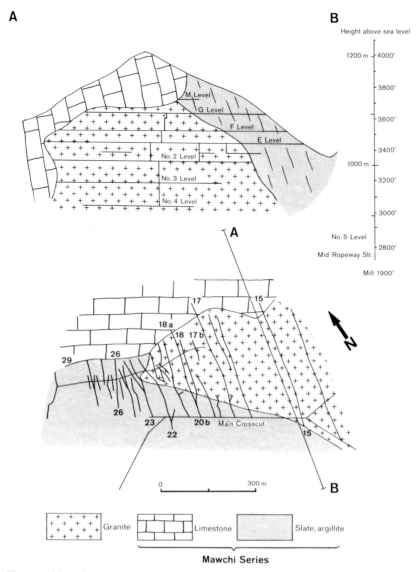

Fig. 82. Tungsten/tin mine Mawchi, part of No. 2 level; the numbers mark drives following the ore lodes. Section A–B lies in lode 15 (sketch after DENYER & HEATH 1940).

5.23 Iron and metals for steel alloys

5.231 Iron

Altogether 22 iron ore occurrences have so far been observed in Burma; this figure does not include numerous other very small ones and occurrences of pyrite. None of these has so far acquired major economic importance. They consist of limonite-hematite ore (13), magnetite ore (3), lateritic iron ore (5) and siderite ore (1).

Residual deposits are found in particular above Early Mesozoic and Paleozoic limestone and dolomitic limestone in the Southern Shan State, and less frequently in the form of lateritic iron ores and crusts above basic igneous rocks at the E edge of the Arakan Yoma, and above and in the vicinity of basalts in Central Burma.

J. C. BROWN (1916) described the iron ore deposit at Twinngé between Maymyo and Mandalay (21° 57′ N, 96° 22′ E) as a typical deposit occurring above limestone or dolomitic limestone of the Shan Plateau, which has presumably been exposed to continental weathering since the end of the Mesozoic. At this location the "Plateau limestone" is overlain by 6–9 m thick red earth ("Indian Red Clay"). At its base an approximately 1 m thick iron ore horizon is developed consisting of limonite and limonite/hematite nodules and concretions. The amount of ore is estimated to be approximately 275,000 tons with Fe contents between 56% and 60%.

In the iron ore deposit at Kya Twinye (Kyetwinye; 21° 53′ N, 96° 32′ 20″ E) about 20 km S of Maymyo, up to 22 m thick limonite/hematite roll ores cover Devonian dolomite and sandstone. The reserves amount to about 3 million tons with an average content of approximately 54% Fe. A further occurrence of this type of deposit is found at Inya (21° 59′ 40″ N, 92° 32′ 30″ E) about 9 km SE of Maymyo. It contains brecciated ores which merge laterally and vertically into lateritic iron ore. About 4.5 million tons of Fe ore with an average Fe content of 35.2% have been discovered in this area; the mean TiO_2 content is 1.5% (KRUPP-ROHSTOFFE 1962).

Although these iron ores are favourably located for open-cast mining, they have so far not been economically exploited because they are difficult to dress (intergrowth of ore/laterite/bauxite), they are not easy to smelt (large amount of slag formation, high Ti contents), and the Fe contents are usually low.

Hematite/limonite concretions occur in the region around Mt. Popa (20° 55′ N, 95° 15′ E) above and in the Miocene-Quaternary Irrawaddy Group. They were used for iron production probably up until the Late Middle Ages, as many small smelting furnaces show.

The largest iron ore deposit so far discovered in Burma is at Pang Pet (20° 44′ 30″ N, 97° 06′ 30″ E) about 10 km SE of Taunggyi. According to NEUBAUER (1965), this deposit is a completely mylonitized primary hematite mineralization which is bound to 2 regional fault systems in the Plateau Limestone, and it also includes an impregnation-type sulphide mineralization in an acid volcanic rock, probably rhyolite. The latter is almost completely weathered and has formed a limonitic gossan. According to KRUPP-ROHSTOFFE (1962), the hematite reserves amount to approximately 10 million tons (average Fe content 56.4%) and the limonite reserves to approximately 70 million tons (average Fe content 42.6%). Most of the iron ore can be open-cast mined. The relatively high content of arsenic (on average 0.011% As) is a disadvantage. In more recent studies, copper contents up to 1.5% were found in addition to small amounts of uranium. Directly NE of Pang Pet is located the occurrence of Kaw Loilaw containing approximately 10,000 tons of low-grade iron ore. It is likely that further Fe-Cu-U occurrences bound to weathered acid volcanic rocks exist in the Taunggyi-Heho-Bawsaing area.

Some magnetite occurrences and their weathering products have been mentioned from the districts of Thaton, Tavoy and Mergui (UN-ECAFE 1952). An iron ore deposit of this type is located on the island of Kho Kyun (10° 23′ N, 98° 27′ E) about 6 km off the Tenasserim coast. It contains a total of 3 million tons of ore with an average Fe content of 46%. A similar occurrence is found on the island of Mah Puteh, approximately 20 km N of Kho Kyun, 1 km

W of the Tenasserim coast. These are limonitic block ores above lense- and vein-like magnetite ore bodies. The reserves are given as 1.3 million tons of Fe ore with contents between 54%–55% Fe.

The only siderite ore occurrence so far found in Burma is located N of Putao on the upper reaches of the Khalaw Ti ("Nogmung"; 27° 30′ N/97° 50′ E). The mountainous, almost inaccessible area is situated at an altitude of about 1,800 m above sea-level. Nevertheless, iron ore was mined here in earlier centuries. According to JURKOVIC & ZALOKAR (1960) the ore body is of unknown shape and extent, although it is known to be "of considerable thickness"; it is in contact with phyllitic slate and sandstone. It consists essentially of siderite, and also of magnetite, hematite, limonite, pyrite, marcasite und psilomelane. Pyrrhotite, arsenopyrite and lepidocrocite occur as accessory minerals. An analysis of an ore sample from Nogmung yielded the following values (JURKOVIC & ZALOKAR 1960, p. 319):

SiO_2	0.2%	CuO	0.8%
Al_2O_3	–	MgO	3.4%
Fe_2O_3	8.5%	SO_3	3.6%
FeO	45.8%	P_2O_5	–
Cr_2O_3	–	S	1.4%
NiO	–		
MnO	7.8%		

5.232 Metals for steel alloys

Manganese, chromium, nickel, cobalt, molybdenum and titanium are discussed here as "metals for steel alloys".

Manganese ores are found in about 10 occurrences, none of which has so far acquired economic significance. Several thousand tons of ore containing between 36% and 56% Mn were mined in the area of Tagaung Taung (23° 34′ 25″ N/96° 10′ 56″ E; Fig. 83). In this area, pyrolusite and psilomelane are interbedded in thicknesses of up to 250 cm in a sequence of reddish chert. Nothing is known about the extent of the deposit or its reserves.

Manganese ore was also mined in small quantities up until 1956 near Hopoing (Kon Nui; 20° 44′ N/97° 15′ E). The ore consists of psilomelane, pyrolusite, rhodochrosite together with limonite, hematite and magnetite interbedded between sandy shale below and limestone above. The ore attains thicknesses of up to 120 cm and contents of about 45% Mn (GOOSSENS 1978, p. 487, from unpubl. reports).

Manganese compounds frequently occur in the deeply weathered tin- and tungsten-bearing pegmatites and dykes of S Burma which contain huebnerite. Manganese oxide and manganese hydroxide formed in the course of the weathering (SOMMERLATTE 1948, p. 135).

The residual Fe deposits that occur above Mesozoic and Paleozoic carbonate series in the Shan States, the lateritic iron ores and the crusts on top of basic magmatites at the E edge of the Arakan-Chin mountains and the Fe concretions and crusts in the Irrawaddy Group (p. 102) contain also manganese – up to 28% Mn_2O_3 in the case of the Irrawaddy Group. CHHIBBER (1934b, p. 238) also mentioned manganese ore occurrences in the Mergui Archipelago, particularly on Gna Islet, in Padaw Bay and on King Island (12° 29′ N/98° 25′ E).

Chromium ore occurrences are widespread in Burma, which is to be expected because they are associated with the likewise widely distributed ultrabasic rocks (p. 174). Nevertheless, none of them have so far been systematically examined.

Chromite has been found at numerous locations in the S–N-striking ophiolite belt that is more than 1,000 km long in the E part of the Indo-Burman Ranges and along the tectonic contact between the Indo-Burman Ranges and the Inner-Burman Tertiary Basin:

- CHHIBBER (1927, pp. 193–194) mentioned chromite in the serpentinites from the districts of Bassein and Henzada in the S section of the ophiolite belt of the Indo-Burman Ranges. In this locality, chromite detritus occurs near Legonywa (17° 40′ N/94° 59′ E), near Shwelaungyin (17° 33′ N/94° 57′ 30″ E) and near Zinbinkwin (17° 36′ 30″ N/95° 00′ E)
- further to the N, chromite was found in the serpentinite NW of Kadaing in the Thayetmyo District (19° 25′ N/94° 37′ E) and in Mon Chaung (20° 46′ N/94° 07′ E) (CHHIBBER 1934b, p. 227, SOMMERLATTE 1948, p. 137)
- still further N, chromite occurs in the peridotite, saxonite (harzburgite) and dunite of the Mwetaung area (23° 25′ N/94° E), locally with contents above 50% Cr_2O_3; it was discovered here in the course of prospecting for nickel (garnierite) ore (p. 198).

Chromite is also found in the ultrabasic rock complexes of the Wuntho Massif (p. 136) and of the Jade Mines Area (Fig. 88).

About 170 km N of Mandalay and directly to the E of the Irrawaddy River, the ultrabasic rock complex of Tagaung Taung (23° 34′ 25″ N/96° 10′ 56″ E; Fig. 83) is exposed, which covers about 100 km². It consists essentially of presumably Upper Cretaceous-Eocene dunite, saxonite, harzburgite, lherzolite, wehrlite, pyroxenite and serpentinite (GEOISTRAZIVANJA 1958) and at many points it contains chromite. The chromite ore crops out over areas measuring a few square metres in extent or it is found in boulders and debris at the bottom of slopes.

From the Jade Mines Area (Myitkyina District) CHHIBBER (1934b, pp. 227, 228) mentioned chromite in peridotite and serpentinite in boulders up to 1 m in diameter about 4 km NE of Tawmaw (25° 42′ 13″ N/96° 15′ 28″ E) and in the boulder conglomerate in the Uyu River. He further mentioned the following chromite locations:

- about 1.5 km WNW of Mahok (25° 44′ 8″ N/96° 22′ 53″ E)
- in the old jadeite mines at Pangmaw (25° 44′ 53″ N/96° 20′ 54″ E)
- along the road from Namshamaw (25° 45′ 31″ N/96° 22′ 28″ E) to Mawsitsit and near Wayutmaw (25° 45′ 42″ N/96° 21′ 38″ E)
- N of Sanhka (25° 41′ 8″ N/96° 20′ 57″ E) where chromite boulders are associated with a volcanic breccia
- in Tertiary conglomerates near Pangmawmaw, E of Kansi (25° 47′ 1″ N/96° 22′ 48″ E).

Nickel ores are also much more widespread than had previously been assumed. Ni is so far produced only from the Bawdwin lead-zinc-silver deposit (p. 177). In this deposit it is associated with the Cu-Ni-Co paragenesis which followed after the main Pb-Zn mineralization (CARPENTER 1964, SOE WIN 1968). It appears in the form of gersdorffite (NiAsS) and, in the oxidation zone as annabergite ($Ni_3(AsO_4)_2 \cdot 8 H_2O$). In 1936, the average high-grade ore production contained 0.23% Ni (DUNN 1937). In 1938, 3,345 tons, in 1960 327 tons and in 1975 only 75 tons of nickel speiss were produced in the Namtu smelter.

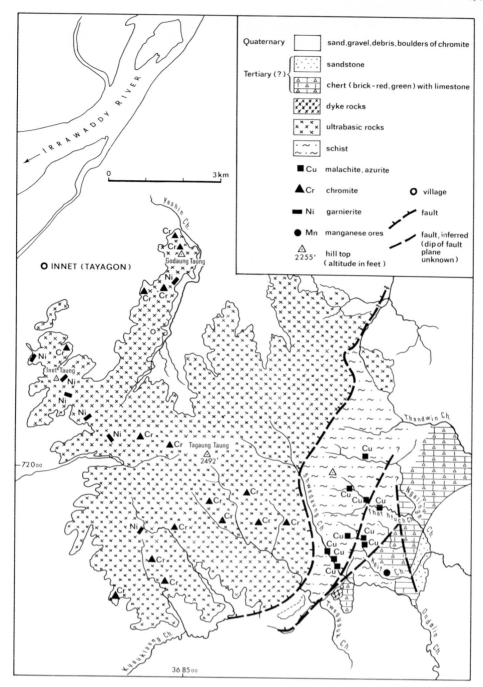

Fig. 83. Geological sketch map of Tagaung Taung area (modified after Geoistrazivanja 1958). (Limestone from Tagaung Taung contains *Orbitolina birmanica* and fragments of Hippuritaceans, proving Albian to Cenomanian age).

In recent years, silicate nickel ores (garnierite) were found as weathering product above ultrabasic rocks in many places, e.g. in the belt of ultrabasic rocks at the E edge of the Indo-Burman Ranges (p. 130), above the ultrabasites in the Jade Mines Area (Fig. 88), and above the ultrabasic rocks at the E edge of the Inner-Burman Tertiary Basin.

In the approximately 100 km long and up to 15 km wide zone of ultrabasites at the E edge of the Indo-Burman Ranges to the W of Kalemyo, garnierite was discovered on the crests of the Bhopi Vum, the Webula Hill, the Muallup and the Mwetaung.

In the Mwetaung area ($23°25'$ N/$94°01'$ E; approximately 60 km^2) the source rocks for the nickel enrichment zone are peridotites, which consist of a serpentinized olivine ground mass with pyroxenes in the cm-size range. Transitions to saxonite (harzburgite) and dunite exist. A three-section profile was formed in the crest area of the Mwetaung, which can be regarded as typical of tropical-humid weathering of ultrabasic rocks in these latitudes:
- deep brown-red lateritic zone, up to 5 m thick
- yellowish-green decomposed zone, approximately 30 m to 60 m thick
- primary ultrabasites.

In the vertical zonation, the maximum nickel contents (up to 3%) are attained in the transition between laterite and the decomposed zone (Fig. 84). The enrichment factor compared with the primary ultrabasite (max. 0.3% Ni) is 10 and above. The Fe_2O_3 content in the laterite varies around 25% to 40% and in the decomposed zone between 5% and 10%; the MgO content of the laterite is usually below 5%, in the decomposed zone it attains up to 30%; the Cu content is 100 ppm in the laterite and less than 30 ppm in the decomposed zone; the Co content in the laterite fluctuates from 500 ppm to 700 ppm and in the decomposed zone it is below 300 ppm. This silicate nickel ore occurrence corresponds to the nickel laterite occurrences of the Philippines and New Caledonia. The probable reserves in the Mwetaung area are currently put at around 10 million tons with contents of $>1.19\%$ Ni.

In the Jade Mines Area (Myitkyina District), garnierite associated with serpentinite was mentioned from near Sanlaik ($25°09'$ N/$98°10'$ E; GOOSSENS 1978, p. 487, from unpubl. reports).

Garnierite occurrences were also observed at the E edge of the Inner-Burman Tertiary Basin, namely in the ultrabasic rock complex of Tagaung Taung ($23°34'25''$ N/$96°10'56''$ E; Fig. 83, p. 197) which also contains chromite. According to GEOISTRAZIVANJA (1958) and FERENCIC (1961, pp. 23–30) the silicate nickel ore occurs in the decomposed zone of serpentinized peridotites or as cement in brecciated serpentinite. The brecciated zones follow the directions of fault zones. Local enrichments of up to more than 10% NiO are found here. The current estimate of the potential ore reserves in the Tagaung Taung area is around 1.1 to 1.3 million tons with Ni contents above 1.5%. About 1,400 tons of this ore were mined in the early fifties.

Cobalt is obtained from the lead-zinc-silver ore deposit at Bawdwin (p. 177), where it appears as cobaltite (CoAsS) and in the oxidation zone as cobalt-annabergite $(CO_3(AsO_4) \cdot 8\,H_2O, Ni_3(AsO_4)_2 \cdot 8\,H_2O)$. The cobalt content of the high-grade ore of Bawdwin was 0.075% (average output in 1936; SOMMERLATTE 1948, p. 186). During smelting, the cobalt is taken up in the nickel speiss which, in the years between 1930 and 1940, contained between 50 and 150 tons of cobalt per annum. As the production of ore from Bawdwin declined, the production of Co dropped in the seventies to approx. 10 tons/annum.

Small amounts of cobalt (up to 700 ppm) are contained in the silicate nickel ores of the Mwetaung area (p. 196) and in those of the Tagaung Taung area (around 600 ppm; p. 197).

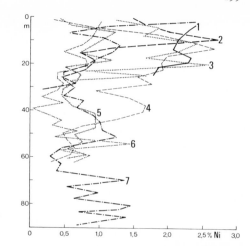

Fig. 84. Vertical distribution of nickel in Mwe-taung area, Chin State.

GOOSSENS (1978, p. 480) mentioned an occurrence of an earthy Co mineral with Mn minerals near Heinza (14° 29′ N/98° 12′ E).

Molybdenum occurs as molybdenite in the primary tin/tungsten deposits of S Burma. It is found as an accessory mineral in pegmatites and dykes, occasionally also in the greisen zone around granite, usually together with cassiterite and wolframite. It also appears in enriched form in the selvages of the dykes where it is frequently intergrown with muscovite. In addition to cassiterite and wolframite, other accompanying minerals are fluorite, bismuth and bismuthinite, arsenopyrite, pyrite, chalcopyrite, and occasionally also beryllium, topaz and tourmaline. Molybdenum occurrences of this type were mentioned by CHHIBBER (1934 b) and SOMMERLATTE (1948) in the Sn/W mining districts of Byingyi, Yamethin, Tavoy and Mergui, e.g. at Yadanabon (11° 17′ 05″ N/99° 24′ E), So Nein-Wagon North mines (14° 11′ N/98° 24′ E), Kadangtaung (14° 04′ N/98° 26′ E) and Zinba (14° 38′ N/98° 12′ E).

Molybdenum was also found in the granodioritic rocks of the Wuntho Massif (p. 136) in N Burma, e.g. in the Shangalon area (23° 42′ N/95° 30′ E). It is associated here with Cu and Au (ZAW PE 1976).

So far none of the molybdenum occurrences have been systematically prospected.

Primary deposits of **titanium ores** have not yet been observed in Burma; occasionally ilmenite and rutile are found in river sands and beach placers in the Tenasserim region. Ilmenite and rutile are mostly associated here with magnetite, zirconium and monazite, e.g.

– in river sediments of the Shwe Du and the Lamawpyin Chaung (near Mergui; HERON 1917, p. 179)
– in river sediments in the Tavoy District (J. C. BROWN & HERON 1919)
– heavy mineral beach sands on the coast of Moulmein.

5.24 Light metals

The "light metals" discussed here are aluminium, magnesium, lithium and beryllium.

Aluminium is produced almost exclusively from **bauxite**. Genetically, there is no difference between bauxite and laterite, which is widely distributed throughout Burma

(p. 223). High-grade bauxite should contain $> 50\%$ Al_2O_3 and $< 4\%$ SiO_2. A distinction is made between silicate bauxite, which forms when silicate source rocks become weathered, and carbonate (calcareous, karst) bauxite which is the weathering product of carbonate source rocks. According to SCHELLMANN (1977, p. 127), the most important conditions under which bauxite may form may be summarized as follows:

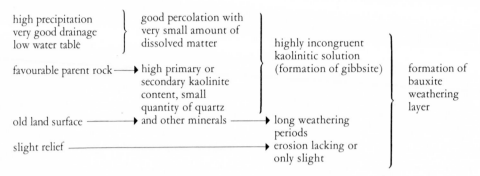

Minable bauxite occurrences have so far not been reported, but there are some areas where the above genetic aspects apply and where bauxite may have formed:

- above argillaceous carbonate rocks, kaolinitic shale, granite, granitic gneiss and other catametamorphics in the Sino-Burman Ranges, particularly in those areas which have been a land surface since the Upper Cretaceous and which exhibit a low relief
- above alkaline, low-quartz or quartz-free igneous rocks in the Inner-Burman Tertiary Basin (syenite, rhyolite, andesite, diorite, anorthosite)
- above the greenschists at the E edge of the Indo-Burman Ranges.

Since bauxite may have formed over almost all types of parent rocks, with the exception of sandstone and ultrabasites, it is worth carrying out prospecting also in all other areas where deep lateritic weathering has taken place over a long period of time.

Raw materials for the production of magnesium have so far not been discovered. FERENCIC (1961, p. 23) mentioned small quantities of magnesite from the Tagaung Taung area (p. 197). Since magnesite is frequently associated with chlorite and talc schists and occurs metasomatically in limestone and dolomite, magnesite prospecting could be carried out in areas having such rock associations.

Amblygonite $(LiAl[(F,OH)PO_4])$, spodumen $(LiAl[Si_2O_6])$ and lepidolite $(K(Li,Al)_3$-$[(F,OH,O)_2/AlSi_3O_{10}])$ are important raw materials of **lithium**. So far only one occurrence of lepidolite has been found in Burma in a pegmatite dyke near Sakangyi/Katha District (J. C. BROWN & DEY 1955, p. 271). Since these lithium minerals appear chiefly in pegmatites, granites, granite dykes and greisen zones, it is very likely that further discoveries of lithium will be made in the Tenasserim region, but also in the extensive granite occurrences in the central and N Sino-Burman Ranges.

Beryllium raw materials occur in the form of beryllium $(3 BeO \cdot Al_2O_3 \cdot 6 SiO_2)$, aquamarine (light blue-green beryllium), emerald (green beryllium), chrysoberyllium $(Be(AlO_2)_2)$, phenacite (Be_2SiO_4) and in the silicates euclase and gadolinite. Apart from the beryllium-bearing gemstone placers of Mogok (p. 207), primary deposits of beryllium occur in pegmatites, chiefly in feldspar pegmatites, e.g. in granite pegmatites of the Sakangyi Mine $(22° 54' N/$

96° 21′ 50″ E), in tungsten-bearing veins of the Byingyi Mine (19° 38′ N/96° 28′ E) and the Sinbo Mine (14° 22′ N/98° 22′ E) (CHHIBBER 1934 b, p. 21).

5.25 Special metals

The "special metals" dealt with here are antimony, arsenic, bismuth, mercury and cadmium.

CRIPER (1885), FERMOR (1906) and HERON (1921) described **antimony** ores from the Moulmein and Amherst districts and from the Northern Shan State. In the meantime, more than 30 occurrences have been discovered in the Sino-Burman Ranges approximately in between 23° 07′ N (Bawdwin) and 15° 31′ N (Thabyu; Fig. 72, folder).

The Pb-Zn deposit at Bawdwin (p. 177) also contains sulphoantimonites such as bournonite, boulangerite and pyrargyrite. The Sb content of the high-grade ore was 1.21% Sb in 1937, which was extracted during smelting as lead-antimony containing 20.9% Sb, 76.8% Pb and approximately 100 g Ag/ton (SOMMERLATTE 1948, p. 142).

Almost all the other Sb occurrences and indications that have been observed so far occur as vein-like or lenticular stibnite in the Late Paleozoic clastic Mergui Series of S Burma or in the carbonate equivalents in the Shan States.

These epithermal precipitates of stibnite are frequently associated with pyrite, quartz, calcite and traces of arsenic; according to SOMMERLATTE (1948), valentinite, cervantite and stibiconite are found in the oxidation zone.

According to unpublished information compiled by GOOSSENS (1978) the occurrences at Tagundaing (16° 10′ 30″ N/97° 47′ 30″ E), Talechaung (16° 26′ N/97° 40′ 20″ E) and Lebyin (20° 41′ N/96° 26′ E) are vein fillings that run transverse to the strike of the sediments, whereas another occurrence at 17° 05′ 20″ N/97° 58′ 50″ E is said to be strata-bound.

Near Natsan, approximately 10 km SE of Moulmein (97° 41′ E/16° 24′ N) stibnite is found in a NW-striking, steeply E-dipping, locally silicified fault zone in folded quartzite and quartzitic schist of the Mergui Series. The fault zone contains up to 150 cm thick stibnite lenses, smaller pockets and stringers of ore. Stibnite also occurs as a dark grey impregnation in the quartzitic country rock.

JONES (1921) and SOMMERLATTE (1948) mentioned a stibnite vein at Naking (Mon State) on the W bank of the Nam Pang in Jurassic sandstone (Kalaw Beds). Apart from massive stibnite, traces of Pb, As, Cu and of noble metals were observed here.

Since the First World War, a cumulative total of only a few thousand tons of stibnite has been produced from the occurrences mentioned above.

Arsenic occurs in the form of arsenopyrite, loellingite, nickel-cobalt arsenide and sulpharsenide such as gersdorffite, cobaltite, smaltite and erythrite in the lead-zinc deposit at Bawdwin. The mean As content of the high-grade ores from Bawdwin was about 0.64% in 1937 (SOMMERLATTE 1948).

Arsenopyrite is also an accessory mineral in many tin/tungsten ore occurrences in S Burma, for example near Maliwun (10° 14′ N/98° 39′ E), Tagu (12° 16′ N/99° 04′ E) and Champang (10° 12′ 30″ N/98° 31′ 30″ E) (GOOSSENS 1978).

Small quantities of **bismuth** are associated with the primary tin/tungsten ores in S Burma and it also occurs in Sn/W placers. HERON (1921 b, p. 81) described it at the Kanbauk Mine (14° 36′ N/98° 03′ E) and at further localities in Tenasserim. GOOSSENS (1978, p. 479) compiled the following list of additional Bi occurrences from unpublished reports:

- Palaw (12° 55′ N/98° 40′ E) in Sn/W placers
- Yadanabon mine (11° 16′ 05″ N/99° 17′ E) Sn/W quartz veins and greisen, with traces of Cu, Mo, F and U
- Byauk Chaung (14° 17′ N/98° 15′ E) in Sn/W placers
- in the Pagaye, Paungdaw and Zinba mines (14° 06′ N/98° 19′ E) together with Sn/W and sometimes in small nodules in the gemstone-bearing detritus of the ruby mines at Mogok (22° 54′ N/96° 30′ E).

Mercury has so far been found in only one place (Pwedaung, 20° 57′ N/96° 39′ 30″ E) where it occurs in traces as cinnabar in Ag-bearing galena. **Cadmium** occurs in sphalerite ores at Bawdwin.

So far, no arsenic, bismuth, mercury or cadmium have been mined in Burma.

5.3 Non-metallic raw materials

5.31 Industrial minerals

Asbestos, barite, corundum, feldspar, fluorite, graphite, gypsum, kyanite, magnesite, mica, ochre, phosphoritic rocks, quartz, rock salt, steatite, sulphur and talc are dealt with in this chapter. Only barite, feldspar, gypsum and rock salt are mined in relatively small amounts to meet local needs. So far none of the occurrences of the industrial minerals have been studied in detail. With the exception of the barite deposits at Maymyo, no systematic prospecting, exploration or evaluation of the industrial minerals has yet been carried out.

Asbestos has been found at an unspecified location in the serpentinite of the Sagaing District (CHHIBBER 1934 b, p. 225). Given the wide distribution of such rocks both at the E edge of the Indo-Burman Ranges as well as in N Burma and at the E edge of the Inner-Burman Tertiary Basin, there is a good chance that further occurrences of asbestos will be discovered.

Barite is widespread, chiefly in strata-bound or vein-like form, in the Paleozoic platform carbonates of the Sino-Burman Ranges. Only a few of the more important occurrences are mentioned here (from N to S):

In Bawdwin (p. 177), barite occurs in a number of veins about 1 m thick which transect the Ordovician volcanoclastic sediments.

According to SEIN MYINT (1968, pp. 457–464), in the area of Maymyo near Taunggyun, Pein-ne-gon, Sitha and Dattaw, ± vertical, N–S-striking barite veins are exposed in the Lower Paleozoic limestone and shale. The veins attain maximum thicknesses of 6 m and can be recognized for up to 350 m at the surface. An analysis of a barite sample from Taunggyun (SEIN MYINT 1968, p. 463) gave the following figures (in percent):

Moisture	0.52	MgO	0.45
SiO_2 insol. matter	1.00	ZnO	0.20
$BaSO_4$	91.52	Pb	0.05
Fe_2O_3	0.64	SO_3	2.14
Al_2O_3	0.20	CO_2	2.32
CaO	trace	Sp.gr.	4.19

The proven barite reserves are put at around 213,000 tons, and the probable reserves are given as about 475,000 tons. In recent years, several thousand tons/annum have been mined in the Maymyo area.

A number of strata-bound and vein-like barite deposits are found in the Middle Ordovician limestones (Wunbye Formation, p. 59) in the area of Bawsaing (Mawson), as was mentioned by CHHIBBER (1934b, p. 226): in the vicinity of Mawson (20° 57′ N/96° 50′ E), E of Kon Lean (20° 48′ N/96° 39′ E) and Myenigon (20° 51′ N/96° 49′ E). In his compilation of unpublished data, GOOSSENS (1978, p. 479) listed "Bawsaing, Hill 5018" (21° 03′ N/ 96° 56′ E) and "Bawetaung" (20° 57′ N/96° 46′ E) as further barite deposits in this region.

Corundum is found in the gemstone placers of the Mogok area and near Nanyaseik in the Myitkyina District. Before the Second World War it was mined together with the gemstones and sold as an abrasive.

Feldspar occurs in possibly minable amounts in pegmatites and albite-quartz aplites in the Mogok area, also in the mica pegmatites of Kansi (25° 47′ N/96° 22′ 18″ E) and in the albite-jadeite rocks in the Jade Mines Area (Fig. 88).

Fluorite was mentioned from a few points and in small quantities in the Sino-Burman Ranges: Yadanabon (11° 6′ 05″ N/99° 17′ E), Bawhningon (20° 31′ N/96° 37′ E; an abandoned quarry), Gwet (14° 10′ N/98° 21′ E) (GOOSSENS 1978, from unpubl. data). Fluorite was quarried in 1972 in Permian limestone near the road from Aungban–Pinlaung at about 21° 33′ N/96° 37′ E in the Southern Shan State.

Graphite is widely distributed in the Precambrian metasediments at the W edge of the Sino-Burman Ranges and occurs locally also in the greenschist at the E edge of the Indo-Burman Ranges.

CHHIBBER (1927, pp. 195, 196; 1934b, pp. 234, 235) mentioned the following graphite occurrences:

- Graphite schist with high graphite contents near Lawa Htensa (25° 33′ N/96° 49′ E); near Tathum (25° 36′ N/96° 50′ E); near Mupaw (25° 51′ N/96° 53′ E); between Nphum (25° 45′ N/96° 51′ E) and Warawng (25° 41′ N/96° 50′ E) (Myitkyina District)
- Ruby-bearing marble, calciphyre and scapolite gneiss with high graphite contents 6 km SW of Wabyudaung (22° 52′ N/96° 9′ E); approximately 8 km NE of Kyaukgyi (22° 59′ N/ 96° 9′ E) (Mogok area)
- about 30 km NE of Taungoo, on the Kanni River (Taungoo District)
- about 3 km SW of Wadawkin, lenticular graphite and quartz accumulations in chlorite schist; about 1.5 km S of Kyibin, graphite and quartz lenses in outcrops in the Thitya Chaung (Henzada District). CHHIBBER (1934b, p. 235) interpreted the formation of this graphite as the result of contact metamorphism of a coal-bearing shale caused by the intrusion of peridotite.

PIDDINGTON (1847) described coarsely foliated graphite under the name "Tremenheerite" from the region of Therabwin (12° 18′ N/99° 3′ E) on the Great Tenasserim River (Mergui District).

Gypsum occurs in thin layers and bands in the Pegu Series (Pegu Group, p. 89) in the Inner-Burman Tertiary Basin, for example in the districts of Thayetmyo, Minbu, Pakokku, Magwe and Myingyan. These occurrences are of minor or no economic value. Because of the lack of other sources, gypsum is gathered in the badlands in these districts to be used as an aggregate material in the cement industry. According to GRIFFITH (1956, p. 17), gypsum occurrences more than 50 m thick are found near Hsipaw, about 60 km SW of Lashio in Northern Shan State. This thickness has been confirmed by several boreholes which revealed that the gypsum is underlain by anhydrite. These evaporites probably belong to the Upper Triassic-Jurassic Namyau Group (p. 83). Several thousand tons of gypsum per annum are mined in this area.

Kyanite was found by CHHIBBER (1934b, p. 29) in kyanite schist approximately 1 km SE of Saingmaw (25° 35′ N/96° 17′ 30″ E) in the Jade Mines Area. The schist contains bright green blades and sheaf-like aggregates of kyanite set in a talcose matrix. Nothing is known about the quality and quantity of this kyanite occurrence.

FERENCIC (1961, p. 23) mentioned an occurrence of **magnesite** in the Tagaung Taung area (Fig. 83). It is probably also found in the Jade Mines Area and, in association with the ultrabasites, at the E edge of the Arakan Yoma.

BARNARD (1933, quoted in FERMOR 1934), discovered **mica** occurrences in pegmatites approximately 15 km E of Fort Hertz in N Burma (Putao subdivision of Myitkyina District). HOLLAND (1902, p. 54) mentioned leases granted for mica mining near Ye-nya-u (Thabeikkyin Township, Mogok area) in the area between Sakaw and Nanyaseik (Myitkyina District), 12 km E of Manwe on the Indaw River and at Shwedaung Gyi Hill at the S end of Indawgi Lake. CHHIBBER (1934b, p. 239) mentioned further localities where locally coarsely foliated, colourless "books" of mica are found:

- NE slopes of the Bumraung Bum, approximately 3 km NW of Hkumgahtaung (25° 31′ 42″ N/96° 36′ 3″ E)
- Kathan Hka near Sakaw (25° 29′ N/96° 38′ E)
- near Nweyon, Mandalay District
- near Chaunggyi, Katha District.

In all cases mica occurs in granite pegmatites which transect the complex of crystalline rocks, i.e. granite, gneiss, chlorite and mica schists and marble.

According to CHHIBBER (1934b, p.241), yellow **ochre** is found in thicknesses up to 10 m near Panbe (20° 50′ N/95° 5′ E) in the Myingyan District; red ochre is said by MASON (1850) to occur near Ka-lein-aung (14° 22′ N/98° 12′ E) on the Tavoy River in the Tavoy District.

So far no rocks containing economically extractable **phosphorite** have been found. Prospective facies conditions (calcareous sediments containing large percentages of chert deposited in a shallow marine depositional environment) have not developed in Burma. Samples of marly limestone of the Cenomanian which GRAMANN collected to the W and N of Kyauktu (e.g. on the Kyi Chaung upstream from Kyi) contained 1.54% P_2O_5, 1.51% K_2O and 23.5% Ca (BGR Lab. No. RF 28 574, 1980). This rock could be crushed and possibly used directly as a fertilizer for soils deficient in Ca, P and K.

The numerous karst caves of the Sino-Burman Ranges, particularly in the area of the Shan Plateau, may be a further possibility for prospecting for phosphorite-bearing sediments.

In addition, apatite-bearing igneous rocks, such as occur in the Wuntho Massif (p. 136), in the Mogok area and in the Jade Mines Area (Fig. 88), may be worth investigating.

Quartz occurs in quartz veins and pegmatites, in lenses, bands and nodules in practically all parts of the country where igneous and metamorphic rocks are found. So far it has not been used as a raw material, and nothing is known about economically interesting occurrences. Rock crystal has occasionally been mined in the Mogok area (Sakangyi).

Solid **rock salt** is not exposed anywhere, but it is likely that it is present in the evaporites of the Upper Triassic-Jurassic Namyau Group (Pangno Evaporites, p. 83) in the Northern Shan State. The water from numerous mineral springs with high NaCl contents is evaporated for producing sodium chloride, e.g. near Bawgyo ($22°35'$ N/$97°16'$ E). Salt springs in the Myitkyina District (e.g. near Lama: $25°41'$ N/$96°21'$ E) and in the Hukawng Basin, where a number of such springs occur along a fault line (CHHIBBER 1934b, p. 213) are used for the same purpose. NOETLING (1891) reported a series of saline springs along the W side of the Maingthon Hill tract in Wuntho, which probably are also arranged along a fault line.

By leaching saline soil and by evaporating NaCl-bearing brines, rock salt was obtained in the Dry Zone of Central Burma (p. 217) and further to the S in the districts of Thayetmyo, Prome and Henzada. In this area THEOBALD (1873) counted 79 saline springs arranged along three approximately parallel lines at the E side of the Arakan Yoma (e.g. Sahdwingyi: $18°2'$ N/$95°9'$ E).

A high percentage of the sodium chloride produced in Burma is obtained by evaporating sea water on both the Arakan and the Tenasserim coasts, and in the Irrawaddy Delta Region (i.e. Chaunggwa).

Because of its association with serpentinized ultrabasites, **steatite** is almost always found wherever such ultrabasites occur. According to MALLET (1889) steatite was mined from veins 20 cm to 30 cm thick which transect the serpentinite in the area about 45 km W of Hpa-aing ($20°15'$ N/$94°23'$ E) (Kyaukpyu District). THEOBALD (1871) described steatite from a hill about 5 km NW of Sandoway ($18°28'$ N/$94°24'$ E) from a quartz vein in shale and conglomerate. HAYDEN (1896) reported steatite in serpentinite near Senlau ($19°54'$ N/ $23°30'$ E; Minbu District). Steatite has also been observed in the Yamethin District (approximately 3 km E of Taungbotha), in the Prome District (Shinbaian Hill: $18°57'$ N/ $94°56'$ E) and in the serpentinite occurrences in the Henzada and Bassein Districts (CHHIBBER 1934b, p. 245); furthermore N of Kalemyo.

Small quantities of native **sulphur** have been found in deposits of H_2S-bearing springs N of Kyin in the Pakokku District.

Pyrite is an accompanying mineral in many metal ores. It occurs as the dominant mineral

- together with Cu carbonates near Pu Khwan Chai ($34°32'20''$ N/$98°57'20''$ E)
- together with Au, Cu minerals, galena, franklinite and altaite near Banmauk ($24°23'$ N/ $95°51'$ E)
- together with auriferous quartz at Nimk In-hill ($24°06'$ N/$95°54'$ E)
- disseminated in stringers and veinlets in an approximately 2 m thick black-grey shale near Hungwe ($23°07'$ N/$97°11'$ E)
- together with traces of Ni near Singu ($22°29'$ N/$96°08'$ E)
- in black shale near Thegon ($20°57'$ N/$96°46'$ E)
- near Yebok ($20°49'$ N/$96°35'$ E)
- in a vein (max. thickness 12 m, average thickness 2 m) in serpentinite about 7 km SW of Kwingyi ($17°28'$ N/$94°39'$ E); according to CHHIBBER (1934 b, p. 112) an ore sample yielded the following analysis:

Fe	42.84%
Cu	1.68%
S	45.42%
Si	10.08%

– together with arsenopyrite near Champang ($10° 12' 30'' N/98° 31' 30'' E$).

Talc was mentioned by CHHIBBER (1934b, p. 29) in kyanite schists from Saingmaw ($25° 35' 0'' N/96° 17' 30'' E$) in the Jade Mines Area and by FERENCIC (1961, p. 23) from the Tagaung Taung area (p. 197).

5.32 Stones and earths

Bentonite, building stones, cement raw materials, glass sand, kaolin (pottery clay), laterite, tuff, sand and gravel are the materials that are dealt with here under the heading "stones and earths". None of these near-surface raw materials has so far been systematically explored and assessed in Burma. Only building stones, cement raw materials, pottery clay and sand/gravel are being mined in significant quantities.

Bentonite, which consists of aluminium silicates – chiefly montmorillonite – occurs usually in conjunction with decomposed volcanic tuffs. It is very likely present in the extensive areas of volcanic rocks, particularly in the central and N section of the Inner Volcanic Arc (cf. kaolin, pottery clay).

Limestone of the Shan Dolomite Group (p. 72) in the Sino-Burman Ranges (Zebingyi Beds, limestones of the Napeng Formation) as well as hard, fine-grained sandstone from the Namyau Series (Jurassic) are regarded as good **building stones.** In the Myitkyina District, Tertiary sandstone and serpentinite are used for building purposes. Sandstone from the Pegu Series is used throughout the Inner-Burman Tertiary Basin (Fig. 48), and laterite is used as building stone throughout the country. Unweathered, fine-grained to medium-grained granite of the type that occurs in the districts of Thaton, Amherst, Tavoy and Mergui, is probably suited as ornamental building stone (CHHIBBER 1934b, p. 308). Good quality white marble is mined in the Sagyin Hills, about 35 km N of Mandalay. Similar marble is exposed near Kyaukse. The marble from Loikaw is widely used as ornamental stone. CHHIBBER (1934b, p. 309) mentioned further outcrops of good quality marble in the Kamaing subdivision in the Myitkyina District.

Porphyry, basalt, quartzite, serpentinite, diorite, syenite, granite, quartz-rich gneiss, limestone and dolomitic limestone suitable for roadway and railway roadbed construction are found practically everywhere in the Sino-Burman Ranges and in parts of the Inner Volcanic Arc as well as of the Indo-Burman Ranges.

Raw materials for the production of **Portland cement** should contain between 25% and 60% $CaCO_3$ and between 40% and 75% non-carbonates ("clay"), while for the production of hydraulic lime and Roman lime the contents should be between 60% and 75% $CaCO_3$ and 25% and 40% non-carbonates. Such limestone and marl are found in the Middle Devonian Zebingyi Beds (p. 65), in the Triassic Napeng Formation (p. 82) or the Natteik Limestone Formation (p. 82) and in the Jurassic Namyau Formation in the central and N Sino-Burman Ranges. Only the $CaCO_3$-rich portions of these sequences are used here to calcine limestone in lime kilns (e.g. near Zebingyi, Mandalay).

At the W edge of the Prome Embayment (p. 185) foraminiferal limestone of the Oligocene-Miocene crops out occasionally in thicknesses up to 100 m, for example W of Thayetmyo (approximately 60 km N of Prome), near Alechaung ($94° 52' E/19° 22' N$, about 47 km NW of Thayetmyo) and near Kyanjin (about 60 km SW of Prome). This limestone is mined near Thayetmyo for cement production. About 410,000 tons of limestone/a,

100,000 tons of marl and about 16,500 tons of gypsum are required for a planned production of 300,000 tons of cement/a. Adequate quantities or marl of the desired quality overlie the limestone, and the gypsum is mined manually from strata of the Pegu Series N of Thayetmyo (p. 89).

CHHIBBER (1934b, p. 233) found **glass sand** of probably usable quality and in practically unlimited quantities on the coast of some islands in the Mergui Archipelago (e.g. about 3.5 km N of the S end of Thamihta Island). The white sands are weathering products of pure quartzitic sandstone and quartzite of the Mergui Series (p. 70). The latter and further occurrences in the Bassein area are also suitable as a raw material for glass manufacture and are used in the production of sheet glass (Bassein).

Kaolin, like bentonite, also consists of aluminium silicates, chiefly kaolinite ($Al_2(OH)_4Si_2O_5$), which is usually a weathering product of decomposed volcanic rocks. Kaolin of more or less high quality (pottery clay) is found in the Shwebo District where it constitutes the raw material for the centuries-old pottery industry, e.g. in Nwenyein, Shwegun, Shwedaik and Kyaungmyaung ($22° 30' N/95° 57' E$). The pottery clay appears here within the Irrawaddy Group (p. 99).

Pottery clay, as a lenticular intercalation in the Irrawaddy Group, is found E of Yedwet ($95° 14' E/20° 29' N$) and is widespread (up to 15 m thick) at the base of the Irrawaddy Group near Pakokku. According to CHHIBBER (1934b, p. 236) this clay is of high quality and could be economically exploited on a large scale.

Light-grey to white high-quality kaolin is also found about 15 km W of Yamethin and near Tayetpin ($20° 21' N/96° 6' E$), and as a weathering product of granite and pegmatites near Yennyein in the Thaton District, furthermore near Indaingon ($17° 59' N/95° 10' E$) in the Henzada District and on the Tenasserim River about 70 km upstream from Tenasserim in the Mergui District (HELFER 1839, p. 35, THEOBALD 1873, p. 341, BALL 1881, p. 567, CHHIBBER 1934b, p. 237).

Laterite is widespread in Burma. It is used practically throughout the country for road construction and in many places for building purposes. When freshly quarried, it is easy to cut into blocks and it subsequently hardens when exposed to air. Laterite is quarried in particular in the regions of Thaton, Bassein, Prome, Tavoy and Amherst.

The extensive **tuff** occurrences of acid, intermediate and basic volcanics, which are widespread mainly in the central part of the Inner Volcanic Arc (Mt. Popa-Lower Chindwin area; p. 131) have not so far been examined for their industrial suitability. Thick pumice tuffs are exposed in the vicinity of some explosive craters on the Lower Chindwin, near Monywa.

Practically unlimited quantities of **sand and gravel** for construction purposes and for building roads and railways are found throughout the Inner-Burman Tertiary Basin and are extracted at numerous localities. The most suitable occurrences are located mainly in the terraces of the Irrawaddy and its tributaries and consist of reworked sands and gravels of the Irrawaddy Group.

5.33 Precious stones and semi-precious stones

Most of the **rubies** and **sapphires** are found in the "Mogok Stone Tract" which is built up of Precambrian gneiss, migmatite and marble (p. 43). Precious stones are also found in the Sagyin Hills, about 25 km N of Mandalay and in the Nanyaseik Stone Tract/Myitkyina

District (CHHIBBER 1934 b, pp. 12 et seqq.). The ruby and sapphire occurrences, as well as those of the less valuable **spinels,** are bound to marbles and calc-silicates and are closely connected genetically with aluminium-rich melts of magmatic rocks which transect the metasediments. According to IYER (1953, p. 37) **corundum** and spinel formed as a function of the richness of the intrusions and limestone in alumina. SEARLE & BA THAN HAQ (1964) assumed that the necessary genetical conditions were satisfied only by the reaction of alaskitic melts with the carbonates to form these precious stones. With their suite of hybrid rocks, the probably Precambrian alaskites (p. 124) document the capability of geochemical reactions with their country rock, while the intrusion of Tertiary granites, e.g. of the "Kabaing Granite" (p. 142) in the Mogok area, took place without any major chemical reaction in the country rock. According to SEARLE & BA THAN HAQ (1964, p. 143), the alaskitic melts were rich in water and volatile elements such as chloride, fluorine and boron. In addition, high temperatures ensured a fluid state in the contact zone and thus provided the necessary conditions for the geochemical exchange processes to take place. As a result, the varieties of corundum formed during reactions of alaskitic melts with marble and the more frequently occurring spinels formed as a result of melt reactions with Mg-containing carbonates, e.g. dolomite.

Rubies, sapphires and spinels are mined in the Mogok area from marbles in simple drift mines, but they are chiefly obtained from placers in eluvial and alluvial sediments. Apart from a few open-cast mines operated with monitors and gravel pumps (Figs. 85–87), the precious stones are extracted by primitive manual mining methods. The alluvial placers are uncovered by digging holes up to 15 m deep and approximately 1 m in diameter. The gemstone-bearing strata ("byon") which are reached in this way are followed in a horizontal direction for several tens of metres ("Twinlon System"). The mined product is washed in rivulets to recover the precious stones. Because of the high groundwater table, manual mining in the alluvial deposits is restricted to the dry season. On the other hand, the eluvial placers are mined during the rainy season when the water is brought into the individual mining fields through channels and passed over bamboo mats. This loosens the gemstone-bearing debris and it is then passed through pits and sluices in which the heavy minerals are gravitationally enriched and examined for their content of precious stones ("Hmyawdwins Method"). Particularly rich finds have been made in karst caves and karst fissures in the marbles which the miners have followed down to depths of up to 70 m ("Loodwins Method").

IYER (1953, p. 82) gave production figures of the Burma Ruby Mines Ltd. as over 100,000 carats of rubies/a in the years 1924–1925 and over 200,000 carats/a in the years 1938 and 1939. From 1970 to 1975, the total production of precious stones was between 30,000 and 120,000 carats/a. Exact production statistics are lacking.

Together with rubies, sapphires and spinels, the contact zones of the acid intrusives also contain **zirconium,** which is mostly golden yellow and very occasionally green in colour; also, **iolites** and **garnet** of polishable quality are found. **Moonstone,** as microperthitic potash feldspar, forms one of the mineral components of alaskite. Of the pegmatite minerals, colourless **topaz** ("Mogok diamond") is mined. In addition, **amethyst, aquamarine** and **tourmaline** are found. **Lapis-lazuli** occurs as a pneumatolytic mineral in marbles and calc-silicates. **Chrysolite** is found in peridotite near Bernardmyo (about 10 km N of Mogok).

NOETLING (1893 a, 1896), BAUER (1895) and BLEECK (1907) reported on the **jade** occurrences. Jade (jadeite), of the finely felty pyroxene variety ($NaAl-Si_2O_6$), is found in the Lonkin area about 150 km SSW of Myitkyina (Fig. 88) (emerald green [chromite-bearing],

Fig. 85. Mogok Valley, view from N. Photo: F. BENDER.

Fig. 86. Gemstone mining in eluvial and alluvial placer deposit, approx. 3 km N of Mogok. Photo: F. BENDER.

Fig. 87. Gravelpump lifting gemstone-bearing gravel and mud from open castmine; ca. 4 km SW of Kyatpin near Mogok. Photo: F. BENDER.

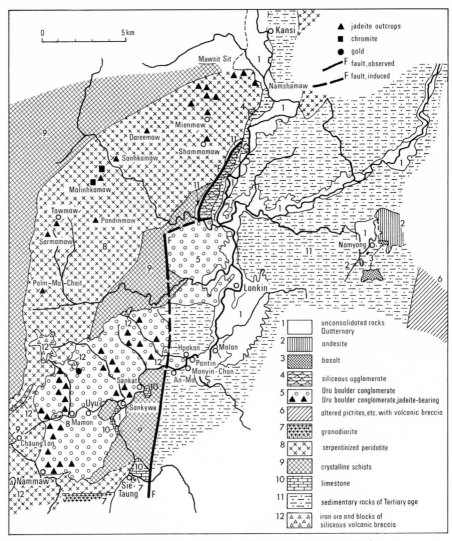

Fig. 88. Geological sketch map of Jade Mines Area (modified after CHHIBBER 1934 b).

apple green [ferruginous] and lavender blue [manganiferous] coloured types). They are extracted from primary deposits by simple drift mining and by diggings from Tertiary to Recent fluviatile gravel sheets.

According to CHHIBBER (1934b, p. 44) the primary occurrences are composed of jadeite-albite dykes or sills which are "intrusive into peridotites and serpentinites" and are pre-Tertiary in age. Four of such "intrusions" are known. The largest is the one at Tawmaw (25° 42′ 13″ N/96° 15′ 28″ E) which is 30–40 m long. The shallow-dipping occurrence consists (from top to bottom) of serpentinized peridotite, light green chlorite schist, silicified serpentinite, amphibolite, amphibolite-albitite rock and lenticular albitite with jadeite intercalations up to 2 m thick.

Fig. 89. Boulder of jadeite approx. 180 cm in diameter, from Jade Mines Area. Exhibited at Myanma Gem Corporation, Rangoon. Photo: F. BENDER.

An interpretation of the jadeite-albitite occurrence as an intrusive rock is unlikely. More probably the jadeite formed under high pressure caused by regional metamorphism of the epizone (WINKLER 1974, p. 67; lawsonite-jadeite-glaucophane facies). The joint occurrence of jadeite and albite together with silica-rich rocks matches the experimentally verified reaction in which jadeite is produced from "de-silicified" albite. The jadeite occurrences at Tawmaw may therefore have been formed by such high pressures, possibly in tectonically severely affected zones within the serpentinized basic intrusive body.

A large percentage of the jade is extracted from fluviatile gravels. The Tertiary occurrences include those at Kansi ($25°47'1''$N/$96°22'18''$E) and Hwehka ($25°29'3''$N/$96°16'43''$E), which fill broad channels in the drainage system of serpentinite massifs and which are composed of sandstone, shale and conglomerate. The diameters of individual boulders, which can be in the metre size range, indicate that the rock material was deposited in torrential streams (CHHIBBER 1934 b, p. 53). Mining is also being carried out in the gravels of the Uyu Chaung and its tributaries, for example near Hpakan ($25°36'38''$N/$96°18'40''$E) and Haungpa ($25°30'$N/$96°6'15''$E). The gravels are overlain by m-thick alluvial sands. Detritus containing good quality jade is found in sandy layers directly above the bedrock; these sands also contain some gold (p. 175).

Jade production in 1970–1975 is given as 3,900 kg – 16,300 kg/a.

Burmese **amber** (burmite, NOETLING 1893 b), has been known in China for almost 2,000 years. The major occurrences are located in the Hukawng Valley near Maingkwan and Shingban ($26°16'$N/$96°35'$E). The amber-bearing strata consist of a tightly folded sequence of dark blue-greyish shale and sandstone of Eocene age. The numerous genera of insects which COCKERELL (1920–1921) identified in the amber are also Eocene in age. Amber occurs mainly in cm-thick coal seams which are intercalated in the shale and sandstone. The burmite, which exists in 14 different colours, is generally somewhat harder than the amber from other deposits. Its fluorescence makes it resemble the Sicilian type of amber known as simetite (HELM 1893, p. 63).

The amber occurrences in the Hukawng Valley are mined by primitive manual methods in shafts up to 15 m deep. The amber production in 1929 was about 1,000 kg, and in 1930 about 100 kg (CHHIBBER 1934 b, p. 5). Further occurrences of low quality are known in areas where Eocene sediments crop out, namely near Mantha/Shwebo District (22° 55′ N/96° 1′ E), Seikkwa/Pakokku District (21° 8′ N/94° 51′ E) in the Mibauk/Thayetmyo District (19° 27′ N/94° 53′ E) and in the N Chindwin Basin.

5.4 Water and soil

5.41 Surface water

The orientation of the **river systems** is governed by the general N–S-strike of the major geotectonic features and thus also of the physiographic units (Fig. 7, folder).

With a length of about 2,010 km and a catchment area covering 415,700 km², the Irrawaddy is the most important river in Burma. Both its headwater streams, the Nmai Kha and Mali Kha, rise in the E Himalayas. The E (main) tributary, the Nmai Kha, rises in the Languela Glacier N of Putao. Both head streams join up about 50 km N of Myitkyina. The Irrwwaddy passes through 3 important defiles: the highest one is below Sinbo, the second one is below Bhamo, and the third starts at Thabeikkyin and ends near Kyaukmyaung in the Shwebo District. According to STAMP (1940, p. 346), the total average discharge of the Irrawaddy at Saiktha, before it flows into the delta, is 398,912,000,000 m³/a (mean value calculated from the discharge measurements made by GORDON 1869–1879, 1885). VOLKER (1966, p. 373), mentioned that no other measurements had been carried out since that time. The average monthly discharge in m³/sec in the period from 1869 to 1879 is shown in Fig. 90. The total annual load of undissolved material carried by the Irrawaddy is in the order of 261,000,000 tons. Thus, the Irrawaddy Delta is building up in a seawards direction at a maximum rate of 4.5 km/100 years (BA KYI 1964, p. 73).

Above the first defile near Sinbo, the **Mogaung River,** coming from the W, and the **Taping** and **Shweli Rivers,** coming from the E, flow into the Irrawaddy.

The **Mu River** rises near Mansi in the Katha District and drains a catchment area of about 18,900 km² along its meandering course to the S before it joins up with the Irrawaddy on about the same latitude as Mandalay.

A little further S, about 15 km NE of Pakokku, the Irrawaddy joins up with the **Chindwin,** its main tributary. Its catchment area covers about 114,000 km². One of its principal sources is the Noungyaung Lake (27° 13′ N/96° 11′ E). SARIN (1968) examined the composition of the channel sediments of the Chindwin River. He observed a distinct increase in the quartz component (2 mm–4 mm) downstream and a decline in the less resistant sediment components. He believed that the different composition of the channel sediments was caused by the progressive downstream sorting due to grain shape and selective abrasion.

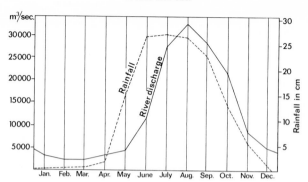

Fig. 90. Average monthly discharge of Irrawaddy near head of delta and rainfall 1869–1879 (modified after STAMP 1940).

From Hkamti to the confluence of the Chindwin in the Irrawaddy S of Monywa, a distance of 750 km, the feldspar content in the 0.5 mm and 0.125 mm grain fractions declined by only 3% and 2% respectively.

The **Uyu River** is an important tributary of the Chindwin, and it flows into the latter from the E about 6 km below Homalin.

The Irrawaddy is navigable all the year round as far as to Bhamo (1,040 km from Rangoon by water). The Chindwin is navigable to Hkamti.

To the E of the Pegu Yoma, the **Pegu River** drains a catchment area of about 5,440 km² before it joins up with the Rangoon River near Rangoon.

Further to the E, the **Sittang River** flows into the Gulf of Martaban. Along its approximately 320 km long course, it drains a catchment area of about 34,450 km².

On about the same latitude as Bhamo, the **Salween River** enters Burmese territory and flows through the Sino-Burman Ranges over a distance of about 1,000 km approximately parallel to the Irrawaddy. It has cut narrow gorges locally more than 1,000 m deep and, as CHHIBBER (1933, p. 33) pointed out, there is no rift, defile or canyon on earth that exceeds them in length. In the Shan States the average difference between high and low water levels of the Salween is between 20 m and 30 m. The river divides up into two arms to the W and S of Moulmein and flows into the Gulf of Martaban.

In the Tenasserim region the **Tavoy River** and, further to the S, the **Great Tenasserim River**, drain into the E Gulf of Martaban.

In the Arakan Coastal Area the **Kaladan, Lemro, Mayu** and **Naaf Rivers** flow into the Bay of Bengal. The largest of these rivers is the Kaladan, which rises in the Chin Hills, and then flows first to the S and then to the N before it turns to the W, enters the Arakan Yoma and then flows S again. After flowing for about 480 km, it discharges into the Bay of Bengal near Akyab.

As in the case of the rivers, no modern studies have been published on the **lakes** in Burma.

The following lakes are located in the region of the Inner-Burman Tertiary Basin (from S to N):

– the Engma (Imma) Lake in the Prome District, which is about 15 km long and up to 6 km wide and up to 4.50 m deep in the rainy season; in the dry season it is a marshy area
– the smaller **Htoo** and **Doora lakes**, both in the Henzada District

- the **Daga Lake** (about 4.5 km long, 15 km wide and up to 12 m deep) in the Bassein District
- the salt lakes in the Dry Zone (p. 217) in the Shwebo and Sagaing districts, e.g. the Kadu, Halin, Mahanand, Thamantha and Ye-myet lakes. These lakes, which are only a few metres deep and up to a maximum of 30 km² in size, shrink in the dry season to an area of just a few km² or change completely into salt marshes
- the crater lakes in the area of the Lower Chindwin River (Fig. 61):
  Twin Taung crater lake (Figs. 62, 63). The rim of the crater, which is composed of tuff, is at an altitude of 210 m above sea level and is thus about 90 m higher than the surrounding plains (approx. 120 m above sea level); the surface of the crater lake, which is about 0.9 km² in size, is at about 90 m and its bottom at about 65 m above sea level; its water is highly alkaline and not suitable for irrigation
  the craters of the Taungbyauk (about 1 km², approx. 60 m deep), of the Twinywa and 6 other smaller craters in this area also contain lakes, some of them with fresh-water
- the **Indaw Lake** W of Katha (about 15 km²)
- the largest lake in Burma, the **Indawgyi Lake** in the Myitkyina District; with an SSW–NNE extent of about 26 km and a width of up to 11 km; it covers an area of about 210 km²; it is situated in a depression between chains of hills in the E, S and W with an outflow to the NE, the Indaw River; CHHIBBER (1933, p. 48) assumed that it was created as a result of tectonic subsidence.

The **Noungyaung Lake** (Nawng Yang, Nong Yang; 27° 13′ N/96° 11′ E) in the border area between Burma and India is about 700 m above sea level, about 3 km long and up to 1.5 km wide (JENKINS 1869, PEAL 1881).

A great number of intramontane basins and numerous smaller depressions, which usually contain Late Tertiary-Quaternary lacustrine sediments are located in the area of the Sino-Burman Ranges, especially in the area of the Shan Plateau. A regular lake country must have existed here, probably until the Pleistocene. With the exception of the **Inle Lake,** these lakes disappeared as a result of karstification and subsurface run-off in the carbonate rocks, or locally also as a result of peat formation and silting-up processes.

The Inle Lake is situated at an altitude of 910 m above sea level in a N–S-striking graben zone more than 100 km long (Fig. 33). The N–S-striking mountain chains in the W (E of Heho) rise about 270 m and those in the E (Taunggyi) about 400 m above the level of the Inle Lake. It is about 23 km long and 6.4 km wide. Its depth varies considerably between the dry season (up to 4 m) and the rainy season (up to 7 m). The mean surface temperature of the very clear water was 21.7 °C towards the end of the dry season, and the bottom temperature was 20.0 °C (ANNANDALE 1923). It is not possible to determine the area covered by the lake exactly because it increases in size during the rainy season and it merges at the margins into floating islands of living and decaying vegetation which in turn grow together further towards the land and then merge into marshy areas.

5.42 Thermal springs

The first and so far the only identification of thermal springs in Burma goes back to OLDHAM (1883). A systematic hydrogeological, hydrological and hydrochemical inventory of the springs still has not been made.

Thermal springs are numerous and widespread throughout Burma. They are concentrated in the Tenasserim region more or less along an SSE–NNW-oriented zone, the NNW-extension of which is located in the E parallel to the Sittang River, and the N-extension runs along the Shan Boundary Fault Zone. Such springs are also found in the Sino-Burman Ranges between Inle Lake and the area around Lashio, in the NE Myitkyina District and at the central E edge of the Arakan Yoma. Some of the more important thermal springs are listed below.

Sino-Burman Ranges, Tenasserim Region – Southern Shan State (from S to N)

- Palauk (13° 13′ N/98° 40′ E); group of 30–40 small, hot springs
- Pai (13° 26′ N/98° 33′ E); series of hot springs at the edge of the granite range near Pai, along a fault; temperature about 198 °F; accompanied by a jet of steam (STEVENSON 1863)
- Taungbyauk (13° 33′ N/98° 40′ E)
- Maung Mayan (14° 09′ N/98° 09′ E)
- Kaukyen (14° 12′ N/98° 25′ E)
- Myitha (14° 13′ N/98° 33′ E)
- Nat Gyi Zin (14° 55′ N/98° 00′ E)
- Thalan Khoung (15° 10′ N/98° 03′ E)
- Myan Khoung (15° 13′ N/98° 07′ E)
- Ataran (16° 06′ N/98° 02′ E); group of 10 hot springs, temperature 130 °F, much CO_2 (HELFER 1839)
- Bonet (16° 27′ N/97° 37′ E)
- Damathat (16° 33′ N/97° 52′ E)
- Yebu (16° 34′ N/98° 09′ E)
- Poung (16° 09′ N/98° 14′ E)
- Mya-waddi (16° 41′ N/98° 30′ E)
- Kyoung Chaung (17° 35′ N/97° 02′ E)
- Maiting (17° 55′ N/97° 04′ E)
- Koon Pai (17° 55′ N/97° 01′ E)
- Vadai Chaung (17° 56′ N/97° 12′ E)
- Hteepahtoh (17° 56′ N/97° 03′ E)
- Sair-O-Khan (18° 04′ N/97° 04′ E)
- Mai Pouk (18° 19′ N/96° 54′ E)
- Bin Byai (18° 24′ N/96° 44′ E)
- Hlayloo-myoung chaung (18° 33′ N/96° 51′ E); "very hot spring" acc. to CHHIBBER (1934 a, p. 74)
- Chaungna-nay (18° 44′ N/96° 46′ E), temperature 108 °F (CHHIBBER 1934 a, p. 74)
- Kayenchaung (19° 10′ N/96° 36′ E)
- Lepanbewchaung (19° 16′ N/99° 36′ E)

Sino-Burman Ranges, Northern Shan State

- Khaungdaing (20° 35′ N/96° 53′ E), approximately 20 km SE of Heho, karst spring originating in Ordovician limestone near the W edge fault of the Inle Lake graben;

temperature 68–70 °C; slight radioactivity: background 70–75 cps, at mouth of spring 110–115 cps (analysis BGR, Lab. No. 25 845, 1980):

pH value: 7.90
conductivity: 928 microsiemens/cm (20 °C)
carbonate hardness: 21.7 degrees
total hardness: 6.1 degrees

	ppm	mval	mval%
Cations			
K^+	17.0	0.43	4.8
Na^+	150.0	6.5	71.3
Mg^{2+}	5.0	0.41	4.5
Ca^{2+}	35.6	1.78	19.4
Fe^{2+}	< 0.1	–	–
Mn^{2+}	< 0.05	< 0.002	–
U^{6+}	0.0003	–	–
Pb^{2+}	< 0.002	–	–
Cu^{2+}	< 0.0005	–	–
Zn^{2+}	0.007	–	–
Anions			
Cl^-	16.6	0.47	5.1
SO_4^{2-}	47.0	0.98	10.6
HCO_3^-	473.0	7.8	84.3
NO_3^-	< 1.0	–	–

This is a soft $Na\text{-}HCO_3$ water

– on the Namtu river near Hsipaw, near Pengwai and a series of hot springs between Nakang and Loinmauk, also near Namon and Lashio; CHHIBBER (1934a, p. 71) mentioned that these thermal springs are associated with fault lines, also the important thermal spring at Lashio (22° 56′ N/97° 47′ E).

1 litre of the water (48°–50 °C, pH 7.5) from the Lashio thermal spring contained (BRAEUNING 1975, p. 33):

	mg/l	mval	mval%
Insoluble residue 378			
Cations			
Na^+	2	0.09	1.6
K^+	2	0.05	0.9
Ca^{2+}	64	3.19	57.6
Mg^{2+}	27	2.21	39.6
Fe^{2+}	–	–	–
Mn^{2+}	–	–	–
Anions			
Cl^-	4	0.11	2.1
SO_4^{2-}	74	1.54	29.0
HCO_3^-	220	3.61	68.0
NO_3^-	3	0.05	0.9

Sino-Burman Ranges, NE Myitkyina District

– Yinchyingpa (25° 56' N/98° 23' E); hot water with hot steam from a group of 8 springs issuing from fissures in the granite
– Munglang Hka (25° 58' N/98° 29' E); hot water with hot steam from a group of springs issuing from granite conglomerates on the right bank of the Munglang Hka; odour of H_2S, spring deposits consisting of Fe-oxides and Fe-sulphate (CHHIBBER 1934 a, p. 76)

Inner-Burman Tertiary Basin and Arakan Yoma

– hot water flowing into the Bulay (Bule) river (19° 15' N/95° 16' E; THEOBALD 1873, pp. 352–353)
– thermal spring in the source area of the Sandoway river (18° 06' N/94° 54' E) emerging from shales, at a rate of approximately 1½ gallons/min at a temperature of 110 °F (CHHIBBER 1934 a, p. 77)
– Halin (22° 27' N/95° 49' E), several hot springs emerge along an N–S-oriented line, temperatures between 53 °C and 73.5 °C, southernmost spring: 64 °C. NaCl 0.68%, $MgCl_2$ 0.02%, $CaCl_2$ 0.09%, $CaSO_4$ 0.01% (CHHIBBER 1934 a, p. 76).

5.43 Groundwater

The contours of mean annual rainfall (Fig. 14) clearly indicate the "Dry Zone" in the Inner-Burman Tertiary Basin which is approximately demarcated by the 1,250 mm line. The water supply in this region is problematical, especially for agriculture. Growing urbanization and increasing industrialization coupled with more intensive farming have created similar problems also in areas with higher precipitation. It should be possible to solve some of the increasing water supply problems by judiciously drawing on the groundwater. This solution should be more economical, particularly in the Dry Zone, than the very expensive step of making more direct use of the surface water (Irrawaddy, flash flood runoff). In the following, a brief summary is given of the hydrogeological conditions in areas where the water supply or parts of it could be obtained from the groundwater reserves.

Dry Zone

The area of the Dry Zone is located between about 19° to 23° N and about 94° to 96° E (Fig. 91). It receives less than 1,250 mm of precipitation and its central parts less than 500 mm/a. Most of the precipitation falls in May–June and from August to October (Fig. 15). Because of the high air temperatures and very low relative humidity throughout the year, the area is classified as semi-arid. The near-surface rocks are composed essentially of Quaternary to Middle Miocene sediments (Irrawaddy Group, p. 99), alluvia and, to a lesser extent, of shale and sandstone of the Tertiary, and of volcanic rocks. The surface runoff is estimated to be as much as 30% (TAHAL WATER PLANNING 1963 a, p. 16). Intensive erosion occurs and has frequently created a "badland topography". In the flat areas with predominantly sandy soils, there is little if any runoff. The Dry Zone is part of the

Irrawaddy-Chindwin River Basin and is situated at altitudes of about 45 m to 200 m above sea level, if one disregards the Mt. Popa Massif. The Dry Zone region is broken up by a number of chains of hills consisting of Tertiary sediments and by mountains made up of volcanic rocks (Mt. Popa, Lower Chindwin volcanic area, Shinmataung area). In the W it is bounded by the Arakan Yoma and Chin Hills, in the E by the Shan Highlands. The Irrawaddy is the base level for the runoff of the entire Dry Zone.

A number of potentially good sandstone aquifers, separated by argillaceous aquicludes, exist in the **pre-Irrawaddian Tertiary sediments.** These rocks are exposed along the W side of the Dry Zone (Western Outcrops; Fig. 27, folder), so that they form a groundwater recharge area in a zone which receives much more precipitation and where there is considerable surface runoff from the Arakan Yoma. Therefore, good opportunities for groundwater exploitation exist in the pre-Irrawaddian sediments in the Dry Zone W of the Irrawaddy, particularly where there is only a thin cover of sediments of the Irrawaddy Group or no cover at all. Most of the oil wells, however, which were drilled to the E of the Irrawaddy, found saline groundwater in the aquifers of the pre-Irrawaddian sediments. To the E of the Inner Volcanic Arc (Fig. 16), which is characterized here by the Mt. Popa Volcanics, the chance of tapping fresh groundwater from the pre-Irrawaddian Tertiary sediments seems to be better because these sediments are here exposed over large areas at the E edge of the Dry Zone and thus – as in the W – constitute a good recharge area.

In the Dry Zone, the fluviatile sediments of the **Irrawaddy Group** attain thicknesses of more than 2,000 m. In keeping with the conditions under which they were laid down, the facies changes very rapidly and irregularly in both the lateral and vertical directions from coarse clastic to fine clastic. As a result, the properties of the reservoir rock also vary so that no extensive and continuous aquifers or aquicludes have formed. Consequently, the groundwater regime is not uniform over large areas and this makes it difficult to prospect for groundwater reserves. In addition, although in general a certain amount of permeability exists, rocks with good reservoir properties have formed relatively infrequently in the Irrawaddy Group. Near-surface groundwater of good quality is encountered locally, but the quantities are small; deeper-lying groundwater in the Irrawaddy Group is frequently saline. Large quantities of groundwater probably exist in the areas of the groundwater runoff from the Irrawaddy River and its tributaries and possibly also in local depressions. In each case, major drilling campaigns to find groundwater should be preceded by systematic hydrogeological inventories and geoelectric surveys.

The widespread Pleistocene and Holocene alluvia in the countless meanders of former river courses offer better chances of finding groundwater than the Irrawaddy Group. Drilling has revealed that they are frequently more than 80 m thick, consist of clay and more or less argillaceous fine-grained, medium-grained and coarse-grained sands with occasional gravel beds. Their surface is mostly smooth and flat ("alluvial plains") and almost always covered by fine sandy soil. The pervious clastic sediments predominate; in general, they overlie the less pervious sediments of the Irrawaddy Group and are partially or entirely surrounded by the latter. This creates a favourable recharge situation: direct infiltration from rainfall (pervious soils), groundwater runoff from perennial and seasonal streams, water inflow from the surrounding less pervious sediments of the Irrawaddy Group. Test wells sunk for groundwater (TAHAL WATER PLANNING 1963a) in the alluvial deposits (e.g. in the Pagan alluvial plain) have shown that there are good chances of finding large quantities of groundwater in these sediments.

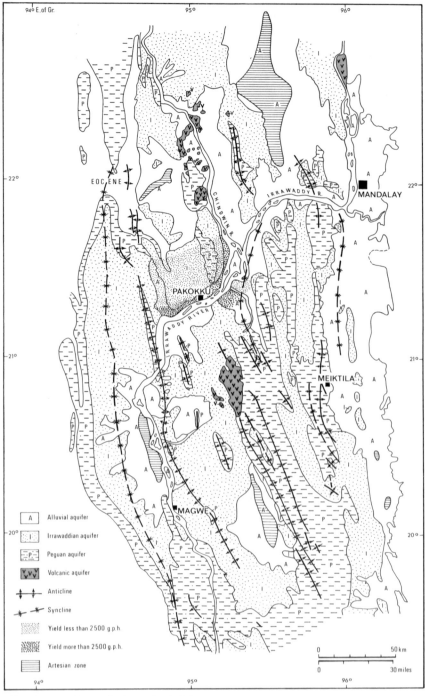

Fig. 91 Sketch map of some hydrogeological features in the Dry Zone (simplified and redrafted after Social. Rep. of Union of Burma UNICEF Colon Assignmt. report, 1980).

15 *

There is also some possibility of obtaining groundwater from the **volcanic rocks** in the Dry Zone. In the W part of the zone the volcanic rocks of the Lower Chindwin area occur together with tuff and basalt lava sheets (Fig. 61) which are locally interbedded as permeable rocks between the less permeable sediments of the Irrawaddy Group. As a result they form potentially good aquifers. The Mt. Popa volcanic rocks (p. 131) are located in the central E part of the Dry Zone. Freshwater springs occur on their slopes where permeable lava sheets overlie less permeable tuffs and impermeable sediments of the Irrawaddy Group. Since no springs emerge from the lower slopes of Mt. Popa, but the lava sheets and tuffs of the Irrawaddy Group extend into the plains far to the S and W, it is probable that further chances of finding groundwater exist here. The hydrogeological conditions in certain areas of the Dry Zone were recently studied (UNICEF 1980) and compiled in a hydrogeological map (Fig. 91).

Arakan Coastal Area

This area receives > 3,000 mm/a of precipitation. The large amount of surface runoff drains to the W from the Arakan Yoma Range, which runs parallel to the coast, into the coastal area of the Bay of Bengal. Fluviatile clastic sediments and marine sands overlie more or less structurally deformed alternating sequences of Tertiary shale and sandstone. The thick Miocene sandstone is a particularly good potential aquifer. At many points, e.g. in Akyab, groundwater is produced from the Recent and sub-Recent sands where fresh groundwater occurs above saline groundwater. As always in comparable hydrogeological situations, there is the danger that if too much groundwater is produced, salinization will occur. Once the boundaries between the saline and the fresh water have been determined by geoelectric surveys, a large part of the region's water requirements could be met by producing groundwater from these sandy aquifers.

Bassein-Rangoon-Irrawaddy Delta Area

This area receives > 2,500 mm/a of precipitation. It is covered by thick alluvia and by clastics of the Irrawaddy Group. W of Bassein, faulted and folded sequences of the pre-Irrawaddian plunge to the E beneath the younger sediments. A spur of the Pegu Anticline, with Miocene shale and sandstone, extends to the S as far as Rangoon, where it also disappears under the Irrawaddian further to the S. According to LEICESTER (1932), in the Bassein area, the older alluvia of the Lower Irrawaddy Delta Area, with several gravel-sand horizons, are good aquifers. The deeper aquifers contain brackish water. As in the other coastal areas of Burma and in the Dry Zone, the subterranean salt water/freshwater boundaries must be identified in order to be certain about the safe yield. Also, Recent or sub-Recent redeposited river sands in the groundwater runoff from the delta arms yield high-quality groundwater. According to OLDHAM (1893, pp. 58–63), the gravel layers in the older alluvia in the Rangoon area are to a large extent continuous, but their permeability varies rapidly and irregularly. Even in those days, the problem of salinization of the groundwater brought about by excessive production rates played a role.

As is the case in the Dry Zone, there are no extensive and uninterrupted aquifers and aquicludes in the fluviatile, thick deposits of the Irrawaddy Group in the delta area either. Locally, however, lenticular and "shoe-string" aquifers may have formed, and they may be potential sources of groundwater.

The Tertiary sequences beneath the Irrawaddy Group contain sandy aquifers which come to the surface N of Rangoon and W of Bassein and thus are in a favourable position for recharge. These aquifers have not yet been tested, although they are located just N of Rangoon and at the E edge of the Arakan Yoma just W of Bassein at depths which can be reached by shallow drilling.

Tenasserim Coastal Area

On average more than 4,000 mm of precipitation falls per annum in the Tenasserim coastal area between Moulmein in the N and Mergui in the S. Nevertheless, it is increasingly necessary to draw on the groundwater reserves of this region. The main reason for this is that the expanding coastal towns are situated on rivers which, as a result of the tidal inflows of sea water, contain brackish water only as far as 60 km inland (Salween River).

Close to the coast in the Tenasserim coastal area, Recent and sub-Recent fluviatile sediments lie directly on top of Paleozoic schist, argillite, greywacke, sandstone, quartzite and limestone (Mergui Series, Moulmein Limestone; p. 74) or above granitic rocks. Depending on the conditions under which they were laid down, the fluviatile sediments can be very argillaceous and thus yield little groundwater, as for example in the town of Moulmein (TAHAL WATER PLANNING 1963b). In other places (Tavoy, Mergui), fluviatile clastic accretions with good reservoir properties are found in the areas of river mouths. When large amounts of groundwater are withdrawn, it is naturally important here also to demarcate the salt water/freshwater boundaries, particularly since salinization can occur not only from the sea but also as a result of the groundwater runoff from the brackish rivers.

Among the Paleozoic rocks, only the usually karstified, massive limestone can be considered as a potential aquifer.

5.44 Soils and soil utilization

A number of authors have reported on the soils found in individual areas in Burma: ROMANIS (1881, p. 227), MCKERRAL (1910, pp. 5–7; 1911, pp. 4–7). LA TOUCHE (1913, pp. 322–325), WARTH (1916), WARTH & PO SHIN (1918, pp. 132–156), WARTH & PO SAW (1919, pp. 157–172), STAMP (1923, pp. 136–138), LORD (1924, pp. 4–8), CLARK (1930), BARRINGTON (1931, pp. 5–12), CHARLTON (1931, 1932). CHHIBBER (1934b, pp. 247–268) summarized the sporadic knowledge that had been gathered up to that time.

ROZANOV (1965) prepared the first schematic soil map of Burma (scale 1: 253,000) on the basis of a soil survey conducted in the early sixties and covering about 40% of the country (KARMANOV 1960, ROZANOV & ROZANOVA 1961, 1962).

General descriptions of the soils of Burma are also to be found in the "World Atlas of Agriculture" (International Association of Agricultural Economists/Istituto Geografico de Agostini, Novara, 1969–1976) and in the "Soil Map of the World" (UNESCO, Paris 1971).

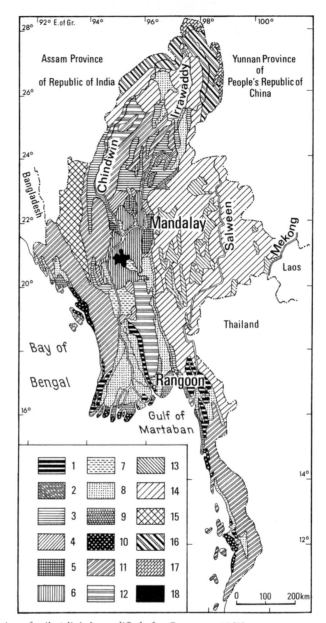

Fig. 92. Distribution of soils (slightly modified after ROZANOV 1965).

Plain area: 1: Lateritic Soils and Laterites, 2: Red-brown Forest Soils of tropical evergreen forests, 3: Yellow-brown Forest Soils of tropical monsoon forests, 4: Cinnamon-brown Soils of tropical dry shrub stands, 5: Dark Compact Soils of tropical dry savannas, 6: Red-brown Soils of tropical dry savannas, 7: Primitive Gravelly Soils of tropical dry savannas.

River valleys and lowlands: 8: Complex of Meadow, Bog, and Alluvial Soils of river valleys, 9: Salinized Soils of maritime lowlands, 10: Salty muds of tropical mangrove forests.

Mountain areas: 11: Mountain Red-brown Forest Soils of tropical evergreen forests, 12: Yellow-brown Mountain Forest Soils of tropical monsoon forests, 13: Red Earths and Yellow Earths, 14: Mountain Red Earths, 15: Complex of Mountain Sod, Meadow, and Red-Brown Forest Soils, 16: Brown Mountain-Forest Soils of temperate coniferous-broadleaf forests, 17: Mountain-Meadow Alpine Soils, 18: Mountain Sod Soils.

ROZANOV (1965) distinguished between 7 soil types in plain areas, 3 in river valleys and lowlands and 8 in mountain areas (Fig. 92). Depending on the main climate zones and the height above sea level, the following are the most important soils according to ROZANOV's classification:

5.441 Soils of the Wet Tropical Monsoon Zone

In the evergreen jungle, **Red-brown Forest Soils** have developed mostly at altitudes between 300 m and 1,300 m above sea level. Their profiles attain thicknesses up to about 2 m. In the uppermost part they are mostly of dark reddish brown colour due to the high humus content, while in the lower portions the colour is light red to orange. The moderately to highly loamy, frequently stony soils merge into the parent material without clear signs of an illuviation process (B horizon). Usually they have an average humus content of only 2%. The exchange capacity is 10–15 meq/100 g, gradually decreasing with depth; the pH value is about 5.5–6.5.

Because of the usually steep slopes on which they occur, these soils are not very suited for agricultural purposes; on the other hand, they are well suited for silviculture. In less steep locations they are used for horticulture and plantations.

Yellow-brown Forest Soils are widespread at altitudes between about 100 m and 450 m above sea level, and in this zone they replace the Red-brown Forest Soils that occur at higher altitudes. The Yellow-brown Forest Soils occur in the lower hills of the Pegu Yoma and in the foothills of the Arakan Yoma, of the Tenasserim Yoma and in hilly parts of Central Burma to about 25° N. They are up to 2 m thick, friable and well structured. The clay content increases with depth and attains its maximum in the B horizon; the humus content is on average between 2 and 4%. In general the soils are light textured, but also loamy to clayey, and they occur over marl and mudstone. The sum of exchangeable bases ranges from 5 to 15 meq/100 g, and the pH values from 5.0 to 6.5. The water-holding capacity is 30–35%, and this together with the high porosity (40–50%) results in deep moisture penetration during the rainy season.

The Yellow-brown Forest Soils are best suited for forestry purposes, but in topographically favourable locations they are also used for horticulture and plantations.

Lateritic Soils are found in Lower Burma to the S of the Pegu Yoma and in the Tenasserim area, particularly near Moulmein, mostly at altitudes below 100 m above sea level. They are of a reddish-yellow colour, up to several metres thick and gradually merge with depth into the laterite.

Because of its high humus content (1.5–3%), the A horizon is somewhat darker in colour and relatively rich in nitrogenous compounds. Its C : N ratio varies from 9 to 12 and becomes more narrow with depth.

In the B horizon, which shows no signs of an illuviation process, the humus content drops abruptly. The underlying horizon is rich in small, rounded concretions and sesquioxides. Weathered, mostly vesicular-textured mottled laterite, which may be up to 15 m thick, leads over into the original laterite. When soil erosion occurs, the weathered laterite is frequently exposed and can be relatively easily excavated. When it is exposed to atmospheric weathering, it becomes hard and is then used for road construction and as building material (p. 207). The lateritic soils are very poor in available phosphorus but usually moderately high in potassium content. The sum of exchangeable bases is only in the range of 2 to 5 meq/100 g; pH values range from 4.5–5.0, slightly increasing with depth.

These soils are well suited for rubber tree plantations, fruit growing and horticulture.

5.442 Soils of the Dry Tropical Monsoon Zone

Cinnamon Soils are restricted to a relatively narrow belt surrounding the Dry Zone (p. 15). A distinction can be made between Dark Cinnamon Soils, Cinnamon Soils and Light Cinnamon Soils, and it is found that the darker colours develop at relatively wet sites. The humus content of these slightly to moderately loamy, and in the upper part often sandy, soils varies between 1% (Light Cinnamon Soil) to 3–5% (Dark Cinnamon Soil). The C : N ratio may exceed 20, and the available phosphorus and potassium contents are fairly high. The sum of exchangeable bases varies from 2–5 meq/100 g in light clay soils to 10–15 meq/100 g in dark ones; pH ranges from 5.5 to 6.5.

These fertile soils are suited for agricultural purposes when properly irrigated.

Red-brown Savanna Soils are typical of the Dry Zone (precipitation < 1,250 mm/a) in Central Burma (p. 217). The colours of this soil complex vary between dark-chestnut – red – brown-red and orange. There are frequent enrichments of carbonate within the soil profile; iron-calcareous concretions are found in the lower portions. The humus contents are below 2%; nitrogen and available phosphorus compounds are low. The base exchange capacity varies between 10 and 30 meq/100 g, and the pH between 5.0 and 7.0 (in extreme cases it may reach 9.0).

Red-brown Savanna Soils are well suited for dry farming. They are particularly endangered by soil erosion.

Dark Compact Savanna Soils are found chiefly on flat, even terrain and above alluvia in the area of the Dry Zone of Central Burma. They resemble the "Black Cotton Soils" in India. Despite the black to dark-grey-brown colours, their humus content is barely 1%. The heavy, clayey soils are calcareous and show horizons with carbonate enrichments. They possess high base exchange capacities. When they dry out, these soils become very hard and exhibit deep mud cracks; rain turns them into mud.

They are not very suited for farming; given suitable irrigation they can be used to grow paddy rice (lowland rice) and chilly.

The regions with Dark Compact Savanna Soils also contain small areas with **Solonetz** (alkaline) and **Solonchak** (saline) soils.

In the Dry Zone, **"Primitive Crushed Stone Soils"** are also found; they consist essentially of stone fragments with small residues of humus soils and are mostly covered by calcareous and limonitic concretions. These soils are not suitable for agriculture.

5.443 Soils of the Subtropical Monsoon Zone

Large parts of the Shan Limestone Plateau are covered by **Red Earths** at altitudes around 1,000 m ab.s.l. They extend to the N to the mountainous area N of Myitkyina. They are occasionally up to 15 m thick; their texture is friable and easily penetrable for water and air. In the light Red Earths the humus content is between 2 and 4% and in dark Red Earths it may attain up to 8%. The C : N ratio varies between 10 and 15. The amounts of available nitrogen, phosphorus and potassium vary greatly, but they are higher than those in the soils of the tropical zone. The base exchange capacity is between 2 and 5 meq/100 g, the pH varies between 6.0 and 7.0.

These soils are well suited for growing wheat, maize, soybean, groundnut, potatoe, also for tea, coffee and pineapple plantations, in addition for the cultivation of dry paddy rice (upland rice) (diversified farming).

Yellow Earths occur on level surfaces on slopes at lower altitudes on the Shan Plateau. They are bright yellow to reddish-yellow in colour, less structured and more compact than Red Earths. They are slightly less suited for agriculture than the Red Earths.

5.444 Soils of the Temperate High Mountain Zone

The soils at sites > 3,000 m ab.s.l. have as yet nowhere been studied or classified. According to observations made by KINGDON-WARD (1949), the Red Earths are gradually replaced at higher elevations by **Brown Mountain Forest Soils** and by **Brown Mountain Meadow Soils** in the alpine zone.

5.445 Azonal soils

Two different soil complexes exist along the coast and in the areas of the river deltas, namely soils which are affected by the daily tide and soils which are affected only by high tides. The former type consists of saline mud, which offers a good substrate for certain Rhizophora. The second type can be used for cultivating paddy rice if they are protected from the high tides.

5.446 Intrazonal soils

Under this heading ROZANOV (1965) grouped the complex of the Meadow, Swampy and Alluvial Soils of river valleys.

The following soils have formed in Lower Burma and in the Arakan and Tenasserim coastal areas: Meadow Alluvial Soils, Meadow Soils, Meadow Gley Soils, Meadow Saline Soils, Meadow Degraded Soils, Meadow Swampy Soils and Swampy Soils. They are usually very gleyized, impermeable and acid to extremely acid. Most of these soils are structureless, compact and low in humus and nutrients. They are used almost exclusively for the cultivation of paddy rice.

In the Dry Zone, this complex of soils consists of Alluvial Soils, Meadow Alluvial Soils, Meadow Gley Soils, Meadow Alkaline Soils and Meadow Saline Soils and they are all more light textured and less impervious here than in the coastal area. They are only weakly acid to neutral, and locally even alkaline. With the exception of the Meadow Alluvial Soils they are also low in humus content and nutrients. They are better suited for diversified agriculture than the Intrazonal Soils in the coastal area.

The Intrazonal Soils in the area of the Shan Plateau and in Upper Burma can be subdivided into Meadow Soils, Meadow Swampy Soils, Meadow Swampy Carbonate Soils and Swampy Soils. These mostly black, heavy soils are usually well structured; their humus content is high. When suitably meliorated, agricultural utilization is possible.

Bibliography

ABDEL-GAWAD, M. (1971): Wrench movements in the Baluchistan Arc and relation to Himalayan – Indian Ocean tectonics. – Geol. Soc. Amer. Bull. **82**: 1235–1250.

ADAM, J. (1832): Account of Barren Island in the Bay of Bengal. – J. Asiat. Soc. Bengal, **1**: 128–131.

ADAM, J. W. H. (1960): On the geology of the primary tin-ore deposits in the sedimentary formation of Billiton. – Geol. en Mijnb., **39**, 10: 405–423.

ADAMS, F. D. (1926): A visit to the gem districts of Ceylon and Burma. – Bull. Canad. Inst. Miner. Met. **39**: 213–246.

ADAMS, F. D. & R. P. D. GRAHAM (1926): On some minerals from the Ruby Mining District of Mogok, Upper Burma. – McGill Univ. Publ., 5 Ser., No. 13. – In: Transact. Roy. Soc. Canada, Sect. 4: 113–136, 6 figs., Ottawa.

ADAMS, T. (1977): The mineral industry of Burma. – U.S. Bur. Mines, Miner. Yearb., **3**, 207–213.

AHLFELD, F. (1938): Die Erzlagerstätten Chinas. – Met. Erz, **35**: 245–252, Halle.

AHMAD, I. (1965): A new genus and species of Micrelytrinae from Upper Burma (Hemiptera: Alylidae). – Proc. Roy. Entomol. Soc. London, Ser. B, **34**, 11–12: 137–140, 7 figs.

D'AMATO, G. (1833): Short description of the mines of precious stones, in the District of Kyatpyen in the Kingdom of Ava. – (Transl.), J. Asiat. Soc. Bengal, **2**: 75–76..

AMOS, B. J. (1975): Stratigraphy of some of the Upper Palaeozoic and Mesozoic carbonate rocks of the Eastern Highlands, Burma. – Newsl. Stratigr., **4**, 1: 49–70, 6 figs., 1 pl., 2 tables.

ANDERSON, J. (1870): The Irrawaddy and its sources. – J. R. Geogr. Soc., **60**: 286–303.

– (1871): A report on the Expedition to Western Yunnan via Bhamo. – 458 pp., Calcutta.

– (1876): Mandalay to Momein, a Narrative of the Two Expeditions to Western China of 1868 and 1875. – 479 pp., London.

ANDERSON, M. M., BOUCOT, A. J. & J. G. JOHNSON (1969): Eifelian Brachiopods from Padaukpin, Northern Shan States, Burma. – Bull. Brit. Mus. Nat. Hist. Geol., **18**, 4: 107–163, 10 figs., 10 pls.

ANNANDALE, N. (1919): The Gastropod fauna of Old Lake Beds in Upper Burma. – Rec. Geol. Surv. India, **50**, 3: 209–240, 3 pls.

– (1923): An introductory account of the Inle Lake. – Rec. Indian Mus., **14**: 1–7.

– (1924): Fossil molluscs from the oil measures of the Dawna Hills, Tenasserim. – Rec. Geol. Surv. India, **55**, 2: 97–104, 2 pls.

ANNANDALE, N., BROWN, J. C. & F. H. GRAVELY (1913): The limestone caves of Burma and the Malay Peninsula. – J. Asiat. Soc. Bengal, N.S., **9**: 391–423.

Anonymus (1972): Burma's tin potential outline. – Min. J. (G.B.), **279**, 7158: 332, London.

Anonymus (1974): Nickel in Burma. – Min. J. (G.B.), **282**, 7233: 259, London.

Anonymus (1974): Copper mines for Burma? – Min. J. (G.B.), **283**, 7261: 340, London.

ANTHONY, H. (1941): Mammals collected by the Vernay Cutting Burma Expedition. – Zool. ser. Field Manuals. Nat. Hist., **27**: 37–121.

ASAMA, K. (1966): Geology and Paleontology of Southeast Asia, ed. by KOGAYASHI, T. & R. RARIVANA. – 2, 197 pp., Univ. of Tokyo Press.

– (1968): Permian plants from Phetchabun, Thailand, and problems of floral migration from Gondwanaland. – Contr. to the Geol. and Paleont. of Southeast Asia, XXV. – Nat. Sci. Mus., Tokyo, Bull., **9**, 2: 171–209.

AUBERT, L. (1914): The petroleum wells of Yenangyat. – J. Soc. Burma Res. Soc., **4**, 2.

AUNG KHIN (1966): The petroleum industry in Burma. – 1st, Burma Res. Congr., Rangoon.

AUNG KHIN, AUNG TIN U, AUNG SUE & KHIN HAN (1970): A study on the gravity indication of the Shan Scarp Fault. – Union of Burma J. Sci. Technol. (3rd Burma Res. Congr.; 1968), **3**, 91–113, 10 figs., Rangoon.

AUNG KHIN & KYAW WIN (1969a): Preliminary studies of the paleogeography of Burma during the Cenozoic. – Union of Burma J. Sci. Technol. (2nd Burma Res. Congr.; 1967), 1, 2: 241–251, 4 figs., Rangoon.

– – (1969b): Geology and hydrocarbon prospects of the Burma Tertiary Geosyncline. – Union of Burma J. Sci. Technol. (1st Burma Res. Congr.; 1966), 2, 1: 53–81, 5 figs., 4 tables, Rangoon.

AUNG NYUN (1955): Report on the drilling for gypsum at Hkaleng near Hsipaw, Northern Shan States. – Unpubl. Rep., Petrol. Min. Resourc. Develop. Co. Burma, Rangoon.

– (1959): The geology of Heinda Mine, Tavoy District. – Unpubl. Rep., Burma Geol. Dep., Rangoon.

BA KYI (1964): Hydrological work in Burma. – In: Methods Hydrol. Forecasting Utiliz. Wat. Resources, Trans. Inter.-region Seminar, Bangkok, Thailand, 1964. – S.L. United Nations, No. 27: 73–79, 1 fig.

BA MAW (1971): Occurrence of the marine fauna in the formation of Pegu Group from the Kyaukka-Thazi Area, Monywa Districts. – J. Sci. Technol. Burma, 4, 1: 43, Rangoon.

BA MAW, BO SAN, ROSS, J. R. P. & R. L. CIOCHON (1976): The Ordovician Bryozoan (Ectoproct) *Diplotrypa* from Central Burma. – Geol. Mag. (G.B.), 113, 6: 515–518, 1 pl.

BA MAW, CICHON, R. L. & D. E. SAVAGE, (1979): Late Eocene of Burma yields earliest anthropoid primate, *Pondaungia cotteri*. – Nature, 282, 5734; 65–67.

BA THAN HAQ (1966): The geology and economic possibilities of the area East of Pyinmana (in Burmese). – 1st, Burma Res. Congr. 1966, Rangoon.

– (1967): Geology and structures of the Eastern margin of the Burma Tertiary Basin. – 2nd Burma Res. Congr., Rangoon.

– (1970): Metallogenic provinces of Burma. – 3rd, Burma Res. Congr., Rangoon.

– (1972): Tectonics and metallogenic provinces of Burma. – Geol. Soc. Malaysia, Newsletter Malays. (Reg. Conf. Geol. Southeast Asia), Abstr., No. 34: 3, Kuala Lumpur, Malaysia.

BA THAN HAQ & D. L. SEARLE (1961): The geology and economic possibilities of the area between the Pawn Chaung and Salween River, South East of Loikaw, Kayah State. – Burma Res. Soc., 44, 1: 13–23, Rangoon.

BA THAN HAQ et al. (1970): Geology and economic possibilities of the Myogi-Yeywa Area, East of Kyaukse. – 3rd Burma Res. Congr., Abstr.: 100, Rangoon.

BALASUNDARAM, M. S. & M. N. BALASUBRAHMANYAN, (1973): Geochronology of the Indian Precambrian. – Geol. Soc. Malaysia, Bull., 6: 213–226.

BALL, V. (1870): Brief notes on the geology and on the fauna in the neighbourhood of Nancowry Harbour, Nicobar Island. – J. Asiat. Soc. Bengal, 39: 25–37.

– (1873): Barren Islands and Narcondam. – Rec. Geol. Surv. India, 6, 4: 81–90.

– (1875): On some stone implements of the Burmese type found in Pargana Dalbhum, District of Songhbum, Chota-Nagpur Division. – Proc. Asiat. Soc. Bengal: 118–125.

– (1879): On the volcanoes of the Bay of Bengal. – Geol. Mag., 6: 16–27.

– (1881): A Manual of the Geology of India, Part 3, Economic Geology. – 663 pp., Calcutta.

– (1886): The mineral resources of India and Burmah, being a lecture delivered at the Colonial and Indian Exhibition on the 5th June 1886. – Min. J., 56: 674–675.

– (1893): On the volcanoes and hot springs of India, and the folklore connected there-with. – Proc. Roy. Irish Acad.: 151.

BALL, V. & R. R. SIMPSON (1913): The coal fields of India. – Mem., Geol. Surv. India, 46: 1–47.

BANNERT, D. & D. HELMCKE (1981): The evolution of the Asian Plate in Burma. – Geol. Rdsch., 70, Stuttgart.

BARAZANGI, M. & J. DORMAN (1969): World seismicity maps compiled from E.S.S.A. Coast and Geodetic Survey Epicenter Data for 1961–1967. – Bull. Seismol. Soc. Amer., 59, 1: 369–380, 7 figs., Baltimore.

BARBER, C. T. (1935): The natural gas resources of Burma. – Mem. Geol. Surv. India, 66 1: 1–172, pl. 14, Calcutta.

– (1936): The Tertiary igneous rocks of the Pakokku District and the Salingyi Township of the Lower Chindwin District, Burma, with special reference to the determination of the feldspars by the Fedoroff Method. – Mem. Geol. Surv. India, 68, 2: 121–292, pl. 11, 30 figs., 13 tables, Calcutta.

BARRINGTON, A. H. M. (1931): Forest soil and vegetation in the Hlaing Forest Circle, Burma. – Burma Forest Bull., 25: 5–12, Rangoon.

BASTIN, H., v. BRAUN, E., HESS, A., KOCH, K. E., STEIN, V., STOPPEL, D. & R. WOLFART (1970): Silurian and Early Devonian biostratigraphy in Northwest Thailand. – Newsl. Stratigr., 1, 2: 25–32.

– – – – – – – (1977): The Silurian-Devonian Boundary, Thailand. – IUGS Ser. A, No. 5: 238–244.

BATESON, J. H., MITCHELL, A. H. G. & D. A. CLARKE, (1972): Geological and geochemical reconnaissance of Seikpudaeng-Padatgyaung area of Central Burma. – Inst. Geol. Sci. London, unpubl. Rep. No. 25.

BAUER, M. (1895): On the jadeite and other rocks from Tawman in Upper Burma. – Rec. Geol. Surv. India, 28, 3 (transl. by F. NOETLING and H. H. HAYDEN): 91–105.

– (1896): Über das Vorkommen der Rubine in Burma. – N. Jb. Miner. Geol. Petrefact., 2: 197–238.

– (1906): Weitere Mitteilungen über den Jadeit von Ober-Birma. – Centralbl. Miner., 7: 97–112.

BAUM, F., v. BRAUN, E., HAHN, L., HESS A., KOCH, K. E., KRUSE, G., QUARCH, H. & M. SIEBENHÜNER (1970): On the geology of Northern Thailand. – Beih. geol. Jb., 102, 23 pp., 1 geol. map, Hannover.

BAUM, F. & K. E. KOCH (1968): Ein Beitrag zur stratigraphischen Neuordnung des Paläozoikums in Süd-Thailand. – Geol. Jb., 86: 879–884.

BEADON, C. (1924): in G. P. DE COTTER: Mineral Deposits of Burma. – Govt. Rangoon: Printer.

BECHSTÄDT, T. & H. MOSTLER, (1976): Riff-Becken-Entwicklung in der Mitteltrias der westlichen Nördlichen Kalkalpen. – Z. dt. geol. Ges., 127: 271–289, 6 figs., Hannover.

BEDFORD, H. (1831): Extract from the Journal of Apothecary H. BEDFORD, Deputy to Yenangyoung in Ava, search of fossil remains. – Gleaning in Sci., Calcutta, 3: 168–170.

BEETS, C. (1949): On the occurrence of *Biplanispira* in the Uppermost Eocene (Kyet-u-bok Band) of Burma. – Geol. en Mijnb., N.S., 11, 7: 229–232, 1 fig.

BENDER, F. (1974): Entwicklungshilfe und Rohstoffversorgung am Beispiel Birma. – Echo aus Deutschland, 11, 4: 16–17, Mainz (Verl. Echo); Nachr. dtsch. Geol. Ges. (126. Jahrestagung der DGG, Bonn, 1974), 11: 6–8, Hannover.

BENTHAM, R. (1966): Oil in Burma – a study of exploration and development. – Inst. Petrol. Rev., London 20: 247–249.

BEQUAERT, J. (1943): Fresh-water shells from cave deposits in the Southern Shan States, Burma. – Trans. Amer. Phil. Soc., N.S., 32: 431–436, pl. 33.

BERRY, W. B. N. & A. J. BOUCOT (1972): Correlation of the Southeast Asian and Near Eastern Silurian rocks. – Geol. Soc. Amer., Spec. Pap. 137: 65 pp.

BGR = Bundesanstalt für Geowissenschaften und Rohstoffe (1976–1981): Laboratory analyses. – Hannover, Fed. Rep. of Germany

BION, H. S. (1913): The gold-bearing alluvium of the Chindwin River and tributaries. – Rec. Geol. Surv. India, 43, 4: 241–263, 3 pls.

BLANFORD, W. T. (1882): Account of a visit to Puppa Doung, an extinct volcano in Upper Burma. – J. Asiat. Soc. Bengal, 1862, 31: 215–216 (Rep. of the Brit. Ass. for the Advance of Sci., 1860, 2: 69–70) reprinted in Geol. Pap. on Burma.

– (1895): The Burmese chipped flints Pliocene, not Miocene. – Nature, 51: 608.

BLEECK, A. W. G. (1907): Die Jadeitlagerstätte in Upper Burma. – Z. prakt. Geol., 15: 341–365.

– (1908a): Jadeite in the Kachin Hills, Upper Burma. – Rec. Geol. Surv. India, 36, 4: 254–285.

– (1908b): Rubies in the Kachin Hills, Upper Burma. – Rec. Geol. Surv. India, 36: 164–170.

– (1913): On some occurrences of wolframite lodes and deposits in the Tavoy District of Lower Burma. – Rec. Geol. Surv. India, 43, 48–73.

BOGLE, A. (1814): Note of petroleum in Ramree Island. – Calcutta J. Nat. Hist. 1: 562.

– (1844a): On the appearance of a new volcanic island on the Arakan Coast. – Calcutta J. Nat. Hist., 4: 455.

– (1844b): Account of the eruption of False Island, Ramree. – J. Asiat. Soc. Bengal, 13, Proc.: 35.

– (1845): Note on a supposed submarine eruption off the Coast of Arracan. – J. Asiat. Soc. Bengal, 14, Proc.: 28–29.

BOSE, P. N. (1893a): Note on granite in the Districts of Tavoy and Mergui. – Rec. Geol. Surv. India, 26, 3: 102–103.

– (1893b): Notes on the geology of a part of the Tenasserim valley with special reference to the Tendau-Kamapying coal-field. – Rec. Geol. Surv. India, 26, 4: 148–164, Calcutta.

BOUCOT, A. J., JOHNSON, J. G. & J. A. TALENT (1967): Lower and Middle Devonian faunal provinces based on Brachiopoda. – Internat. Symp. on the Devonian System, Calgary, Alberta 1967, 2: 1239–1254. – Alberta Soc. Petrol. Geol., Calgary.

BRAEUNING, E. (1975): Die Thermalquelle von Lashio/Nordbirma. – Heilbad Kurort, 27, 1: 32–33, 2 figs., Bonn.

v. BRAUN, E., BESANG, C., EBERLE, W., HARRE, W., KREUZER, H., LENZ, H., MÜLLER, D. & I. WENDT (1976): Radiometric age determinations of granites in Northern Thailand. – Geol. Jb., B 21: 171–204, Hannover.

v. BRAUN, E. & R. JORDAN (1976): The stratigraphy and paleontology of the Mesozoic Sequence in the Mae Sot Area in Western Thailand. – Geol. Jb. B 21: 5–21, 7 figs., 1 table, 7 pls., Hannover.

BREWSTER, D. (1834): Notice respecting a remarkable specimen of amber from Ava. – Rep. Brit. Assoc., 4: 574; Proc. Roy. Soc. London.

BRINCKMANN, J. & C. HINZE (1976): Concluding report on lead-zink-silver-ore exploration in the vicinity of the Bawdwin Mine, Shan State (NE Burma). – Unpubl. rep. BGR Hannover.

BRINCKMANN, J. & C. HINZE (1981): On the geology of the Bawdwin Lead-Zinc Mine, Northern Shan State – NE-Burma. – Geol. Jb., D 43: 7–46, 1 table, 3 pls., 2 figs., Hannover.

BRÖNNIMANN, P., WHITTAKER, J. E. & L. ZANINETTI (1975): Triassic foraminiferal biostratigraphy of the Kyaukme-Langtawkno area, Northern Shan States, Burma. – Riv. Ital. Paleont., 81, 1: 1–30.

BROMLEY, A. H. (1896): Notes upon gold mining in Burma. – Trans. Inst. Min. Engin., 12: 506–514.

BROWN, C. B. & J. W. JUDD (1895): The rubies of Burma and associated minerals; their mode of occurrence, origin and metamorphoses, a contribution to the history of corundum. – Phil. Trans. Roy. Soc. London, Ser. A, 187: 151–228, London.

BROWN, G. F., BURAVAS, F., CHARAIJAVANAPHET, J., JALICHANDRA, N., JOHNSTON, W. D., STRESTHRAPURA, V. & G. C., TAYLOR (1951): Geological reconnaissance of the mineral deposits of Thailand. – U.S. Geol. Surv. Bull. 984, Roy. Dep. Mines Geol. Surv. Thail., Mem. 1, 183 pp., 20 pls., 38 figs., 4 tables.

BROWN, J. C. (1909a): Recent accounts of the mud volcanoes of the Arakan Coast, Burma. – Rec. Geol. Surv. India, 37, 3: 264–279.

– (1909b): Stone implements from the Tenyueh District, Yunnan Province, Western China. – J. Asiat. Soc. Bengal, N.S., 5: 299–305.

– (1912a): Report on certain gold-bearing deposits on Mong-Long, Hsipaw State, Northern Shan States, Burma. – Rec. Geol. Surv. India, 42, 1: 37–51.

– (1912b): Fiery eruption of a mud volcano on Foul Island, Arakan Coast, Burma. – Rec. Geol. Surv. India, 42, 4: 279.

– (1912c): Eruption of a submarine mud volcano off Sandoway, Arakan Coast, Burma. – Rec. Geol. Surv. India, 42, 4: 278.

– (1916a): Contributions to the geology of the Province of Yunnan in Western China: 5. Geology of parts of the Salween and Mekong Valleys. – Rec. Geol. Surv. India, 47, 4: 205–266.

– (1916b): A note on the iron ore deposits of Tweingang, Northern Shan States. – Rec. Geol. Surv. India, 47: 137–141.

– (1917): Solubility of tungsten minerals. – Min. Sci. Press, 115: 302–303, San Francisco.

– (1918a): Geology and ore deposits of Bawdwin Mines. – Rec. Geol. Surv. India, 48, 3: 121–178.

– (1918b): The cassiterite deposits of Tavoy. – Rec. Geol. Surv. India, 49, 1: 23–33.

– (1919a): Tungsten ore deposits of Burma. – Min. Mag. (G.B.), 22, 4.

– (1919b): The genesis of tungsten ores. – Geol. Mag., 6: 44–46.

– (1923a): Contributions to the geology of the Province of Yunnan in Western China: 6. Traverses between Tali Fu and Yunnan Fu. – Rec. Geol. Surv. India, 54, 1: 68–86.

– (1923b): Contributions to the geology of the Province of Yunnan in Western China: 7. Reconnaissance surveys between Shunning Fu, Puerh Fu, Ching-tung Ting and Ta-li-Fu. – Rec. Geol. Surv. India, 54, 3: 296–323.

– (1923c): Contributions to the geology of the Province of Yunnan in Western China: 8. A traverse down the Yang-tzechiang Valley from Chin-Chiang-Kai to Hi-li-Chou. – Geol. Surv. of India, 54: 324–336.

– (1923d): The mines and mineral resources of Yunnan, with short accounts of its agricultural products and trade. – Mem. Geol. Surv. India, 47: 1–201.

– (1924a): A geographical classification of the mineral deposits of Burma. – Rec. Geol. Surv. India, 56, 1: 65–108.

– (1924b): Submarine mud eruptions of the Arakan Coast, Burma. – Rec. Geol. Surv. India, 56, 3: 250–256.

- (1929): The iron ore deposits of the Northern Shan States. – Rec. Geol. Surv. India, **61**, 2: 180–195.
- (1932): The geology and lead ore deposits of Mawson, Federated Shan States. – Rec. Geol. Surv. India, **65**, 3: 394–433.
- (1935): Gold in Burma and the Shan States. – Min. Mag., **52**: 9–20, 82–92.
- (1936a): India's Mineral Wealth. – Calcutta.
- (1936b): Contributions to the geology of the Province of Yunnan in Western China: 9. The Brachiopod Beds of Liuwun and related formations in the Shan States and Indo-China. – Rec. Geol. Surv. India, **71**, 2: 170–216.
- (1938): Contributions to the geology of the Province of Yunnan in Western China: 10. The distribution, age and relationships of the Red Beds. – Rec. Geol. Surv. India, **73**, 2: 514–578.
- (1944): Ruby mining in Upper Burma. – Min. Mag., **68**, 6: 329–340.
- (1950: Contributions to the geology of the Province of Yunnan in Western China: 11. The regional relationships of the Ordovician faunas. – Rec. Geol. Surv. India, **81**: 321–376.
BROWN, J. C. & A. K. BENERI (1931–32): General report for 1930. – Rec. Geol. Surv. India, **65**, 5.
BROWN, J. C. & A. K. DEY (1955): India's Mineral Wealth. – 3rd Ed., Oxford Univ. Press, 761 pp.
- – (1975): The mineral and nuclear fuels of the Indian Subcontinent and Burma. – 536 pp., 75 figs., 19 pls., Delhi: Oxford Univ. Press.
BROWN, J. C. & A. M. HERON (1919): The distribution of ores of tungsten and tin in Burma. – Rec. Geol. Surv. India, **50**, 2: 101–121.
- – (1923a): The geology and ore deposits of the Tavoy District. – Mem., Geol. Surv. India, **44**, 2: 167–354, pl. 8.
- – (1923b): The Northern extension of the wolframite-bearing zone in Burma. – Rec. Geol. Surv. India, **54**, 2: 235–237.
BROWN, J. C. & F. R. C. REED (1913a): Contributions to the geology of the Provinces of Yunnan and Western China: 1. The Bhamo-Teng Yuen Area. – Rec. Geol. Surv. India, **43**, 3: 173–205.
- – (1913b): Contributions to the geology of the Province of Yunnan: 2. Notes on the stratigraphy of the Ordovician and Silurian Beds of Western Yunnan. – Rec. Geol. Surv. India, **43**, 4: 327–334.
- – (1913c): The origin of the wolfram-bearing veins of the Tavoy District. (Summary only). – J. Asiat. Soc. Bengal, N.S., **13**, 2: 202–203.
BROWN, J. C. & V. P. SONDHI (1933a): Geological reconnaissance in the Southern Shan States. – Rec. Geol. Surv. India, **67**: 135–165.
- – (1933b): The geology of the country between Kalaw and Taunggyi, Southern Shan States. – Rec. Geol. Surv. India, **67**: 166–248.
BRUNNSCHWEILER, R. O. (1966): On the geology of the Indo-Burma Ranges. – Geol. Soc. Australia, **13**, 1: 127–194, 17 figs., Adelaide.
- (1970): Contributions to the post-Silurian geology of Burma (Northern Shan States and Karen State). – Geol. Soc. Australia, **17**: 59–79, 10 figs., Sydney.
- (1974): Indo-Burman Ranges. – In: SPENCER, A. M.: Mesozoic-Cenozoic Orogenic Belts. – 809 pp., Edingburgh: Scottish Academic Press.
BRUNNSCHWEILER, R. O. & K. S. RODOLFO (1970): Geology of Andaman Basin and Burma, Discussion and reply. – Geol. Soc. Amer., Bull., **81**, 6: 1847–1854.
BUCKLAND, W. (1928a): Geological account of a series of animal and vegetable remains, and of rocks, collected by J. CRAWFURD Esq., on a Voyage up the Irrawaddi to Ava. – Trans. Geol. Soc. London, ser. 2, **2**: 377–392.
- (1928b): Supplementary remarks of the supposed power of the waters of the Irrawaddi, to convert wood to stone. – Geol. Trans., ser. 2, **2**: 403–404.
BUCKMAN, S. S. (1915): The Brachiopoda of the Namyau Beds of Burma. – Preliminary notice. – Rec. Geol. Surv. India, **45**, 1: 75–81.
- (1917): The Brachiopoda of the Namyau Beds, Northern Shan States, Burma. – Pal. Ind., N.S., **3**, 2: 254 pp., 21 pls., 6 figs.
BUFFETAUT, E. (1977): Données nouvelles sur les Crocodiliens Paleogènes du Pakistan et de Birmanie. – C. R. Acad. Sci. Paris, D, **285**, 8: 869–872.
- (1978): A Dyrosaurid (Crocodylia, Mesosuchia) from the Upper Eocene of Burma. – N. Jb. Geol. Pal., Mh., **5**: 273–281, 2 figs.
BUNDESANSTALT FÜR BODENFORSCHUNG (BfB) (1970–1975): Unpubl. reports and laboratory analyses. – Hannover, Fed. Rep. of Germany/Rangoon, Soc. Rep. of the Union of Burma.

Bundesanstalt für Geowissenschaften und Rohstoffe (BGR) (1976–1981): Unpubl. reports and laboratory analyses. – Hannover, Fed. Rep. of Germany/Rangoon, Soc. Rep. of the Union of Burma.

Bunopas Sangad & P. Vella (1979): Late Paleozoic and Mesozoic structural evolution of Northern Thailand – A Plate Tectonic Model. – 3rd Reg. Conf. on Geol. and Min. Res. of Southeast Asia Bangkok, Thailand 1978, pp. 133–140.

Burk, C. A. & C. L. Drake (1974): The Geology of Continental Margins. – 1009 pp., Berlin-Heidelberg-New York: Springer.

Burma Meteorological Department (1953): Statement showing the monthly and annual rainfall at rainrecording stations in Burma for the years 1949, 1950, 1951. – Rangoon: Govt. Print. and Stationary.

Burney, H. (1832): Note on platina ore from Ava. – J. Asiat. Soc. Bengal, 1: 16–17.

– (1837): Note on a fossil *Hippopotamus* from Burma. – J. Asiat. Soc. Bengal, 6: 1099.

Burri, C. (1931): Pyroxenolith aus dem Vulkangebiet des Lower Chindwin River, Upper Burma. – Schweiz. Miner. Petrogr. Mitt., 11: 417–418.

Burri, C. & H. Huber (1932): Geologie und Petrographie des jungvulkanischen Gebietes am Lower Chindwin (Upper Burma). – Schweiz. Miner. Petrogr. Mitt., 12: 286–344.

Burton, C. K. (1967a): Graptolite and Tentaculites, correlation and paleogeography of the Silurian and Devonian in the Yunnan-Malayan Geosyncline. – Trans. Proc. Palaeont. Soc. Japan, N.S., 65: 27–46, 1 fig., 4 tables.

– (1967b): The Mahang Formation: a mid-Palaeozoic Euxinic Facies from Malaya – with notes on its conditions of deposition and palaeogeography. – Geol. en Mijnb., 46, 5: 167–187, 13 pls., 3 figs., 4 tables.

Burton, C. K. & J. D. Bignell (1969): Cretaceous – Tertiary events in Southeast Asia. – Geol. Soc. Amer. Bull., 80: 681–688.

Burton, R. C. (1913): Contributions to the geology of the Province of Yunnan in Western China: 2. Petrology of the volcanic rocks of the Teng-Yueh District. – Rec. Geol. Surv. India, 43: 206–228.

Cadell, H. M. (1901): A sail down the Irrawaddy. – Scot. Geogr. Mag., 17: 239–265, Edinburgh.

Calhoun, A. B. (1929): Burma – an important source of precious and semi-precious gems. – Eng. Min., 127, 18: 708–712.

Campbell, J. M. (1919a): Water in rock magmas and veins. – Min. Mag. (G.B.), 21, London.

– (1919b): The ore minerals of Tavoy. – Min. Mag. (G.B.), 20: 76–89, London.

– (1926): Tin in the Mergui District, Lower Burma. – Min. Mag., 35: 155–160.

Campbell, K. V. & P. Nutalaya (1975): Structural elements and deformational events. – In: Stokes, R. B., et al. Proceedings of the Conference on the Geology of Thailand. – Dep. of Geol. Sci. Chiang Mai Univ. Spec. Publ., 1: 155–164.

Carey, S. W. (1975): Tectonic Evolution of Southeast Asia. – Reg. Conf. on Geol. and Min. Res. of Southeast Asia, Jakarta, Indonesia.

Carpenter, A. (1888): The birds-nest of Elephant Island, Mergui Archipelago. – Rec. Geol. Surv. India, 21, 1: 29–30.

Carpenter, R. H. (1964): The Pertinent Geologic Features and Exploration Possibilities of the Bawdwin Mine Area, Northern Shan State, Burma. – United Nations Spec. Fund Proj., Lead – Tin, Burma.

Carter, H. J. (1888): Description of a large variety of *Orbitolites mantelli* Cart., from the West Bank of the River Irrawaddy in the Province of Pegu, Burma, about 36 miles above Prome. – Ann. Mag. Nat. Hist., 6th ser., 2: 342–348.

– (1889): *Ramulina parasitica* a new species of fossil Foraminifera infesting *Orbitolites mantelli* var. *theobaldi* with comparative observations on the process of reproduction in the Myzetozoa, fresh water Rhizopoda and Foraminifera. – Ann. Mag. Nat. Hist., 6th ser. 4: 94–101.

Carter, T. D. (1942): Three new mammals of the genera *Crocidura, Callosciurus* and *Pteromys* from Northern Burma. – Amer. Mus. Novitat., No. 120: 1–2.

CCOP – IOC (1974): Metallogenesis, Hydrocarbons and Tectonic Patterns in Eastern Asia. – 158 pp., UNDP (CCOP), Bangkok.

CCOP (1974): The offshore hydrocarbon potential East Asia. – Comm. for Coordin. of Joint Prosp. for Min. Res. in Asian Offsh. Areas; review, 67 pp., Bangkok.

CCOP (1976): A decade of offshore hydrocarbon exploration in the CCOP region. – CCOP-Newsletter 3: 1–9, Bangkok.

CECIL, G. (1928): Ruby mining in Upper Burma. – Eng. Min., 126, 8: 294.

CHARLTON, J. (1931 a): Soil Survey Pegu District, Burma. – Agric. Surv., No. 13. – Dep. of Agriculture, Rangoon.

– (1931 b): A note on soils regarding their suitability for irrigation engineering works exposed to water. – Agric. and Livestock in India, 1, 11: 119–127.

– (1932 a): A further note on soils regarding their suitability for making irrigation works exposed to water. – Agric. and Livestock in India, 2, 3: 290–296.

– (1932 b): Soil survey of the Mandalay Canal Area, Burma. – Agric. Surv., No. 15. – Dep. of Agriculture, Rangoon.

CHAUDHURY, H. M. (1973): Earthquake occurrences on the Himalayan region and the new tectonics. – Proc. Sem. Geodyn. Himalayan Region, Natural Geophys. Res. Inst., Hyderabad: 59–71.

CHHIBBER, H. L. (1926): The rhyolites and rhyolite tuffs of Thaton District, Lower Burma. – J. Burma Res. Soc., 56, 3: 166–175.

– (1927 a): Some intrusive rocks of the Pegu-Yoma. – Trans. Min. Geol. Inst. India, 21: 338–363.

– (1927 b): The rhyolites and rhyolite tuffs of Thaton District, Lower Burma. – Proc. 14th Ind. Sci. Congr.: 239 (Abstr.).

– (1927 c): Geography of South Tenasserim and the Mergui Archipelago. – J. Burma Res. Soc., 17: 127–156.

– (1927 d): The extinct iron industry of the neighbourhood of the Mount Popa, etc. – J. Asiat. Soc. Bengal, N.S. 22: 219–233.

– (1927 e): The origin and mineral constitution of the Late Tertiary fossil wood of Burma. – J. Asiat. Soc. Bengal, N.S., 23, 13–26.

– (1927 f): The ancient volcanoes of Burma. (Abstr.). – J. Burma Res. Soc., 27: 169–173.

– (1927 g): Rhythmic banding of ferric oxide in silicified rhyolite tuff. – Geol. Mag., 64: 7–9.

– (1927 h): An estimation of the age of Mount Popa, deduced from the amount of erosion of the Plateau on the West. – J. Burma Res. Soc., 17.

– (1927 i): The serpentines and associated minerals of Henzada and Bassein Districts, Burma. – Proc. 14th Ind. Sci. Congr.: 239–240 (Abstr.). – J. Burma Res. Soc. 26: 176–199, Rangoon.

– (1927 k): The serpentines and associated minerals of the Henzada and Bassein Districts. – J. Burma. Res. Soc. 16: 195–196.

– (1927 l): The igneous rocks of the Mount Popa Region Burma. – Trans. Min. Geol. Inst. India, 21: 226–310, Calcutta.

– (1928 a): The hornblende lamprophyres and associated rocks of Mokpalin Quarries, Thaton District, Lower Burma. – Proc. 15th Ind. Sci. Congr.: 290, Calcutta.

– (1928 b): A note on the limestone caves in the neighbourhood of Nyaungbinzeik, Kyikmaraw Township, Amherst District, Lower Burma. – J. Burma Res. Soc., 18: 124–131.

– (1929): The salt industry of Amherst District, Lower Burma. – J. Burma Res. Soc., 19, 2.

– (1930): Myitkyina District. – Rec. Geol. Surv. India, 43: 97–102.

– (1931): Hukawng Valley. – Rec. Geol. Surv. India, 65, 1: 77–79.

– (1932): Myitkyina District geological traverses. – Rec. Geol. Surv. India, 66, 1: 85–91.

– (1933): The Physiography of Burma. – 148 pp., London-NewYork-Toronto: Longmans, Green & Co.

– (1934 a): The Geology of Burma. – 538 pp., London: Mac Millan.

– (1934 b): The Mineral Resources of Burma. – 320 pp., London: Mac Millan.

CHHIBBER, H. L. & M. M. WADHWANA (1928): The volcanic rocks of the Irrawaddy Delta, Bassein and Myaungmya District. – Proc. 15th Ind. Sci. Congr.: 290, Calcutta.

CHIT SAING (1969): Planktonic Foraminifera from Kyangin Tondaung Bore Hole, No. 1, Burma. – J. Sci. Technol. (Burma) (3rd Burma Res. Congr.; 1968) 2, 3: 447–486, 8 figs., 5 pls.

CHO CHO, TALMADGE, D. & SOE WWIN (1970): A simple field and laboratory method for the determination of copper in soils. – J. Sci. Technol. (Burma) (4th Burma Res. Congr., 1969): 3, 2: 307–317, 10 figs.

CHOWDHURY, K. A. & K. N. TANDON (1964): A fossil wood of *Terminalia tomentosa* W. and A. from the Tertiary of Burma. – Ann. Bot. (G.B.), 28, 111: 445–450, 1 pl.

CLARK, W. M. (1930): Manuring of paddy in Upper Burma. –Bull. Dept. of Agriculture, 27, Rangoon.

CLEGG, E. L. G. (1923): Note on the Konghka and Mammal-Lang iron ore deposit, Northern Shan State, Burma. – Rec. Geol. Surv. India, 54: 431–435.

- (1927): Notes on the geological traverse in the Yuzalin Valley. – Rec. Geol. Surv. India, **60**, 3: 292–302.
- (1933): Note on an overlap in the Ngapa Area, Minbu District. – Rec. Geol. Surv. India, **66**, 2: 250–254.
- (1936a): Notes on the geology of the second defile of the Irrawaddy River. – Rec. Geol. Surv. India, **71**, 4: 350–358.
- (1936b): Notes on the rocks in the vicinity of Kyaukse, Burma. – Rec. Geol. Surv. India, **71**, 4: 376–378.
- (1938): The geology of parts of the Minbu and Thayetmo Districts, Burma. – Mem. Geol. Surv. India, **72**, 2: 137–317, pl. 6.
- (1941a): The Cretaceous and associated rocks of Burma. Mem. Geol. Surv. India, **74**, 1: 1–102, pl. 1.
- (1941b): Introduction to HOBSON, G. V.: A geological survey in parts of Karenni and the Southern Shan States. – Mem. Geol. Surv. India, **72**, 2.
- (1941c): A note on the Bawdwin mines, Burma. – Rec. Geol. Surv. India, **75**, 13: 1–14.
- (1944a): Notes on tin and wolfram in Burma and India. – Rec. Geol. Surv. India, **76**, Econom. Bull., **15**: 1–168.
- (1944b): The Mineral Deposits of Burma. – 38 pp., Bombay: Times of India Press.
- (1953): The Mergui, Moulmein and Mawchi Series with a note on the evidence for the Permian age of the Moulmein System by H. B. WHITTINGTON. – Rec. Geol. Surv. India, **78**, 2: 157–194, 2 pls., 1 map.
- (1954): A traverse from Padaung to the Taungup Pass in the Prome District, Burma. – Rec. Geol. Surv. India, **78**, 2: 195–202, 1 map, Calcutta.
- (1974): The Mineral Deposits of Burma. – Repr. by the Geol. Soc. of Burma.

CLIFT, W. (1829): On the fossil remains of two new species of *Mastodon,* and of other vertebrated animals found on the left bank of the Irrawadi. – Trans. Geol. Soc. London, ser. 2, **2**: 369–375.

CLOSS, H., NARAIN, H. & S. C. GARDE (1974): Continental margins of India. – In: BURK, C. A. & C. L. DRAKE (eds.): The Geology of Continental Margins. – 629–639, Berlin, Heidelberg, New York: Springer.

COCKERELL, T. D. A. (1917): Insects in Burmese amber. – Ann. Entomol. Soc. Amer., **10**: 323–329, 10 figs.
- (1917a): Arthropods in Burmese amber. – Psyche, **24**: 40–45, 6 figs.
- (1917b): Descriptions of fossil insects. – Proc. Biol. Soc. Washington, **30**: 79–82, 4 figs.
- (1917c): Fossil insects. – Ann. Entomol. Soc. Amer., **10**: 1–22, 6 figs.
- (1920–1921): Fossil Arthropods in the British Museum. – Ann. Mag. Nat. Hist. (Ser. 9), **5**: 274, 276–277, **6**: 211–214, **8**: 541, 544–545.
- (1922): Fossils in Burmese amber. – Nature, **109**: 713–714.

COGGIN BROWN, J.: see BROWN, J. C.

COLBERT, E. H. (1935): Siwalik mammals in the American Museum of Natural History. – Trans. Amer. Phil. Soc., N.S., **26**: 1–401, 198 figs.
- (1938): Fossil mammals from Burma in the American Museum of Natural History. – Bull. Amer. Mus. Nat. Hist. 74.

COLEMAN, R. & W. IRWIN (1975): Ophiolites and ancient continental margins. – The geology of continental margins: 921–931, 6 figs., Berlin.

COLQUHOUN, A. (1931): Bawdwin Mine – Ore reserves calculations. – Min. Mag.

Committee for Co-Ordination of Joint Prospecting for Mineral Resources in Asian Offshore Areas (CCOP) & Intergovernmental Oceanographic Commission UNESCO (IOC) (1974): Metallogenesis, Hydrocarbons and Tectonic Patterns in Eastern Asia – A Programme of Research. – UNDP/CCOP Bangkok, 158 pp.

COOKE, C. B. (1872): The resources of Tenasserim. – Indian Econom., **3**: 148–149.

COTTER, G. P. de (1908a): The structure and age of the Taungtha Hills, Myingyan District, Upper Burma. – Rec. Geol. Surv. India, **36**, 3: 149–155.
- (1908b): Fossils from the Miocene of Burma. – Rec. Geol. Surv. India, **36**, 2: 131–132.
- (1908c): Note on the Tatakan Area, Blocks 21–26 N Yenangyaung. – Rec. Geol. Surv. India, **36**, 2: 136.

– (1909): The southern part of the Gwegyo Hills, including the Payagyigon-Ngashandaung oil field. – Rec. Geol. Surv. India, 37, 3: 225–234, 2 pls.
– (1910): The northern part of the Yenangyat oil filed. – Rec. Geol. Surv. India, 38, 4: 302–307.
– (1912a): The Pegu-Eocene Succession in the Minbu District, near Ngape. – Rec. Geol. Surv. India, 41, 4: 221–239.
– (1912b): Notes on some Nummulites from the Burma Tertiaries. – Rec. Geol. Surv. India, 41, 4: 322–323.
– (1912c): Note on the limestone hill near Thayetmo. – Rec. Geol. Surv. India, 41: 323.
– (1914): Some newly discovered coal seams near the Yaw River, Pakokku District, Upper Burma. – Rec. Geol. Surv. India, 44, 3: 163–185.
– (1915): Geology of the country near Ngahlaingdwin, Minbu District, Burma. – Rec. Geol. Surv. India, 45, 4.
– (1918a): The geotectonics of the Tertiary Irrawaddy Basin. – J. Asiat. Soc. Bengal, N.S., 14: 409–420.
– (1918b): A revised classification of the Gondwana system. – Rec. Geol. Surv. India, 48, 1.
– (1922): A note on the geology of Thayetmo and neighbourhood, including Padaukpin. – Rec. Geol. Surv. India, 54, 1: 103–116.
– (1923a): Note on the age of the limestone deposit Martaban Railway Station, Thaton District, Burma. – Rec. Geol. Surv. India, 54, 3: 343.
– (1923b): The Lamellibranchiata of the Eocene of Burma. – Paleont. Ind., N.S., 7, Mem. No. 2: 1–24, 7 pls.
– (1923c): A brief sketch of the Tertiary geology of Burma. – Proc. Pan-Pacific Sci. Congr.: 942–949.
– (1924a): The Mineral Deposits of Burma. – Publ. by the Superintendent, Rangoon: Government Printing, (revised 1939).
– (1924b): The oil-shales of Eastern Amherst, Burma, with a sketch of the geology of the neighbourhood. – Rec. Geol. Surv. India, 55, 4: 273–313.
– (1930): Confidential report on the Loian coalfield. – Rec. Geol. Surv. India, 63.
– (1933a): Notes on the geological structure and distribution of the oilbearing rocks of India and Burma. – Proc. World Petr. Congr., pap. 168, 1: 7–14, London.
– (1933b): Note on an Ammonite from Ramri Island. – Rec. Geol. Survey India, 66.
COTTER, G. P. de & E. L. G. CLEGG (1938): Geology of parts of Minbu, Myingyan, Pakokku, and Lower Chindwin Districts, Burma. – Mem. Geol. Surv. India, 72, 1: 1–136.
COTTER, G. P. de & S. JETHU RAMA RAUS (1916): Pakokku and Thayetmo. – Rec. Geol. Surv. India, 47, 1: 32.
COX, H. (1799): An account of the petroleum wells in the Burmha Dominions, extracted from the journal of a voyage from Ranghong up in the river Erai-wuddey to Amarapoorah, the present capital of the Burmha Empire. – Phil. Mag., 9: 226–234; Asiat. Res., 6: 127–136.
CRAWFORD, A. R. (1972): A displaced Tibetan Massif as a possible source of some Malayan Rocks. – Geol. Mag. (G.B.) 109, 6: 483–489, 1 fig.
– (1974): The Indus suture line, the Himalaya, Tibet and Gondwanaland. – Geol. Mag. 111: 369–383.
CRAWFORD, J. (1823): Geological observations made on a voyage from Bengal to Siam and Cochin China. – Geol. Trans., ser. 2, 1: 406–408.
– (1827): Brief narrative of an embassy . . . to the King of Ava, in 1826–1827. – Edinb. New Phil. J., 3: 359–370, and Edinb. J. Sci., 8: 10–25.
CREDNER, W. (1935): Siam, das Land der Tai. – Stuttgart.
CRIPER, W. R. (1885): Notes on some antimony deposits in the Moulmein District. – Rec. Geol. Surv. India, 18, 3: 151–153.
CULLEN, W. (1840): Note on coal on the Tenasserim Rivers. – Madras J. Lit. Sci., 11: 388–389.
CURRAY, J. R. & D. G. MOORE (1971): Growth of the Bengal Deep-Sea Fan and denudation in the Himalayas. – Bull. Geol. Soc. Amer., 82: 563–572.
– – (1974a): The Bengal Geosyncline, from rift to orogeny. – Geology.
– – (1974b): Sedimentary and tectonic processes in the Bengal Deep-Sea Fan and Geosyncline. – In: C. A. BURK & C. L. DRAKE (Eds.): The Geology of Continental Margins. – 617–627, New York: Springer.
CURRAY, J. R., MOORE, D. G., LAWVER, L. A., EMMEL, F. J., RAITT, R. W., HENRY, M. & R. KIECKHEFER (1979): Tectonics of the Andaman Sea and Burma. – AAPG Mem. 29: 189–198, 6 figs., Tulsa.

CURRIE, E. D. (1940): Note on Echinoidea from Burma. – Rec. Geol. Surv. India, 74, 2.

DALTON, L. V. (1908): Notes on the geology of Burma. – Quart. J. Geol. Soc. London, 64: 604–644.

DANA, J. D. (1886): Volcano of Barren Island in the Bay of Bengal. – Amer. J. Sci., 31: 394.

DATTA, P. N. (1900): Note on the geology of the country along the Mandalay – Kaunglon Ferry Route, Upper Burma. – Gen. Rep. Geol. Surv. India (1899–1900): 96–122.

– (1907): Gold-bearing deposits of Loi Twang, Shan States. – Rec. Geol. Surv. India, 35.

DAVE, V. K. S. (1973): Metallic mineral deposits of India and the neighbouring countries, a Review (Abstr.). – In: Symp. on Strategic Minerals of India, Exploration, Sufficiency, Deficiency, Procurement of Strategic Minerals: 34–35, Vikram Univ., Stud. in Geol., Ujjain, India.

DAVIES, T. (1878): Notes on "Jadeite" and "Jade". – Geol. Mag., 5: 192.

DAVIS, J. H. (1960): The forests of Burma. – Rep. Dep. of Bot., Univ. of Florida, Gainesville.

DAY, A. E. (1932): Trans. Min. & Geol. Inst. India, 27.

DEKA, P. J. (1968): Certain aspects of metallogeny of gold and some associated minerals in the region of Burmese Arc. – J. Mines, Metals, Fuels, India, 16, 6: 199–202.

DENYER, A. & K. C. G. HEATH (1940): Mining and milling tin-tungsten ore at Mawchi Mine, Burma. – Bull. Inst. Min. Metall. 426, 1–29, 13 figs.

Dep. of Agriculture, Burma (1922–33): Annual Report of the Agric. Chemist, Burma.

DE TERRA, H. & MOVIUS, H. L. (1943): Research on Early Man in Burma. – Trans. Amer. Phil. Soc., N.S., 32, 3, 267–466, 102 figs., Philadelphia.

DEWEY, J. F. & K. C. A. BURKE (1972): Tibetan, Variscan, und Precambrian basement reactivation – Products of Continental collision. – J. Geol., 81: 683–692, Chicago.

DEY, A. K. & A. K. SAHA (1954): A petrological study of rock types between Dening and Minutang, Mishmi Hills, Assam. – Rec. Geol. Surv. India, 83: 493–500.

DEY, B. P. (1968): Aerial photo interpretation of a major lineament in the Yamethin – Pyawbwe Quadrangle. – Sci. Technol. Burma, 1st Burma Res. Congr. 1966, 1, 3: 431–443, 6 figs., Rangoon.

DIENER, C. (1977): Anthracolithic fossils of the Shan States. – Paleontol. Ind., N.S., 3, Mem. 4: 1–74, pl. 7.

DOYLE, P. (1879): A Contribution to Burman Mineralogy. – 15 pp., Calcutta.

DRESCHER, H. P. & W. GOCHT (1978): Burma's Heinda Tin Mine. – Krupp-Rohstofftechnik, Repr. from "Tin International" London, April 1978: 6 pp., 11 figs., 4 tables.

DRUMMOND, R. R. (1958): A climatic map of Burma. – North Burma Council for Stud. in Geogr., 1: 1–20, Mandalay.

DUNBAR, C. O. (1932): Fusulinids of Lower *Productus* limestone of Salt Range. – Rec. Geol. Surv. India, 81, 2.

DUNN, J. A. (1932a): Reaction minerals in a garnet cordierite gneiss from Mogok. – Rec. Geol. Surv. India, 65, 4: 445–456.

– (1932b): Some ore minerals from Bawdwin, Shan States. – Rec. Geol. Surv. India, 69, 4: 469–474.

– (1937): A microscopical study of the Bawdwin ores, Burma. – Rec. Geol. Surv. India, 72, 3: 333–359.

– (1938a): Tin-tungsten mineralization at Mawchi, Karenni States, Burma. – Rec. Geol. Surv. India, 73, 2: 209–237.

– (1938b): Thin-tungsten mineralization at Hermyingyi, Tavoy District, Burma. – Rec. Geol. Surv. India, 73, 2: 238–246.

DUNN, P. J. (1977): Uvite, a newly classified gem tourmaline. – J. Gemmol. (G.B.), 15, 6: 300–308, 5 figs.

DUTT, A. B. (1942): The mineral resources of the Shan Scarps, included in the Kyaukse, Meiktila and Yamethin Districts and Yengan State. – Rec. Geol. Surv. India, 77, prof. pap. 10: 1–64, pl. 6.

EAMES, F. E. (1950): The Pegu System of Central Burma. – Rec. Geol. Surv. India, 81, 2: 377–388, 1 fig., 4 pls.

Earth Sci. Res. Div., Res. Pol. Dir. Board & Govt. Soc. Rep. of the Union of Burma (1977): The Geological Map of Burma 1 : 1 Mio (with explan. brochure). – Rangoon.

Economic Commission for Asia and the Far East (1952): Coal and iron ore resources of Asia and the Far East. – United Nations Publ., ST/ECAFE/5, Bangkok.

Economic Commission Far East, Miner. Resources Develop. (1970): Geochemical prospecting activities in Burma. – Proc. 2nd Semin. Geochem. Prospect. Methods Techn.; Peradeniya, 38, 25–26.

EGUCHI, T., SEIYA, U. & T. MAKI (1979): Seismotectonics and tectonic history of the Andaman Sea. – Tectonophys., 57: 35–51, Amsterdam.

EPPLER, W. F. (1974): Über einige Einschlüsse im Birma-Rubin. – Z. dtsch. Gemmol. Ges., 23, 2: 102–108, 9 figs.
– (1976): Negative crystals in ruby from Burma. – J. Gemmol., 15, 1: 1–5.
EVANS, P. (1941): The geology of British oilfields. 3. The oilfields of Burma. – Geol. Mag., 78, 321–350.
– (1958): The oilfields of Assam and Burma. – In: E.C.A.F.E. Symp., New Delhi. Bangkok: United Nations.
– (1964): The tectonic framework of Assam. – J. Geol. Soc. India, 5: 80–96.
– (1965): Structure of north-eastern frontier area of Assam. – in: D. N. WADIA commem. Vol.: Indian Min. Geol. & Met. Inst.: 640–646.
EVANS, P. & W. COMPTON (1946): Geological factors in gravity interpretation from India and Burma. – J. Geol. Soc. London, 102, 211–249.
FALCON, N. L. (1967): The geology of the north-east margin of the Arabian basement shield. – Advan. Sci., 24: 1–12.
FALVEY, D. A. (1974): The development of continental margins in plate tectonic theory. – Austr. Petrol. Explor. Assoc. J., 14: 95–106.
FAO (1971): Soil map of the world, 1:5 000 000. – Paris (UNESCO).
FAY LAIN & WIN MAW (1970): Report on the geology of the Arakan Coast and off-shore islands (Kyaukryu and Sandoway Districts). – Unpubl. Myanma Oil Corp. Rep. F.L. 7, W.M. 7, Rangoon.
– – (1971): The circular synclinal structures of Ramree and Cheduba (Manaung) Islands. – Union of Burma J. Sci. Technol. 4: 31–41; Rangoon.
FERENCIC, A. (1961): Nickel ore in Tagaung Taung, Upper Burma. – J. Geol. Soc. India, 2: 23–30, 2 pls., Bangalore.
– (1962): Laterites and iron rich laterites in Southern Burma. – 3 Kongr. Geol. Yougoslavia, 2: 377–397, 11 figs. (in Serbo Croatian; Abstr. in Engl.: 395–397).
FERMOR, L. L. (1906): Ores of antimony, copper and lead from the Northern Shan States. – Rec. Geol. Surv. India, 33, 3: 234.
– (1909): The manganese ore deposits of India. – Mem. Geol. Surv. India, 37, 104: 669–671.
– (1932): General Report of the Geological Survey of India for the year 1930: The Mogok Stone Tract, Katha District. – Rec. Geol. Surv. India, 65, 1: 80–86.
– (1933): General Report of the Geological Survey of India for the year 1931: The Mogok Stone Tract, Katha District. – Rec. Geol. Surv. India, 66, 1: 92–96.
– (1934): General Report of the Geological Survey of India for the year 1932. – Rec. Geol. Surv. India, 67: 1–150.
– (1935): General Report of the Geological Survey of India for the year 1933: The Mogok Stone Tract, Katha District. – Rec. Geol. Surv. India, 68, 1: 50–58.
– (1936): General Report of the Geological Survey of India for the year 1934: The Mogok Stone Tract, Katha District. – Rec. Geol. Surv. India, 69, 1: 50–54.
FITCH, R. (1599): The voyage of M. Ralph FITCH, Marchant of London, by the way of Tripolis in Syria, to Ormus, and so to Goa in the East India, to Cambaia, and all the Kingdome of Zalabdin Echebar the great Mogor, to the mighty river Ganges, and down to Bengala to Bacola, and Chonderi, to Pegu, to Lamhay in the Kingdome of Siam, and back to Pegu and from. – Hakluyt's Principal Navigation, 2: 250–268.
FITCH, T. J. (1970): Earthquake mechanisms in the Himalayan, Burmese and Andaman regions and continental tectonics in central Asia. – Geophys. Res. 75: 417–422.
– (1972): Plate convergence, transcurrent faults, and internal deformation adjacent to Southeast Asia and the Western Pacific. – J. Geophys. Res., 77: 4432–4660.
FOLEY, W. (1833): On the coal from Arracan. – J. Asiat. Soc. Bengal, 2: 368.
– (1834): On fossil shells and coal from Kyauk Phyoo, Ramree. – J. Asiat. Soc. Bengal, 3: 412–413.
– (1835): Journal of a tour through the Island of Ramree with a geological sketch of the country, and a brief account of the customs, etc. of its inhabitants. – J. Asiat. Soc. Bengal, 4: 20–39, 82–95.
– (1836): Notes on the geology of the country in the neighbourhood of Maulayeng (vulg. Moulmein). – J. Asiat. Soc. Bengal, 5: 269–281.
FOSS, K. M. (1904): The occurrence of tin and gold in Lower Burma. – Min. J. (G.B.), 76: 505–506.
FOX, C. S. (1930): On the occurrence of Cretaceous Cephalopods in the "Red Beds" of Kalaw, Southern Shan States, Burma. – Rec. Geol. Surv. India, 63, 1: 182–187.

– (1953): The mineral production of India and Burma during 1939. – Rec. Geol. Surv. India, **78**, 2: 301–405.

FRIEDLANDER, H. (1874): The country of the earthoil in Upper Burma. – Rangoon, Suppl. to the Brit. Burma Gazette, Feb. 14th.

FROMAGET, J. (1929): Note préliminaire sur la stratigraphie des Formations Secondaires et sur l'âge des prouvements majeurs en Indochine. – Bull. Surv. Geol. Indochina, **18**, 5.

FRYAR, M. (1872 a): Report on some mineraliferous localities in Tenasserim. – Indian Econom., **4**: 42–43.

– (1872 b): Coal at Moulmein. – Indian Econom., **1**: 326–328.

– (1873): Reports on minerals in Shwegeen, Toungoo and Pahpoon Districts Tenasserim Division, Burma. – Coll. Guard., **30**: 390.

– (1878): Mineral resources of British Burma. – J. Soc. Arts., **26**: 169.

– (1882): Report on minerals in the Amherst District of the Tenasserim Division. – 450–459, Burma.

GANSSER, A. (1964): Geology of the Himalayas. – 289 pp., London, New York, Sydney: Wiley Interscience.

– (1966): The Indian Ocean and the Himalayas, a geological interpretation. – Eclog. Geol. Helv., **59**: 831–848.

– (1973): Ideas and problems on Himalayan geology. – Geodynamics of the Himalayan region. – India Nat. Geophys. Res. Inst.: 97–103.

– (1976): The Great Suture Zone between Himalaya and Tibet. – A Preliminary Account. – In: Himalaya, Coll. Internat. du C.N.R.S., No. 268 – Ecol. and Geol. de l'Himalaya: 181–191, 1 map.

GARSON, M. S., AMOS, B. J. & A. H. G. MITCHELL (1976): The geology of the country around Neyaungga and Ye-ngan, Southern Shan States, Burma. – Inst. Geol. Sci., Overseas Div., Mem. 2: 70 pp., 10 figs., 1 pl., 1 map, London.

GARSON, M. S., FIMM, B. Y., MITCHELL, A. H. G. & B. A. R. TAIT (1975): The geology of the tin belt in Peninsular Thailand around Phuket, Phangnga and Takua Pa. – Inst. Geol. Sci., Overseas Div., Mem. 1: 1–112, 27 figs., 10 pls., 9 maps, London.

GARSON, M. S. & A. H. G. MITCHELL (1970): Transform faulting in the Thai Peninsula. – Nature, **228**, 5266: 45–47, 2 figs.

GARSON, M. S., MITCHELL, A. H. G., AMOS, B. J., HUTCHINSON, D., KYI SOE, PHONE MEYINT & NGAW CIN PAU (1972): Economic geology and geochemistry of the area around Neyaungga and Ye-ngan, Southern Shan States of Burma. – Inst. Geol. Sci., Overseas Div., Rep. No. 22, London.

GEE, E. R. (1926): The geology of the Andaman and Nicobar Islands, with special reference to Middle Andaman Island. – Rec. Geol. Surv. India, **59**, 2.

Geoistrazivanja (1958): Report on ore occurrences in Tagaung Taung. – Unpubl. Rep., Zagreb/Yugoslavia and Rangoon.

Geological Survey of India (1977): Gondwana Geology – Status, Problems and Possibilities. – IV Internat. Gondwana Symp. Geol. Surv. India, Calcutta.

Geolog. Map of the Soc. Rep. of the Union of Burma 1:1 000 000 (1977): with explanatory brochure. Earth Sci. Res. Div., Rangoon.

Geological Maps of Asia, Scale 1:5 000 000 (1975): Compiled by Acad. Geol. Sci. of China, Beijing (Peking).

GEORGE, E. C. S. (1908): Memorandum on the tourmaline mines of Maingnin. – Rec. Geol. Surv. India, **36**: 233–238.

GLAESSNER, M. F. (1963): Preliminary report on generalized stratigraphic correlation between sedimentary basins in the ECAFE Area. – Proc. 2nd Symp. on the Development of Petroleum Resources of Asia and the Far East, **1**, 18: 139–144.

GOBBETT, D. J. (1972): Carboniferous and Permian stratigraphic correlation in Southeast Asia (Abstr.). – In: Regional Conf. on the Geol. of Southeast Asia. – Geol. Soc. Malaysia, Newsl., **34**: 13.

– (1973): Carboniferous and Permian correlation in Southeast Asia. – In: Regional Conf. on the Geol. of Southeast Asia (Proc.). – Geol. Soc. Malaysia, Bull. 6: 131–142.

GOBBETT, D. J. & C. S. HUTCHINSON (1973): Geology of the Malay Peninsula. – 438 pp., New York, London, Sydney, Toronto: Wiley Interscience.

GOCHT, W. R. A. (1976): Associated heavy minerals in alluvial tin deposits of Southeast Asia (Abstr.). – Internat. Geol. Congr., No. 25, **1**: 211–212.

GOOSSENS, P. J. (1978a): The metallogenic provinces of Burma; their definitions, geologic relationships and extension into China, India and Thailand. – Third Regional Conf. Geol. and Miner. Resources of Southeast Asia, Bangkok: 431–492.

– (1978b): Earth Sciences Bibliography of Burma, Yunnan, and Andaman Islands. – Third Regional Conf. Geol. Miner. Resources Southeast Asia, Bangkok: 495–536.

– (1978c): The Burmese tin and wolfram belt, distribution and types of deposits and presentations of the New Metallogenic Map of Burma. – In: Geology of Tin Deposits, Malaysia, Annex to Warta Geol., 4, 2: 10 (Abstr.).

GORDON, R. (1879): Report on the Irrawaddy River. – Fol., Rangoon.

– (1882): The Irrawaddy and the Snapo. – Proc. Roy. Geogr. Soc. London, N.S., 9: 559–563.

– (1885): The Irrawaddy River. – Proc. Royal Geogr. Soc. London, N.S., 7: 292–325.

– (1888): On the ruby mines near Mogok, Burma. – Proc. Roy. Geogr. Soc. London.; N.S., 10: 261–275.

– (1889): The ruby mines of Burma. – Asiat. Quat. Rev., 7: 410–423 (Abstr. Min. J., 69: 475).

GORSHKOV, G. P. (1961): Problems of the seismotectonics and seismic zoning of the territory of the Union of Burma. – Byull. Sovj. Seism. 12 (in Russ.).

GRAHAM, S. A., DICKINSON, W. R. & R. V. INGERSOLL (1975): Himalayan-Bengal model for flysch dispersal in the Appalachian-Ouachita system. – Geol. Soc. Amer. Bull., 86: 273–286.

GRAMANN, F. (1974): Some paleontological data on the Triassic and Cretaceous of the western part of Burma (Arakan Islands, Arakan Yoma, western outcrops of Central Basin). – Newsl. Stratigr. 3, 4: 277–290, 3 figs. Leiden.

– (1975a): Paläontologische Daten aus dem westlichen Birma in ihrer Bedeutung für die regionale Geologie zwischen indischem Subkontintent und Sundabogen. – In: 45. Jahresvers. Paläontol. Ges., Kurzfass. Vortr., 45: 12, Hannover.

– (1975b): Ostracoda from Tertiary sediments of Burma with reference to living species. – Geol. Jb., B 14: 1–46, 3 figs., Hannover.

GRAMANN, F., LAIN, F. & D. STOPPEL (1972): Paleontological evidence of Triassic age for limestones from the Southern Shan State and Kayah States of Burma. – Geol. Jb., B 1: 1–33, 75 figs., Hannover.

GRANT, F. T. (1832): Mode of extracting the gold dust from the sand of the Ningthee River, on the frontier of Manipur. – J. Asiat. Soc. Bengal, 1: 148–149.

GREGORY, J. W. (1923): The geologic relations of the oil shales of Southern Burma. – Geol. Mag., 60: 152–159, London.

GREGORY, J. W. & W. N. EDWARDS (1923): Tertiary oil shales of Southern Burma. – Geol. Mag. 60.

GREGORY, J. W., REED, F. R. C. & G. L. ELLES (1925): The geology and physical geography of Chinese Tibet and its relations to the mountain system of South Eastern Asia, from observations made during the Percy SLADEN expedition, 1922. – Phil. Trans. B 23: 171–290.

GREGORY, J. W., WEIR, J., TRAUTH, F. & J. PIA (1930): Upper Triassic fossils from the Burma-Siamese frontier; the Thaungyi Trias and description of corals. – Rec. Geol. Surv. India, 63, 1: 155–167.

GREGORY, W. (1834): On the composition of petroleum of Rangoon with remarks on petroleum and naphtha in general. – Trans. Roy. Soc. Edinburgh, 13: 124–140.

GRIESBACH, C. L. (1892): Geological sketch of the country North of Bhamo. – Rec. Geol. Surv. India, 25, 3: 127–130.

– (1924): In C. BROWN: A geographical classification of the mineral deposits of Burma. – Rec. Geol. Surv. India, 56, 1: 65–108.

GRIFFITH, S. V. (1956): The mineral resources of Burma. – Min. Mag. (G.B.), 95, 1: 9–18.

GRIFFITHS, H. D. (1917a): The Kanbauk wolfram mines. – Min. Mag. (G.B.), 17: 211–219.

– (1917b): The wolfram deposits of Burma. – Min. Mag. (G.B.), 17: 60–66.

GRIMES, G. E. (1898): Geology of parts of the Myingyan, Magwe and Pakokku Districts, Burma. – Mem. Geol. Surv. India, 28, 1: 30–71.

GÜBELIN, E. (1969): On the nature of mineral inclusions in gemstones. – J. Gemmol., 11, 5: 149.

GUPTA, H. C. DAS (1930): Lower Chindwin District. – Rec. Geol. Surv. India, 62, 1.

GUPTA, H. K. (1973): Geophysical investigations for the Himalayan region undertaken at the National Geological Research Institute. – In: Geodynamics of the Himalayan region. – India Nat. Geophys. Res. Inst.: 72–96.

HAGEN, D. & E. KEMPER (1976): Geology of the Thong Pha Phum Area (Kanchanaburi Province, Western Thailand). – Geol. Jb., 21, 3: 53–91, Hannover.

HAHN, L. (1976): The stratigraphy and paleogeography of the non-marine Mesozoic deposits in Northern Thailand. – Geol. Jb. **B 21**: 155–169, Hannover.

HALLOWES, K. A. K. (1920): On the coal-seams of the foot hills of the Arakan Yoma between Letpan Yaw in Pakokku and Ngape in Minbu, Upper Burma. – Rec. Geol. Surv. India, **51**, 1: 34–49.

HAMADA, T. (1961): Some Permo-Carboniferous fossils from Thailand. – Sci. Pap. Coll. Gen. Educ. Univ. Tokyo, **10**, 2: 338–361.

– (1964): Some Middle Ordovician brachiopods from Satun, southern Thailand. – Japan. J. Geol. Geogr. **35**: 213–221.

HANDCOCK, A. R. W. (1909): Volcanic upheaval off the Coast of Burma. – Geogr. J., **34**: 690.

HANDLIRSCH, A. (1908): Über einige fossile Insekten. – Verh. Zool. Bot. Ges., Wien, **58**: 205–207.

HANNAK, W. (1972): Die Blei-Silber-Zink-Lagerstätte von Bawdwin (Birma), ein Erzlager. – In: Tag. Montangeol. Clausthal-Zellerfeld: 22.

HARK, H. U. (1975): Die weltweite Erdöl- und Erdgasexploration. Rückschau auf die erdölgeologischen Beiträge des 9. Welt-Erdöl-Kongresses in Tokio. – Erdöl-Erdgas-Z., **91**, 12: 421–434, 14 figs.

HAYDEN, N. H. (1896): Report on the steatite mines, Minbu District, Burma. – Rec. Geol. Surv. India, **29**, 4: 71–76.

HEADLAM, E. J. (1907): A new island in the Bay of Bengal. – Geogr. J., **24**: 430–436.

HEALEY, M. (1908): The fauna of the Napeng Beds or Rhaetic Beds of Upper Burma. – Paleontologia Ind., N.S., **2**, 4: 88 pp., pl. 9.

HEIM, H. & H. HIRSCHI (1939): A section of the Mountain Range of North-western Siam. – Eclog. geol. Helv., **32**, 1: 1–16, 5 figs., 1 pl.

HEINRICH, G. H. (1965): Burmesische Ichneumoninae I. – Entomol. T., Sverige, **86**, 1–2: 74–130, 1 pl.

HELFER, J. W. (1837): Report on Amherst in the Tenasserim Provinces. – 40 pp., Calcutta.

– (1838a): Note on minerals found in Tenasserim. – J. Asiat. Soc. Bengal, 7: 171.

– (1838b): Report on the coal discovered in the Tenasserim. – J. Asiat. Soc. Bengal, 7: 701–706.

– (1839): The provinces of Ye, Tavoy and Mergui on the Tenasserim Coast. – Calcutta.

– (1859): Gedruckte und ungedruckte Schriften über die Tenasserim Provinzen, den Mergui Archipel und die Andaman Inseln. – Mitth. k.k. Geogr. Ges., **3**: 167–390, Wien.

HELM, O. (1892): On a new fossil, amber-like resin occurring in Burma. – Rec. Geol. Surv. India, **25**, 4: 180–181.

– (1893): Further notes on Burmite a new amber-like fossil resin from Upper Burma. – Rec. Geol. Surv. India, **26**, 1: 61–64.

– (1894): Über Birmit, ein in Oberbirma vorkommendes fossiles Harz. – Schr. Naturf. Ges. Danzig, **8**: 63–66.

HELMCKE, D. & S. RITZKOWSKI (in press): The Geology of Ramree and Cheduba Islands (Arakan, Burma).

HENDERSON, J. W. et al. (1971): Area Handbook for Burma. – Foreign Area Studies, Washington: U.S. Governm. Print. Off.

HENDRY, D. (1928): Fertilisers for Paddy (Hmawbi). – Bull. Dep. of Agric., Burma, **25**.

HENNIG, W. (1972): Insektenfossilien aus der Unteren Kreide, IV, Psychodidae (Phebotominae), mit einer kritischen Übersicht über das phytogenetische System der Familie und die bisher beschriebenen Fossilien (Diptera). – Beitr. Naturk., **241**: 1–69, 83 figs., Stuttgart.

HERON, A. M. (1917): Monazite in Mergui and Tavoy. – Rec. Geol. Surv. India, **48**, 3: 179–180.

– (1921a): The antimony deposit of Thabyu, Amherst District, Burma. – Rec. Geol. Surv. India, **53**, 34–43.

– (1921b): Bismuth in Tenasserim. – Rec. Geol. Surv. India, **53**, 1: 81.

– (1936a): General Report of the Geological Survey of India for the year 1935. – Rec. Geol. Surv. India, **71**, 1: 1–104.

– (1936b): Mogok Stone Tract (General Report). – Rec. Geol. Surv. India, **71**, 1: 58–63.

– (1937): Age of Mogok Series (General Report of the Geol. Surv. of India). – Rec. Geol. Surv. India, **72**, 1: 62.

– (1938): (1) Disturbed areas bound Mogok Series on the North; (2) Varying degrees of metamorphism North of Mogok; (3) Sindhi – Southern Shan States (General Report of Geol. Surv. of India, 1937). – Rec. Geol. Surv. India, **73**, 1: 64–65.

HILDE, T. W. C., UYEDA, S. & L. KROENKE (1976): Evolution of the Western Pacific and its margin. – UN ESCAP, CCOP Techn. Bull. **10**: 1–19, Bangkok.

HOBSON, G. V. (1941): Report on a geological survey in part of Karenni and the Southern Shan States. – Mem., Geol. Surv. India, 74, 2: I–XVIII, 103–155, pl. 4.

HOCHSTETTER, F. VON (1866): Geologische Beobachtungen während der Reise der österreichischen Fregatte Novara um die Erde in den Jahren 1857, 1858, 1859. – Wien.

– (1869): Geology and physical geography of Nicobar Islands. – Rec. Geol. Surv. India, 2, 3.

– (1885): Gesammelte Reise-Berichte von der Erdumsegelung der Fregatte Novara. – Die Nikobarischen Inseln: 138–182, Wien.

HOKE MAW (1972): Structural outline of Tenasserim Division, Burma. – Geol. Soc. Malaysia, Newsletter (Malays. Reg. Conf. Geol. Southeast Asia) Abstr., 34: 23, Kuala Lumpur.

HOLDEN, R. (1916): A fossil wood from Burma. – Rec. Geol. Surv. India, 47, 4: 267–272.

HOLLAND, H. (1893): Note on the occurrence of quartz in an Indian basic volcanic rock (Narcondam). – Bull. Microsc. Soc., 2, 6: 3, Calcutta.

HOLLAND, T. H. (1902): The mica deposits of India. – Mem. Geol. Surv. India, 34: 11–121.

– (1904): Tin ore in Burma. – Rec. Geol. Surv. India, 31, 1: 43.

– (1909): General Report of the Geological Survey of India for the year 1908. – Rec. Geol. Surv. India, 38: 1–70.

HOLLAND, T. H., KRISHNAN, M. S. & K. JACOB (1956): Birmanie-Burma. – In: Lexique Stratigraphique International, 3 (Asie), 6d: 61–115.

HOSKING, K. F. G. (1970): The primary tin deposits of Southeast Asia. – Miner. Sci. Engin., 2, 4: 24–50.

HUANG CHI-CHING (T. K. HUANG) (1978): An outline of the tectonic characteristics of China. – Eclog. geol. Helv., 71, 3: 611–635, 7 figs., 4 tables.

HUBER, H. & C. BURRI (1933): Grundzüge der Geologie von Burma mit besonderer Berücksichtigung des jungen Vulkanismus. – Z. Vulkanol., 15, 3: 153–179, Berlin.

HUGHES, T. W. H. (1889): Tin-mining in Mergui District. – Rec. Geol. Surv. India, 22: 188–208.

– (1892): Coal on the Great Tenasserim River, Mergui District, Lower Burma. – Rec. Geol. Surv. India, 25: 161–163.

HUTCHISON, C. H. (1839): Report on the New Tenasserim coal field. – J. Asiat. Soc. Bengal, 8: 390–393.

HUTCHISON, C. S. (1973): Tectonic evolution of Sundaland: A Phanerozoic synthesis. – Geol. Soc. Malaysia, Bull. 6: 61–86.

– (1975): Ophiolite in Southeast Asia. – Geol. Soc. Amer., Bull., 86, 6: 797–806, 4 figs.

ICHIKAWA, M., SRIVASTAVA, H. N. & J. DRAKOPOULOS (1972): Focal mechanism of earthquakes occurring in and around Himalayan and Burmese Mountain Belts. – Pap. Meteorol. Geophys., 23: 149–162, Tokyo.

IGO, H. & T. KOIKE (1967): Ordovician and Silurian Conodonts from the Langkawi Islands. – Geol. Pal. Southeast Asia, 3: 1–29, 4: 1–21.

IYER, L. A. N. (1931): Plants. – Rec. Geol. Surv. India, 65, 1.

– (1934): In FERMOR, L. L.: General Report of the Geological Survey of India for the year 1932, Shwebo District, Burma. – Rec. Geol. Surv. India, 67, 1: 48–50.

– (1938a): Amherst District. – Rec. Geol. Surv. India, 73, 1.

– (1938b): In HERON, A. M.: General Report for 1936, Amherst District. – Rec. Geol. Surv. India, 73, 1: 66–67.

– (1953): The geology and gemstones of the Mogok Tract, Burma. – Mem. Geol. Surv. India, 82: 100 pp., 8 pls., 2 figs., 14 tables.

JACOB, K. (1954): The occurrence of Radiolarian Cherts in association with ultrabasic intrusives in the Andaman Islands, and its significance in sedimentary tectonics. – Rec. Geol. Surv. India, 83, 2.

JAEGER, H., NAKINBODEE, V., MAHAKAPONG, V., v. BRAUN, E., HESS, A., KOCH, K. E. & V. STEIN (1968): Graptolites of the Lower Devonian from Thailand (preliminary results). – N. Jb. Geol. Paläont., Mh., 1968, 12: 728–730.

JAEGER, H., STEIN, V. & R. WOLFART (1969): Fauna (Graptoliten, Brachiopoden) der unterdevonischen Schwarzschiefer Nordthailands mit einem Beitrag von D. STOPPEL. – N. Jb. Geol. Paläont., Abh. 133, 2: 171–190.

JAIN, A. K., THAKUR, V. C. & S. K. TANDON (1974): Stratigraphy and structure of the Siang District, Arunachal, Himalaya. – Himalayan Geol., 4: 28–60.

JAMES (1853): Mean result of three analyses of laterite from Burma. – J. Asiat. Soc. Bengal, 22: 198.

JAVANAPHET, JUMCHET C. (1969): Geological Map of Thailand – 1:1 000 000. – Dept. Min. Res., Ministry of Natl. Development, Bangkok.

JEN CHI-SHUEN & CHU CHING-CHUAN (1970): Geosynclinal Indonesian structure in the Lanping-Weisi Region, Western Yunnan. – Int. Geol. Rev., 12, 4: 447–463.

JENKINS, H. L. (1869): Notes on the Burmese Route from Assam to the Hookoong Valley. – Proc. Asiat. Soc. Bengal: 67–73.

JOB, A. L. (1973): Burma's mines and mineral potential. – World Min. (USA), 26, 1: 34–38.

JOHNSON, B. D., POWELL, McA, C. & J. J. VEEVERS (1976): Spreading history of the eastern Indian Ocean and Greater India's northward flight from Antarctica and Australia. – Bull. Geol. Soc. Amer., 87, 11: 1560–1566.

JOHNSTON, J. H. (1841): Reports on trials of coal from Mergui. – Coal Com. Rep., App. Nos. 2, 12 and 13.

JONES, C. R. (1968): Lower Paleozoic rocks of Malay Peninsula. – Amer. Assoc. Petrol. Geol. Bull., 52, 7: 1259–1278.

– (1970): The geology and mineral resources of the Grik area, Upper Perak. – Geol. Surv. Dep. West Malaysia Mem., 11: 1–144.

JONES, E. G. (1887 a): Notes on Upper Burma. – Rec. Geol. Surv. India, 20, 4: 170–194.

– (1887 b): On the metalliferous mines in the neighbourhood of Kyauktat and Pyingaung in the Shan Hills. – Rec. Geol. Surv. India, 20, 191–194.

JONES, H. C. (1920): Note on an occurrence of Graptolites in the Southern Shan States. – Rec. Geol. Surv. India, 51, 2: 156.

– (1921): Note on some antimony deposits of the Southern Shan State. – Rec. Geol. Surv. India, 53: 44–50.

JONES, W. R. (1917): Tin and wolfram lodes. – Min. Mag. (G.B.), 17: 230.

– (1918 a): The origin of wolfram deposits. – Min. Mag. (G.B.), 18: 319–320.

– (1918 b): Tungsten in manganese ore. – Eng. Min. J., 106: 779.

JUNGWIRTH, J. (1959): Die Bergwirtschaft der Burmesischen Union, derzeitiger Stand und Entwicklungsmöglichkeiten. – Berg- u. Hüttenmänn. Mh. Österr., 104, 7: 143–151, 13 figs., Wien.

JURKOVIC, I. & B. ZALOKAR (1959 a): Notes on the mineral in the Wuntho Region, Burma. – Geol. Vjesn. 12 (1958): 125–134, Zagreb.

– – (1959 b): The ore occurrences of the Shangalon Area, South-West of Kawlin, Upper Burma. – Geol. Vjesn. 12 (1958): 235–266, Zagreb.

– – (1961): Silver-bearing galena and siderite occurrences in the Putao Region, Northern Burma (Asia). – Geol. Vjesn. 14: 311–322, Zagreb.

JURKOVIC, I., ZALOKAR, B. & L. MARIC (1965): Notes on the ore occurrences on the Eastern and South-Eastern slopes of Tagaung-Taung, Katha District, Upper Burma (Asia). – Geol. Vjesn., 18, 1 (1964): 53–60, 2 pls., 1 map (Abstr. in Serbo-Croatian).

KARMANOV, I. I. (1960): Soils of the rice fields of Lower Burma and certain other regions of the Union of Burma (in Russ.). – Pochvovedeniye 8.

KARUNAKARAN, C. (1974): Geology and mineral resources of the States of India, Part IV – Arunachal Pradesh, Assam, Manipur, Meghalaya, Mizoram, Nagaland and Tripura. – Geol. Surv. India, Misc. Publ. No. 30, 124 pp., Delhi.

KARUNAKARAN, C., PAWDE, M. B., RAINA, B. K., RAY, K. K. & S. S. SAHA (1964): Geology of South Andaman Islands, India. – Rep. 22nd Internat. Geol. Congr. India, 11: 79–100.

KARUNAKARAN, C., RAY, K. K. & S. S. SAHA (1964): Sedimentary environment of the formation of Andaman Flysch, Andaman Islands, India. – Rep. 22nd Internat. Geol. Congr. India, 15: 226–232.

– – – (1967): Implications of the East Indian and Australia. – Internat. Symp. on continental Drift. – UNESCO, Montevideo, Uruguay: 1–15.

– – – (1968): Tertiary sedimentation in the Andaman-Nicobar Geosyncline. – J. Geol. Soc. of India, 9, 1: 32–39, Bangalore.

KATILI, J. A. (1974): Metallogenesis, hydrocarbons and tectonic patterns in Eastern Asia. – Rep. of the IDOE Workshop on Tectonic Patterns and Metallogenesis in East and Southeast Asia, Bangkok, Thailand, Sept. 1973, UNESCO CCOP, IOC: 158 pp.

KELTERBORN, P. (1925): Einige Notizen über Gesteine des jungvulkanischen Gebietes am Lower Chindwin, Upper Burma. – Eclog. Geol. Helv., 19, 2: 352–359.

KEMPER, E. (1976): The Foraminifera in the Jurassic Limestone of West Thailand. – Geol. Jb., B 21: 129–153, Hannover.

KEMPER, E., MARONDE, H. D. & D. STOPPEL (1976): Triassic and Jurassic Limestone in the region Northwest and West of Si Sawat (Kanchanaburi Province, Western Thailand). – Geol. Jb., B 21: 93–127, 1 fig., 1 table, 6 pls., Hannover.

KEVAN, D. K. McE., SINGH, A. & S. S. AKBAR (1964): A new genus of Pyrogamarphidae (Orthoptera, Acrictoidea) from Burma. – Nat. Hist., USA, No. 363: 7, 7 figs., 1 pl.

KHIN, A. & K. WIN (1969): Geology and hydrocarbon prospects of the Burma Tertiary geosyncline. – Union of Burma J. Sci. Technol. 2, 1: 53–82.

KHIN MAUNG TAW & TIN WIN (1970): Forecasting of lower Irrawaddy floods due to heavy rainfall in head water regions. – Union of Burma J. Sci. Technol. (5th Burma Res. Congr., 1970), 3, 1: 231–243, 6 figs., Rangoon.

KIM MINS, D. E. (1963): On the Leptocerinae of the Indian Subcontinent and Northeast Burma (Trichoptera). – Bull. Brit. Mus. Nat. Hist. Entomol., 14, 6: 263–316, 164 figs.

KINGDON-WARD, F. (1949): Burma's icy mountains. – London.

KLOOTWIJK, C. T. (1976): The drift of the Indian continent; an interpretation of recent paleomagnetic data. – Geol. Rdsch., 65, 3: 885–909.

KO KO (1970): Choice of effective anomalous values from among the local thresholds in Lower Yeboke area, Southern Shan State, Burma. – Econ. Comm. Asia Far East, Miner. Resources Develop. Ser. (USA) (Proc. 2nd Semin. Geochem. Prospect. Methods Techn.; Peradeniya, 1970), 38: 309–313, 6 figs.

KO KO LAY, HTIN AUNG & TIN AUNG (1969): Mn and Mg contents in laterite of Mingaladon Area. – Union of Burma J. Sci. Technol. (3rd Burma Res. Congr., 1968), 2, 2: 403–413, 10 figs., 10 analys., Rangoon.

KOBAYASHI, T. (1934): The Cambro-Ordovician formations and faunas of south Chosen. Paleontology Part 1 – Middle Ordovician faunas. – J. Fac. Sci. Imp. Univ. Tokyo, sect. 2, 3, pt. 8: 329–419.

– (1957): Upper Cambrian fossils from Peninsula Thailand. – J. Fac. Sci. Univ. Tokyo, sect. 10, 10, pt. 3: 367–382, pls. 4–5.

– (1959): On some Ordovician fossils from Northern Malaya and her adjacence. – J. Fac. Sci. Univ. Tokyo, 31, 2–4: 387–407, 4 pls., 3 figs., 1 table.

– (1960): Notes on the geologic history of Thailand and adjacent territories. – Jap. J. Geol. Geogr., 31: 129–148.

– (1964a): Paleontology of Thailand. – (1916–1962), Ibid, 1: 17–29.

– (1964b): On the orogenies of the Burmese-Malayan Geosyncline. – 22nd Internat. Geol. Congr., Abstr., 11, Proc. Sect. 2: 175, pt. 11: 123–131, New Delhi.

– (1972): The early stage of the Burmese-Malayan Geosyncline. – Geol. Soc. Malays., Newsl. (Malays. Reg. Conf. Geol. Southeast Asia, Abstr.; Kuala Lumpur, Malaysia; 1972), 34: 33.

– (1973a): The early stage of the Burmese-Malayan Geosyncline. – Geol. Soc. Malaysia, Bull., 6: 119–129, Kuala Lumpur.

– (1973b): The Early and Middle Paleozoic history of the Burmese-Malayan Geosyncline. – Geol. Paleont. Southeast Asia, 11: 93–107.

KOBAYASHI, T. & T. HAMADA (1970): A cyclopygid-bearing Ordovician Faunule discovered in Malaysia with a note on the Cyclopygidae. – Geol. Pal. Southeast Asia, 8: 1–18.

KOBAYASHI, T. & HISAYOSHI (1965): On the occurrence of Graptolite shales in North Thailand. – Contr. to the Geol. and Paleont. of Southeast Asia, XVIII. – Jap. J. Geol. Geogr. 36, 2–4: 37–44.

KOCH, K. E. (1973): Geology of the Region Sri Sawat-Thon Pha Phum-Sangkhlaburi (Kanchanabury Province/Thailand). – Geol. Soc. Malaysia, Bull., 6: 177–185.

– (1978): Geological map of Northern Thailand 1 : 250 000, Sheet Thon Pha Phum, Bundesanstalt für Geowissenschaften und Rohstoffe, Hannover.

KOENIGSWALD, G. H. R. VON (1965): Critical observations upon the so-called Higher Primates from the Upper Eocene of Burma. – Kkl. Nederl. Wetensch., Proc. Ser. B, 68, 3: 165–167.

KOMALARJUN, P. & T SATO (1964): Aalenian (Jurassic) Ammonites from Mae Sot, North-western Thailand. – Geol. Palaeont. Southeast Asia, 1: 237–251, 6 figs., 1 pl., Tokyo.

KOTAKA, T. & S. UOZUMI (1962): Variation and dimorphism of Pachymelania (Gastropoda) from the Eocene of Burma. – Trans. Proc. Paleont. Soc. Japan, 47: 301–309, 3 figs., 1 pl. (Abstr. in Japan.).

KREBS, W. (1975): Formation of Southwest Pacific Island Arc-Trench and mountain systems: Plate or global-vertical tectonics? – Amer. Assoc. Petrol. Geol. Bull., **59**, 9: 1639–1666.

KRENNER, J. A. (1883): Über Jadeit. – N. Jb. Miner.: 173–174, Stuttgart.

– (1899): Die wissenschaftlichen Ergebnisse der Reise des Grafen Bela SZECHENYI in Ostasien. – Jadeitstein aus Birma, 3, 3: 345–351 (Abstr. J. Chem. Soc. London, 76: 672–673).

KRISHNAN, M. S. (1949): Geology of India and Burma. – 544 p., Madras: Madras Law Journal Press.

KRISHNAN, M. S. & P. K. GHOSH (1940): Manganese ore in Bamra State. – Rec. Geol. Surv. India, **75**, prof. pap. 8.

KRISHNAN, M. S. & K. JACOB (1957): Burma. – Lexique Stratigraphique International, 3, 8b: 285–328.

KRISL, P. H. (1975): Geology of the NW-part of Monywa copper area (Lower Chindwin District/Burma). – Post-Graduate Training in Mineral Exploration, BUR/71/516, Rangoon Arts and Science Univ., Rangoon.

Krupp – Rohstoffe (1962): Rohstoffversorgung Eisen- und Stahlindustrie Burma, Bd. I: Eisenerzgebiet Taunggyi, Bd. II: Eisenerzgebiet Maymyo, Bd. III: Eisenerzgebiet Mergui. – Unpubl. Rep. Friedr. Krupp Rohstoffe, Essen.

KRUSCH, P. (1933): Burmas Blei-Silber-Zink-Erze als Quelle der deutschen Kobaltproduktion. – Metallwirtsch., **12**: 727–728.

KUDRJAVCEV, G. A. (1969): Geologie von Südostasien, Indochina. – Geol. Jugo-Vostocnoj Azii, Indokitaj, Leningrad, 71–111.

KUDRJAVCEV, G. A., GATINSKIJ, J. G., MISHNA, A. V. & A. N. STROGANOV (1968): Some tectonic features of Burma and the Malacca Peninsula (in Russ.). – Geotektonika, S.S.S.R., 4: 99–113, 1 fig.

KURAKOVA, L. I. (1966): Geomorphological map of Burma: 1:2 000 000 (in Russ.). – Vest. moskov. Univ., Geogr., 2: 122–125, 1 fig.

KUTINA, J. (1969): Ge-Ti-As type of a colloform sphalerite from Katha District, Burma. – Indian Mineral., **10**: 146–151, 2 pls., Mysore.

KYAW NYEIN (1969): Case history of Chauk and Lanywa oilfields. – Union of Burma J. Sci. Technol. (1st Burma Res. Congr., 1966), 2, 2: 243–274, 15 figs., 5 pls., Rangoon.

KYAW WIN & THIT WAI (1971): Geology of the Arakan-Chin Range. – 1971, Burma Res. Congr., Abstr. Vol.: 23.

KYI MAUNG (1970): Biostratigraphy of the Central Burma Basin with special reference to the depositional conditions during Late Oligocene and Early Miocene times. – Union of Burma J. Sci. Technol. (3rd Burma Res. Congr., 1968), 3, 1: 75–90, 4 pls., Rangoon.

KYI MAUNG & MG MG THA (1968): Some aspects of faunal studies of Arakan Coastal and Irrawaddy Delta Areas. – Union of Burma J. Sci. Technol. (1st Burma Res. Congr., 1966), 1, 3: 421–430, 3 figs., Rangoon.

LACROIX, A. (1930): La jadeite en Birmanie; les Roches qu'elle constitue ou qui l'accompagnent, composition et origine. – Bull. Soc. Fr. Miner., **53**: 216–254.

LAMBERT, I. B. & T. SATO (1974): The Kuroko and associated ore deposits of Japan: A review of their features and metallogenesis. – Econ. Geol., 9: 1215–1236.

LANDER, C. H. & F. W. WALKER (1924): Report on the examination of Burmese lignites from Namma, Lashio and Pauk. – Rec. Geol. Surv. India, **61**, 4: 345–351.

LA TOUCHE, T. H. D. (1886): Geology of the Upper Dehing Basin in the Singhpo Hills. – Rec. Geol. Surv. India, **19**: 111–115.

– (1900): Note on the Namya Series. – Rec. Geol. Surv. India 1899–1900: 85.

– (1906a): On recent changes in the course of the Namtu River, Northern Shan States. – Rec. Geol. Surv. India, **33**, 1: 46–48.

– (1906b): Note on the natural bridge in the Gokteik Gorge. – Rec. Geol. Surv. India, **33**, 49–54.

– (1907a): Brine Well at Bawgyo, Northern Shan States. – Rec. Geol. Surv. India, **35**, 2: 97–101.

– (1907b): Report on gold bearing deposits of Loi Twang, Shan States, Burma. – Rec. Geol. Surv. India, **35**, 2: 102–113.

– (1908a): On a volcanic outburst of Late Tertiary age in South Hsenwi, Northern Shan States. – Rec. Geol. Surv. India, **36**, 1: 40–44.

– (1908b): Note on the Plateau Limestone. – Rec. Geol. Surv. India, **37**.

– (1913a): Geology of the Northern Shan States. – Mem. Geol. Surv. India, **39**, 2: 1–379, 28 pls., 3 maps, 11 figs.

– (1913b): Geology of the Northern Shan States. – Mem. Geol. Surv. India, **39**.

LA TOUCHE, T. H. D. & J. C. BROWN (1909): The silver-lead mines of Bawdwin, Northern Shan States. – Rec. Geol. Surv. India, 37, 3: 235–263.

LA TOUCHE, T. H. D. & R. R. SIMPSON (1906): The Lashio coal field, Northern Shan States. – Rec. Geol. Surv. India, 33, 1: 117–124.

LAUFER, B. (1907): Historical joltings on amber in Asia. – Mem. Amer. Anthr. Assoc., 1, 3.

LE FORT, P. (1975): Himalayas: The collided range, present knowledge of the continental arc. – Amer. J. Sci., 275, ser. A: 1–44.

LEHNER, E. J. (1941): Traverse across Taungup Pass. – Unpubl. Rep. Burmah Oil Co. (Digboi).

LEICESTER, P. (1930): Geology of Amherst District. – Rec. Geol. Surv. India, 63, 1 (Dir. Gen. Rep.).

– (1932a): The Eruption of a Mud Volcano off the Arakan Coast. – Rec. Geol. Surv. India, 65, 3: 442–443.

– (1932b): The Geology and Underground Water of Rangoon. – Rangoon: Govt. Printing, 1–78.

LENZ, H. & P. MÜLLER (1981): Rb/Sr age determinations (total rock) of rocks of the Bawdwin Volcanic Formation, Northern Shan State, Burma. – Geol. Jb. D 43: 47–52, 2 tables, 1 fig., Hannover.

LEPPER, G. W. (1933): An outline of the geology of the oil bearing regions of the Chindwin-Irrawaddy Valley of Burma and of Assam-Arakan. – Proc. World Petrol. Congr. pap. No. 169, 1: 15–25.

LORD, L. (1924): Thayetmyo District. – Agric. Surv. No. 4, Dep. of Agriculture, Rangoon.

LOVEMAN, M. H. (1916): The geology of the Bawdwin Mines, Burma. – Bull. Amer. Inst. Min. Engin., No. 120: 2120–2143.

– (1919): A connecting link between the geology of the Northern Shan States and Yunnan. – J. Geol., 27: 204–221.

LOW, J. (1829): Geological appearances and general features of portions of the Malayan Peninsula and the countries lying between it and 18° North Latitude. – Asiat. Res., 18, 1: 128–162; J. Asiat. Soc. Bengal, 3: 305–326.

LUYENDYIK, B. P. (1977): Deep sea drilling on the Ninetyeast Ridge: synthesis and a tectonic model. – In: Indian Ocean Geology and Biostratigraphy (Studies following Deep-Sea Drilling legs 22–29): 165–185; Amer. Geophys. Union, Washington.

LYDEKKER, R. (1876): Notes on the fossil Mammalian faunas of India and Burma. – Rec. Geol. Surv. India, 9, 3: 86–106.

– (1880): Teeth of fossil fishes from Ramri Island and Punjab. – Rec. Geol. Surv. India, 13, 1: 59–61.

LYONS, C. M. (1918): Methods of alluvial mining applicable to Tavoy conditions. – 45–54, Rangoon: Superintendent Government Printing.

MACDONALD, A. S. & S. M. BARR (1979): Tectonic significance of a Late Carboniferous Volcanic Arc in Northern Thailand. – Third Regional Conf. on Geol. and Min. Res. of Southeast Asia, Bangkok, Thailand 1978: 151–156.

MACLAREN, J. M. (1904): The geology of the Upper Assam. – Rec. Geol. Surv. India, 31: 179–204.

– (1907a): The course of the Upper Irrawaddy. – Geogr. J., 30: 507–511.

– (1907b): The auriferous deposits of Burma. – Min. J., 82: 113–114.

MACLEOD, T. E. (1837): Abstract Journal of an Expedition to Kiang Hung on the Chinese Frontier, starting from Moulmein on the 13th December 1836. – J. Asiat. Soc. Bengal, 6: 989–1005.

– (1838): On the hot springs of Palouk River, Tenasserim. – J. Asiat. Soc. Bengal, 7: 466–467.

MALCOLM, H. (1837): Account of the caves near Moulmein. – Asiat. J., N.S., 24, 2: 10.

MALLET, F. R. (1876): On the coal fields of the Naga Hills, bordering the Lakhimpur and Sibsagar Districts, Assam. – Mem. Geol. Surv. India, 12, 2: 269–363.

– (1878a): The mud volcanoes of Ramree and Cheduba. – Rec. Geol. Surv. India, 11, 2: 188–207.

– (1878b): On the mineral resources of Ramree, Cheduba and the adjacent islands. – Rec. Geol. Surv. India, 11, 2: 207–223.

– (1879): Note on recent mud volcano eruption in Ramree Island (Arakan). – Rec. Geol. Surv. India, 12, 1: 70–72.

– (1880): Record of gas and mud eruptions on the Arakan Coast on 12th March 1879, and in June 1843. – Rec. Geol. Surv. India, 13, 206–209.

– (1881): Notice of a mud eruption in the Island of Cheduba. – Rec. Geol. Surv. India, 14, 2: 196–197.

– (1882): Notice of a recent eruption from one of the mud volcanoes in Cheduba. – Rec. Geol. Surv. India, 15: 141–142.

– (1883a): Notice of a fiery eruption from one of the mud volcanoes of Cheduba Island, Arakan. – Rec. Geol. Surv. India, 16, 4: 204–205.

- (1883b): On the native lead from Moulmein and chromite from the Andaman Islands. – Rec. Geol. Surv. India, 16, 4: 203–204.
- (1884a): Notice of a further fiery eruption from the Minbyin Mud Volcano of Cheduba Island, Arakan. – Rec. Geol. Surv. India, 17, 3: 142.
- (1884b): On some of the mineral resources of the Andaman Islands in the neighbourhood of Port Blair. – Rec. Geol. Surv. India, 17: 79–86.
- (1885a): On the alleged tendency of the Arakan Mud Volcanoes to burst into eruption most frequently during the rains. – Rec. Geol. Surv. India, 18, 2: 124–125.
- (1885b): The volcanoes of Barren Island and Narcondam, in the Bay of Bengal: their Topography – Captain J. R. Hobday –, and Geology – F. R. Mallet. – Mem. Geol. Surv. India, 21, 4 (repr. in 1933): 251–286.
- (1887): A Manual of the Geology of India, 4. – In: Mineralogy: 179 pp., Calcutta.
- (1889): Note on Indian steatite. – Rec. Geol. Surv. India, 22: 59–67.
- (1907): A new mud volcano island (Arakan). – Nature, 75: 460.
Mansuy, H. (1916): Faunes Paléozoiques du Tonkin Septentrional. – Mém. Serv. géol. Indoch. Hanoi, 5, 4: 1–71.
Marks, G. N. (1918): Recent progress of wolfram mining in Tavoy. – 23–32, Rangoon: Superintendent Government Printing.
Marr, F. A. (1934): The geology of Arakan. – Unpubl. Rep. Burmah Oil Co. (Digboi).
Mason, F. (1850): The Natural Productions of Burmah, or Notes on the Fauna, Flora and Minerals of the Tenasserim Provinces, and the Burman Empire. – 2 vols., Moulmein.
Mason, F. & W. Theobald (1882): Burmah, its People and Production and Notes on the Fauna, Flora and Minerals of Tenasserim, Pegu and Burma. – 2 vols., Hertford.
Maung Maung Khin, Khin Maung Nyo, Htay Aung, Tha Htay Hla, Kyaw Nyein, Thike, Daw Nyo Nyo, Than Htay & Tin Aye (1970): Geology and copper ore deposit of Sabetaung in Monywa District. – J. Sci. Technol. Burma, 3, 3: 515–532.
Maung Thein (1971): Limestone resources in the Kyaukse Area. – J. Sci. Technol. Burma. 4, 1: 51.
- (1972): The geological evolution of Burma: A preliminary synthesis. – Geol. Soc. Malaysia, Newsl. (Reg. Conf. Geol. Southeast Asia, Abstr., Kuala Lumpur, Malaysia, 1972), 34: 40.
- (1973): A preliminary synthesis of the geological evolution of Burma with reference to the tectonic development of Southeast Asia. – Geol. Soc. Malaysia, Bull. 6: 87–116, Kuala Lumpur.
Maung Thein & Ba Than Haq (1969): The pre-Paleozoic and Paleozoic stratigraphy of Burma: A brief review. – Union of Burma J. Sci. Technol. (2nd Burma Res. Congr., 1967), 2, 2: 275–287.
Maung Thein & Soe Win (1970): The metamorphic petrology, structures and mineral resources of the Shan-Taung-U-Thandwmyet Range, Kyaukse District. – J. Sci. Technol. Burma, 3, 3: 487–514.
Maung Thein et al. (1971): Geology and mineral resources of Myogi South, Kinda-Yechanbyin, Nyaunggyet-Yangan Areas, Kyaukse and Taunggyi Districts. – Burma Res. Congr., Abstr. Vol.: 19–20.
Maxwell-Lefroy, E. (1914): Mining in the Tavoy District in Lower Burma. – Min. Sci. Press, San Francisco, 109: 448.
- (1916): Wolframite mining in the Tavoy District in Lower Burma. – Trans. Inst. Min. Met., London, 25: 83–100.
McDougall, I. & M. McElhinny (1970): The Rajmahal traps of India. – K-Ar ages and palaeomagnetism. – Earth Planet. Sci. Letters 9, 4: 371–378, 2 figs., 3 tables, Amsterdam.
McKenna, J. & F. J. Warth (1911): Sugar cane in Burma. – Bull. 6, Dept. of Agric., Burma.
McKenzie, D. & J. G. Sclater (1971): The evolution of the Indian Ocean since the Late Cretaceous. – Geophys. J. Roy. Astron. Soc., 25: 437–528.
McKerral, A. (1910): Myingyan District. – Agric. Surv. No. 1: 5–7, Rangoon: Superintendent Government Printing.
- (1911): Sagaing District. – Agric. Surv. No. 2: 4–7, Rangoon: Superintendent Government Printing.
Medlicott, H. B., Blanford, W. T., Ball, V. et al. (1959): A Manual of the Geology of India and Burma, Vol. II; 3rd ed., by Sir E. H. Pascoe. – Geol. Surv. India, 1: 485–1345, 17 figs., 17 pls.
Meng, H. M., Chern, K. & T. Ho (1937): Geology of the Kochin tinfield, Yunnan. A preliminary sketch. – Bull. Geol. Soc. China, 16: 421–437.
Metallstatistik (1970–1978): Frankfurt am Main (Metallgesellschaft).
Meyer, R. O. (1955): Report on salt extraction centres in Burma. – Rep. Petrol. Min. Res. Develop. Corp. Burma, Rangoon.

MIDDLEMISS, C. S. (1900): Report on a geological reconnaissance in parts of the Southern Shan States and Karenni. – Gen. Rep. Geol. Surv. India (1899–1900): 122–153.

Mineral Development Corporation (1971): A Résumé of Burma's Tin-Tungsten Mining Industry. – Rapp., 11 pp.

Minerals Yearbook (1970–1978): US Bureau of Mines, Washington D.C.

Mining Annual Review (1970–1978): London.

MISCH, P. (1945): Remarks on the tectonic history of Yunnan with special reference to its relation to the type of Young Orogenic Deformation. – Bull. Geol. Soc. China, 25: 47–153.

MITCHELL, A. H. G. (1973): Metallogenic belts and angle of dip of Benioff Zones. – Nature, 245, 143: 49–52.

– (1976): Tectonic settings for emplacement of subduction-related magmas and associated mineral deposits. – In: D. STRONG: Metallogeny and plate tectonics. – Geol. Assoc. Canada, Spec. Pap. 14: 3–21, Montreal.

– (1977): Tectonic settings for emplacement of Southeast Asian tin granites. – Geol. Soc. Malaysia, 9: 123–140.

MITCHELL, A. H. G. & M. S. GARSON (1972): Relationship of porphyry copper and circum-Pacific tin deposits to paleo-Benioff zones. – Inst. Min. Metallurg. Transact./Sect. B, 81, Bull. 783: B10–B25, London.

MITCHELL, A. H. G., MARSHALL, T. R., SKINNER, A. C., AMOS, B. J. & J. H. BATESON (1977): Geology and exploration, geochemistry of the Yadanatheingyi and Kyaukme-Longtawkno areas, Northern Shan States, Burma. – Overseas Geol. Miner. Res., 51: 35 pp., 2 maps 1:125 000, 18 figs.

MITCHELL, A. H. G. & W. S. McKERROW (1975): Analogous evolution of the Burma Orogen and the Scottish Caledonides. – Geol. Soc. Amer., Bull., 86, 3: 305–315, 10 figs.

MITCHELL, A. H. G., YOUNG, B. & W. JANTARANIPA (1970): The Phuket Group, Peninsular Thailand: a Paleozoic geosynclinal deposit. – Geol. Mag.: 411–428.

MOLDENKE, E. (1921): Geology of the Namma Coal Field, Burma. – Min. Met., 1525: 30–31.

MOORE, D. G., CURRAY, J. R., RAITT, R. W. & F. J. EMMEL (1974): Stratigraphic-seismic section correlations and implications to Bengal Fan History. – In: VON DER VORCH, C. C., J. G. SCLATER et al.: Initial reports of the Deep Sea Drilling Project, XXII. Washington, D.C.: US Govt. Printing Office: 405–412.

MORGAN, A. H. (1904): The Burma ruby mines. – Min. J., 76:4.

MORNAY, S. (1843): Qualitative examination of native copper found on Round Island in the Cheduba Group, S.E. of Ramree. – J. Asiat. Soc. Bengal, 12: 904–906.

MORROW-CAMPBELL, J. (1919): The ore minerals of Tavoy. – Min. Mag. 20: 76–89.

– (1920): Tungsten deposits of Burma and their origin. – Econ. Geol., 15: 511–534.

MOULE, J. (1922): Burma Corporation Ltd. – Proc. Austr. Inst. Min. Met. 46: 82–84.

MOVIUS, Jr., H. L. (1943): The Stone Age in Burma. – Trans. Amer. Phil. Soc., N.S., 32: 341–393.

MÜLLER, K. J. (1967): Devonian of Malaya and Burma. – Internat. Symp. on the Devonian System, Calgary, 1: 565–568, 2 figs.

MÜLLER, P. & T. WEISER (1981): Studies of the petrology and ore mineralogy in the area of the Bawdwin ore deposit, Northern Shan State, Burma. – Geol. Jb., D43: 53–64, 4 tables, Hannover.

MUIR-WOOD, H. M. (1948): Malayan Lower Carboniferous fossils and their bearing on the Visean paleogeography of Asia. – London: Brit. Mus. (Nat. Hist.).

MURPHY, R. W. (1975): Tertiary Basins of Southeast Asia. – Seapex Proc., 2: 1–36.

MURTHY, M. V. N. (1970): Tectonics and mafic igneous activity in northeast India in relation to the Upper Mantle. – Proc. 2nd Symp. Upper Mantle Project, Natl. Geophys. Res. Inst., Hyderabad: 287–304.

MYINT KYAW (1968): Simple method for flood forecasting for the Irrawaddy Lower Reach. – Union of Burma J. Sci. Technol. (2nd Burma Res. Congr., 1967), 1: 155–163, 9 figs.

MYINT LWIN THEIN (1968a): On some Nautiloid Cephalopods from the area East of Kyaukse, Burma. – Union of Burma J. Sci. Technol. (1st Burma Res. Congr., 1966), 1, 1: 67–76, 13 figs.

– (1968b): Composition of indigenous natural gas. – Union of Burma J. Sci. Technol. (1st Burma Res. Congr., 1966), 1, 1: 1–10, 13 figs.

– (1970): On the occurrence of *Daonella* Facies from the Upper Chindwin Area, Western Burma. – Union of Burma J. Sci. Technol. (3rd Burma Res. Congr., 1968), 3, 2: 277–282, 7 figs.

– (1973): The Lower Paleozoic Stratigraphy of Western Part of the Southern Shan States, Burma

(Abstr.). – In: Regional Conf. on the Geology of Southeast Asia, Geol. Soc. Malaysia, Newsl., **34**: 42.

– (1973): The Lower Paleozoic Stratigraphy of Western Part of the Southern Shan States, Burma. – Geol. Soc. Malaysia, **6**: 143–163.

MYINT LWIN THEIN & BA THAN HAQ (1966): The Mesozoic Geology of Burma, a brief review. – Geol. Dep., Arts and Sci. Univ., Rangoon.

– – (1967): The pre-Paleozoic and Paleozoic stratigraphy of Burma, a brief review. – Pap. of 2nd Burma Res. Congr., Geol. Dep. Arts and Sci. Univ., Rangoon.

MYINT LWIN THEIN et al. (1971): Geology and mineral resources of the Pindaya Range, Southern Shan States. – Pap. pres. at 1971, Burma Res. Congr., Abstr. Vol.: 21.

NAGAPPA, Y. (1959): Foraminiferal biostratigraphy of the Cretaceous-Eocene Succession in the India Pakistan Burma Region. – Micropaleontology, **5**: 145–177, 11 figs., 11 pls., 1 table, New York.

NANDY, D. R. (1973): Geology and structural lineaments of the Lohit Himalaya (Arunachal Pradesh) and adjoining areas. – Proc. Sem. Geodyn. Himalayan Region, Natl. Geophys. Res. Inst., Hyderabad: 167–172.

NATAL'IN, N. B. (1960): Soil cultivation in Burma (in Russ.). – Vest. Sel'skokhoz Nauki, U.S.S.R., **6**: 114–118, 6 figs. (Abstr. in English, French, Dutch).

NEUBAUER, W. H. (1965): Möglichkeiten für die Eisen- und Stahlerzeugung in mittel- und fernöstlichen Entwicklungsländern. – Stahl u. Eisen, **85**, 1.

NGAW CIN PAU (1962): Report on a geological reconnaissance in the Naga Hills. – Unpubl. Rep. Geol. Sect. Petrol. Min. Devel. Corp. (Govt. of Burma), Rangoon.

NOETLING, F. (1889): Report on the oilfields of Twingoung and Beme, Burma. – Rec. Geol. Surv. India, **22**, 2: 75–136.

– (1890a): Report on the Upper Chindwin Coal Fields. – Calcutta.

– (1890b): Field notes from the Shan Hills (Upper Burma). – Rec. Geol. Surv. India, **23**, 2: 78–79.

– (1891a): Note on a salt spring near Bawgyo, Thibaw State. – Rec. Geol. Surv. India, **24**: 129–131.

– (1891b): Report on the coal-fields in the Northern Shan States. – Rec. Geol. Surv. India, **24**: 99–119.

– (1891c): Note on the reported Namseka ruby mine in the Mainglon State. – Rec. Geol. Surv. India, **24**, 2: 119–125.

– (1891d): Note on the tourmaline (Schorl) mines in the Mainglon State. – Rec. Geol. Surv. India, **24**: 125–128.

– (1892): Report on the petroleum industry in Upper Burma, from the end of the last century up to the beginning of 1891. – Rangoon, 73 pp., with Appendices Fol. 1892, Prelim. Rep. on the Econom. Resources of the Amber and Jade Mines Area in Upper Burma. – Rec. Geol. Surv. India, **25**, 3: 130–135.

– (1893a): Note on the occurrence of jadeite in Upper Burma. – Rec. Geol. Surv. India, **26**, 1: 26–31.

– (1893b): On the occurrence of Burmite, a new fossil resin from Upper Burma. – Rec. Geol. Surv. India, **26**: 31–40.

– (1893c): Lead deposits of Burma. – J. Soc. Chem. India, **12**: 1075.

– (1893d): Carboniferous fossils from Tenasserim. – Rec. Geol. Surv. India, **26**, 3: 96–100.

– (1894a): Note on the geology of Wuntho in Upper Burma. – Rec. Geol. Surv. India, **27**, 4: 115–124.

– (1894b): On the occurrence of chipped flints in the Upper Miocene of Burma. – Rec. Geol. Surv. India, **27**, 3: 101–103.

– (1894c): Gold in the Mingin Mountains of Northern Burma. – Rec. Geol. Surv. India, **27**: 115–124.

– (1895a): On some marine fossils from the Miocene of Burma. – Mem. Geol. Surv. India, **27**, 1: 1–45.

– (1895b): The development and subdivision of the Tertiary System in Burma. – Rec. Geol. Surv. India, **28**, 2: 59–86.

– (1896): Über das Vorkommen von Jadeit in Ober-Birma. – N. Jb. Miner., **1**: 1–17.

– (1897): The occurrence of petroleum in Burma and its technical exploitation. – Mem. Geol. Surv. India, **27**, 2: 47–272, 17 pls., 1 map.

– (1899): Fauna of Miocene Beds of Burma. – Palaeontologia Ind., N.S., **1**, 3: 182–189.

– (1900): The Miocene of Burma. – Verh. K. Akad. Wetensch. Amsterdam, Sect. 2, 7, 2: 1–131, 1 map, 3 figs.

- (1901): Fauna of the Miocene Beds of Burma. – Palaeontologia Ind. Mem. **3**: 1–378, pl. 25, 5 figs.
- (1910a): Geology and prospects of oil in Western Prome and Kama, Lower Burma (including Namayan, Padaung, Taungbogyi and Ziaing). – Rec. Geol. Surv. India, **38**, 4.
- (1910b): Coal-fields in North-Eastern Assam. – Rec. Geol. Surv. India, **40**.
- (1933/34): Gold near Kalaw, Southern Shan States. – Rec. Geol. Surv. India, **67**: 166–248.
- (1938): The geology of parts of the Minbu and Thayetmyo Districts. – Mem. Geol. Surv. India, **72**, 2.
- (1941): The Cretaceous and associated rocks of Burma. – Mem. Geol. Surv. India, 74, 1.
NUM KOCK & TAUKLIN (1968): Mechanical properties of soil in R.I.T. compound. – Sci. Technol. Burma, **1**, 2: 291–307, 21 figs.
NUM KOCK & TIN MAUNG (1969): A laboratory study of lateritic soil. – J. Sci. Technol. Burma, **2**, 2: 99–119.
NUM KOCK & TIN WIN (1968): Some experiments on the consolidation of lateritic soil. – J. Sci. Technol. Burma (2nd Burma Res. Congr., 1967), **1**, 3: 537–553, 15 figs.
NYI NYI (1964a): The palaeogeographic maps of Burma. – Sunday supplements of the Working People's Daily, Rangoon.
- (1964b): Geological sciences in Burma. – A review. – Contr. to the Symp. **1**: 1271–1295, Rangoon.
OBUKHOV, A. I. (1968a): Minor element content and distribution in soils of the Arid Tropical Zone in Burma (in Russ.). – Pochnovedenie, S.S.S.R., 2: 93–102 (Abstr. in Engl.).
- (1968b): Content and distribution of minor elements in the soils of the Dry Tropical Zone of Burma. – Soviet Soil Sci., Trade U.S. Dep. Pochnovedenie, S.S.S.R., 2: 224–233, 5 tables.
OBUKHOVA, V. A. (1969): Composition of humus from the paddy field soils of the Irrawaddy (in Russ.). – Vest. Moskov. Univ., 6: 89–93, 2 figs., 1 table.
OHN GYAW (1969): Investigation of method for flood fore-casting of Pegu River. – Union of Burma. J. Sci. Technol. Burma (3rd Burma Res. Congr., 1968), **2**, 3: 589–601.
OHN MAUNG, SAN HLA MAUNG & AUNG NAING (1970): A method of forecasting Irrawaddy stages at Nyaung Oo. – Union of Burma J. Sci. Technol. Burma (5th Burma Res. Congr., 1970), **3**, 2: 405–419, 9 figs.
OLDHAM, R. D. (1883): Report on the geology of parts of Manipur and the Naga Hills. – Mem. Geol. Surv. India, **19**, 4: 217–242.
- (1885): Notes on the geology of the Andaman Islands. – Rec. Geol. Surv. India, **18**, 3.
- (1893): Note on the alluvial deposits and subterranean water supply of Rangoon. – Rec. Geol. Surv. India, **26**, 2: 64–70.
- (1895): The alleged Miocene Man in Burma. – Nat. Sci., 7: 201–207.
- (1906): On explosion craters in Lower Chindwin District Burma. – Rec. Geol. Surv. India, **34**: 137–147.
OLDHAM, T. (1853): Memorandum of the results of an examination of specimens of gold dust and gold from Shuy-Gween. – Sel., Rec., Bengal Govt., **13**: 59–62.
- (1856): Notes on the coal field and tin-stone deposits of the Tenasserim Provinces. – Sel., Rec. Geol., Govt. India, **10**: 31–67.
- (1858): Notes on the geological features of the banks of the Irrawaddy and of the country North of Amarapoora. – Yule's Mission to Ava. – Appendix A: 309–351.
- (1872): Discovery of petroleum near Thayetmyo. – Indian Econom., **3**: 191–193.
OLIVER, J., ISACKS, B., BARAZANGI, M. & W. MITRONOVAS (1973): Dynamics of the down-going lithosphere. – Tectonophys., **19**: 133–147, Amsterdam.
O'RILEY, E. (1847): Notes on the geological formations of Amherst Beach, Tenasserim Provinces. – J. Nat. Hist., **8**: 186–189.
- (1849): Remarks on the metalliferous deposits and mineral productions of the Tenasserim Provinces. – J. India Archipelago, **3**: 724–743.
- (1852): Report on minerals in Tenasserim. – Sel. Rec., Bengal Govt., **6**: 21–24.
- (1864): Remarks on the "Lake of the Clear Water" in the District of Bassein, British Burmah. – J. Asiat. Soc. Bengal, **33**: 39–44.
OSSBERGER, R. (1968): Über die Zinnseifen Indonesiens und ihre genetische Gliederung. – Z. dt. geol. Ges., **117**: 749–766.
OTT, E. (1972): Die Kalkalgen-Chronologie der alpinen Mitteltrias in Angleichung an die Ammoniten-Chronologie. – N. Jb. Geol. Paläont., Abh., **141**, 1: 81–115, 2 figs., 1 table.
PAIN, A. C. D. (1943): The gems of Burma. – Gemmologist, **12**: 37–40.

PASCOE, E. H. (1906a): The northern part of the Gwegyo Anticline, Myingyin District, Upper Burma. – Rec. Geol. Surv. India, 34, 2: 261–265.

– (1906b): The asymmetry of the Yenangyat-Singu Anticline, Upper Burma. – Rec. Geol. Surv. India, 34, 4: 253–260.

– (1906c): The Kabat Anticline, near Seiktein, Myingyan District, Upper Burma. – Rec. Geol. Surv. India, 34, 4: 242–252.

– (1907): Fossils in the Upper Miocene of the Yenangyaung oil field, Upper Burma. – Rec. Geol. Surv. India, 35, 2: 120.

– (1908a): Marine fossils in the Yenangyaung oil-field, Upper Burma. – Rec. Geol. Surv. India, 36, 3: 143–146.

– (1908b): The Wetchok-Yedwet Pegu outcrop, Magwe District, Upper Burma. – Rec. Geol. Surv. India, 36, 4: 286–294.

– (1908c): On the occurrence of fresh water shells of the genus *Batissa* in the Yenangyaung oil field, Upper Burma. – Rec. Geol. Surv. India, 36, 3: 143–146.

– (1909): Note on a Pegu Inlier at Ondwe, Magwe District, Upper Burma. – Rec. Geol. Surv. India, 38: 152–153.

– (1912a): A traverse across the Naga Hills of Assam from Dinapur to the neighbourhood of Sarameti Peak. – Rec. Geol. Surv. India, 62, 1: 254–264.

– (1912b) The oil fields of Burma. – Mem. Geol. Surv. India, 40, 1: 1–269, 55 pls., 7 figs.

– (1924): General Report of the Geological Survey of India for the year 1923. – Mergui District, Burma. – Rec. Geol. Surv. India, 55, 1: 31–32.

– (1927): A gas eruption on Ramree Island, off the Arakan Coast of Burma, in July, 1926. – Rec. Geol. Surv. India, 60, 2: 153–156.

– (1929): Myitkyina District (Gen. Rep.). – Rec. Geol. Surv. India, 67: 108–114.

– (1930): General Report of the Geological Survey of India for the year 1929. – Rec. Geol. Surv. India, 63, 1: 1–154.

– (1950, 1959, 1964): Manual of the Geology of India and Burma. – Vol. 1, 3rd ed., 483 pp., Govt. India Press, Calcutta, 1950. – Vol. 2, 3rd ed., 1343 pp., Govt. India Press, Calcutta, 1959. – Vol. 3, 3rd ed., Govt. India Press, Calcutta, 1964.

PASCOE, E. H. & G. P. de COTTER (1908): On new species of *Dendrophyllia* from the Upper Miocene of Burma. – Rec. Geol. Surv. India, 36, 3: 147–148, 1 pl.

PAUL, D. D. & H. M. LIAN (1975): Offshore Tertiary basins of Southeast Asia, Bay of Bengal to South China Sea. – 9th World Petrol. Congr., Panel Discussion 7 (5): 1–15, Tokyo.

PEAL, S. E. (1881): Report on a visit to the Nongyang Lake, on the Burmese frontier, Febr. 1879. – J. Asiat. Soc. Bengal, 1, 2: 1–30.

PEMBERTON, R. B. & S. F. HANNAY (1837): Abstract of the Journal of a route travelled by Capt. S. F. HANNAY, from the Capital of Ava to the amber mines of the Hukong Valley on the Southeast frontier of Assam. – J. Asiat. Soc. Bengal, 6: 245–278.

PENZER, N. M. (1922): The Mineral Resources of Burma. – 176 pp., London: George Routledge & Sons Ltd.

PETER, G., WEEKS, L. A. & R. E. BURNS (1966): A reconnaissance geophysical survey in the Andaman sea and across the Andaman Nicobar island arc. – J. Geophys. Res. 71: 495–509.

PIA, J. (1930): Upper Triassic fossils from the Burma-Siamese frontier; A new Dasycladaceae, *Holosporella siamensis,* nov. gen. spec., with a description of the allied genus *Aciculella* PIA. – Rec. Geol. Surv. India, 63, 1: 177–181.

PIDDINGTON, H. (1831): Analytical examination of a mineral water from the Athan Hills, Tenasserim Province. – Gleaning in Sci., 3: 24–26.

– (1846): Note on gem sands from Ava. – J. Asiat. Soc. Bengal, 15: 61.

– (1847a): On a new king of coal, being volcanic coal from Arracan. – J. Asiat. Soc. Bengal, 16: 371–373.

– (1847b): Notice of Tremenheerite, a new Carbonaceous mineral. – J. Asiat. Soc. Bengal, 16: 369–371.

– (1853): Note on laterite and lateritous clays from Rangoon. – J. Asiat. Soc. Bengal, 22: 206–207.

PILGRIM, G. E. (1904): Fossils from the Yenangyaung oil field, Burma. – Rec. Geol. Surv. India, 31, 2: 103–104.

– (1906): Fossils of the Irrawaddy Series from Rangoon. – Rec. Geol. Surv. India, 33, 2: 157–158.

– (1910): Preliminary note on a revised classification of the Tertiary fresh water deposits of India. – Rec. Geol. Surv. India, 40: 185–205.

– (1913): The correlation of the Siwaliks with Mammal horizons of Europe. – Rec. Geol. Surv. India, 43, 4: 264–326.

– (1925): The Perissodactyla of the Eocene of Burma. – Paleontologia Ind., N.S., 7, Mem. 3: 1–28, 2 pls.

– (1928): The Artiodactyla of the Eocene of Burma. – Paleontologia Ind., N.S., 13: 1–39.

PILGRIM, G. E. & G. P. de COTTER (1916): Some newly discovered Eocene Mammals from Burma. – Rec. Geol. Surv. India, 47, 1: 42–77, 6 pls.

PIMM, A. C. & J. G. SCLATER (1974): Early Tertiary hiatuses in the northeastern Indian Ocean. – Nature, 252: 362–365.

PINFOLD, E. S., DAY, A. E., STAMP, L. D. & H. L. CHHIBBER (1927): Late Tertiary igneous rocks of the Lower Chindwin Region, Burma. – Trans. Min. Geol. Inst. India, 21: 145–225, Calcutta.

PONGPOR ASNACHINDA (1979): Tin mineralization in the Burmese-Malayan Peninsula – A Plate Tectonic Model. – Third Reg. Conf. on Geol. and Min. Res. of Southeast Asia, Bangkok, Thailand, Nov. 1978: 293–299.

PORRO, C. (1915): Geology of the country near Ngahlaingdwin, Minbu District, Burma. – Rec. Geol. Surv. India, 45, 4.

POTONIÉ, R. (1960): Sporologie der eozänen Kohle von Kalewa in Burma. – Senckenbergiana Leth., 41, 1–6: 451–481, 2 pls., 13 figs.

POWELL, C. McA. & P. J. CONAGHAN (1973): Plate tectonics and the Himalayas. – Earth Planet. Sci. Letters, 20: 1–12.

PRAKASH, U. (1965 a): *Dipterocarpoxilon tertiarum* sp. nov., a new fossil wood from the Tertiary of Burma. – Current. Sci., India, 34, 8: 254–255, 3 figs.

– (1965 b): Fossil wood of Dipterocarpacae from the Tertiary of Burma. – Current. Sci., India, 34, 6: 181–182, 2 figs.

– (1971): Fossil woods from the Tertiary of Burma. – Palaeobotanist (India), 20, 1: 48–70, 45 figs., 8 pls.

PRASHAD, B. (1930): On some undescribed fresh water molluscs from various parts of India and Burma. – Rec. Geol. Surv. India, 63, 4: 428–433.

PRETZMANN, G. (1963): Über einige süd- und ostasiatische Potamoniden. – Ann. naturhist. Mus. Wien, 66: 361–372, 4 pls.

PRINSEP, J. (1831 a): Examination of the water of several hot springs on the Arracan Coast. – Gleaning in Sci., Calcutta, 3: 16–18.

– (1831 b): Examination of metallic button, supposed to be platina, from Ava. – Gleaning in Sci., Calcutta, 3: 39–42.

– (1832): Examination of minerals from Ava. – J. Asiat. Soc. Bengal, 1, 14–17: 305.

– (1833 a): Discovery of platina in Ava. – Asiat. Res., 18, 2: 279–284.

– (1833 b): Notes on the coal discovered at Khyuk Phyu in the Arracan District. – J. Asiat. Soc. Bengal, 2: 595–597.

– (1835): Chemical analysis of mineral water from Ava. – J. Asiat. Soc. Bengal, 4: 509.

QUIRING, H. (1945): Antimon. – In: Die metallischen Rohstoffe, H. 7, Stuttgart.

– (1946): Arsen. – In: Die metallischen Rohstoffe, H. 8, Stuttgart.

RAJU, A. T. R. (1968): Geological evolution of Assam and Cambay Tertiary Basins of India. – Bull. Amer. Assoc. Petrol. Geol. 52, 2422–2437.

RAO, M. (1922): Note on the oil shales of Mergui. – Rec. Geol. Surv. India, 54, 3: 342–343.

RAO, S. R. N. (1942): On *Lepidocyclina (Polylepidina) birmanica* sp. nov. and *Pseudophragmina pagoda* s. gen. nov. et sp. nov., from the Yaw stage (Priabonian) of Burma. – Rec. Geol. Surv. India, 77: 1–13, pl. 1–2 (Prof. pap. 12).

RAO, S. R. S. (1930): The geology of the Mergui District. – Mem. Geol. Surv. India, 55, 1: 1–62, 10 pls.

RASTALL, R. H. (1968): The genesis of tungsten ores. – Geol. Mag.: 194–241, 293, 367.

RASTOGI, B. K. (1973): Earthquake focal mechanisms and plate tectonics in the Himalaya-Burma Region. – In: Seminar on Geodynamics Himalayan Reg.: 205–206, Hyderabad (NGRI).

RASTOGI, B. K., SINGH, J. & R. K. VERMA (1973): Earthquake mechanisms and tectonics in the Assam-Burma Region. – Tectonophys., 18: 355–356.

RAU, S. R. (1921): Note on the stratigraphy of the Singu-Yenangyat Area. – Rec. Geol. Surv. India, **53**, 4: 321–330.

RAU, RAO BAHADUR S. SETHU RAMA (1933): The geology of the Mergui District. – Mem. Geol. Surv. India, **55**, 1: 1–62.

RAY, D. K. (1973): Problems of the geology of the Himalaya. – In: Geodynamics of the Himalayan region. – Indian Natl. Geophys. Res. Inst.: 104–119.

RAY, K. K. (1976): Geotectonics of the Circum Indian Meso-Cenozoic mobile belt with special reference to oil and natural gas possibilities. – In: Seminar on tectonics and metallogeny, Southeast Asia and Far East – 1974. – India Geol. Surv. Misc. Publ., **34**: 85–97.

RAY, K. K. & S. K. ACHARYYA (1976): Concealed Mesozoic-Cenozoic Alpine Geosyncline and its petroleum possibilities. – Amer. Assoc. Petrol. Geol. Bull., **60**, 5: 794–808.

REED, F. R. C. (1906): The Lower Paleozoic fossils of the Northern Shan State Upper Burma, with a section on Ordovician *Cystidea* by BATHER, F. A. – Palaeontologia Ind., N.S., **2**, Mem. 3: 1–154.

– (1908): The Devonian faunas of Northern Shan States. – Palaeontologia Ind., N.S., **2**, Mem. 3: 1–183.

– (1913): Further notes on the species *Camerocrinus asiaticus* from Burma. – Rec. Geol. Surv. India, **43**, 4: 335–338, 1 pl.

– (1915): Supplementary memoir on new Ordovician and Silurian fossils from the Northern Shan States. – Palaeontologia Ind. N.S., **6**, Mem. 1: 1–98.

– (1917): Ordovician – Silurian fossils from Yunnan. – Palaeontologia Ind., N.S., **6**, Mem. 3: 1–63.

– (1920): Carboniferous fossils from Siam. – Geol. Mag., **57**: 113–120, 172–178.

– (1924): Provisional list of the Paleozoic and Mesozoic fossils collected by Dr. Coggin BROWN in Yunnan. – Rec. Geol. Surv. India, **55**, 4: 314–326.

– (1927): Paleozoic and Mesozoic fossils from Yunnan. – Palaeontologia Ind., N.S., **10**, Mem. 1: 254–276, 20 pls.

– (1929): New Devonian fossils from Burma. – Rec. Geol. Surv. India, **62**, 2: 229–257.

– (1931): Notes on some Jurassic fossils from the Northern Shan States. – Rec. Geol. Surv. India, **65**, 1: 185–187.

– (1932 a): Note on a specimen of the genus *Maclurites* from the Ordovician of Burma. – Rec. Geol. Surv. India, **66**, 3: 438–440.

– (1932 b): Notes on some Lower Palaeozoic fossils from the Southern Shan States. – Rec. Geol. Surv. India, **66**, 2: 181–211.

– (1933): Anthracolithic faunas of the Southern Shan States, with an appendix by Stanley SMITH describing *Lobhophyllum orientale* nov. sp. – Rec. Geol. Surv. India, **67**, 1: 83–134.

– (1936 a): Jurassic Lamellibranchs from the Namyau Series, Northern Shan States. – Ann. Mag. Nat. Hist. Serv., **18**: 1–28.

– (1936 b): The Lower Paleozoic faunas of the Southern Shan States. – Palaeontologia Ind. N.S., **21**, Mem. 3: 1–130, pl. 7.

– (1954 a): Note on a Nautiloid shell of Permocarboniferous age from Burma. – Rec. Geol. Surv. India, **78**, 2: 298–299, 1 fig.

– (1954 b): Notes on a species of *Lingula* from the Zebingyi Beds of the Northern Shan States. – Rec. Geol. Surv. India, **78**, 2: 300.

REGAN, R. D., AIN, J. C. & W. M. DAVIS (1974): A Global Magnetic Anomaly Map. 1–19, (Greenbelt, Maryland): Goddard Space Flight Center.

REH, H. (1964): Bergwirtschaft und Lagerstättenverhältnisse Burmas. – Z. angew. Geol., **10**, 6: 316–319, 1 fig. (Abstr. in Russ. and Engl.).

REIMANN, K. V. & A. THAUNG (in press): Results of palynostratigraphical investigations of the Tertiary sequence in the Chindwin Basin, Northwestern Burma. – 2 figs., 3 tables, 4 pls.

RICHARDSON, D. (1831): On specimens of coal from Gendah (Kendat), on the Kueendwen (Chindwin). – Gleaning in Sci., **3**: 125, Calcutta.

– (1837): Abstract Journal of an Expedition from Moulmein to Ava through the Karean Country, between December 1836 and June 1837. – J. Asiat. Soc. Bengal, **6**: 1005–1012.

v. RICHTHOFEN, F. (1882) (Ed.): China – Ergebnisse eigener Reisen und darauf gegründeter Studien. – Vols. II and IV, Berlin 1883.

RIDD, M. F. (1971): Faults in Southeast-Asia and the Andaman rhombochasm. – Nat. Phys. Sci. **229**, 2: 51–52, London.

ROBISON, R. A. & J. PANTOJA-ALOR (1968): Tremadocian trilobites from the Nochistlan Region, Oaxaca, Mexico. – J. Palaeont., 42, 3: 767–800.

RODOLFO, K. S. (1966): Marine sedimentation off the Irrawaddy River, Burma. (Abstr.). – Geol. Soc. Amer., Spec. Pap., 101: 179–180.

– (1969): Bathymetry and marine geology of the Andaman Basin, and tectonic implications for Southeast Asia. – Geol. Soc. Amer., Bull. 80: 1203–1230.

ROMANIS, R. (1881 a): Rice soils of Burma. – Chem. News: 227.

– (1881 b): On the hot springs at Natmoo near Moulmein, British Burmah. – Chem. News, 43: 191.

– (1882 a): Analyses of laterite from Haranbee, Pegu. – Trans. Edinb. Geol. Soc., 4: 164.

– (1882 b): Mineral water from Amherst, British Burmah. – Chem. News, 45: 148.

– (1882 c): On the outcrops of coal on Myanoung Division of the Henzada District. – Rec. Geol. Surv. India, 15, 3: 178–181.

– (1882 d): Report on borings for coal at Engsein, British Burma. – Rec. Geol. Surv. India, 15, 2: 138.

– (1884): Report on the Yenanchaung oil wells. – Rangoon.

– (1885): Report on the oil-wells and coal in the Thayetmyo District, British Burma. – Rec. Geol. Surv. India, 18, 3: 149–151.

– (1886 a): Analaysis of gold dust from the Meza River Valley, Upper Burma. – Rec. Geol. Surv. India, 19, 4: 268–270.

– (1886 b): The goldfields of Burma. – Chem. News, 54: 278–279.

– (1886 c): Notes on Upper Burma. – Trans. Edinb. Geol. Soc., 5: 306.

ROSSMAN, G. R. (1974): Lavender jade the optical spectrum of Fe (3+) and Fe (2+)→ Fe (3+) intervalence charge transfer in jadeite from Burma. – Amer. Mineralogist, 59, 7: 868–870, 2 figs.

ROY, S. K. & S. KRISHNASWAMY (1935): Notes on the microscopic characters of Bawdwin ores. – Quart. J. Geol. Min. Met. Soc., India, 7: 59–69.

ROZANOV, B. G. (1961): Ages des Terres Rouges en Birmanie (in Russ.). – Nauch. Dokl. Vyssh. Shkoly biol. Wanki, SSSR, 4: 214–217.

– (1965): Soil map and land resources of Burma. – Nauchn. Dokl. Vyssh. Shkoly biol. Nauki, SSSR, No. 3: 182–188 (transl. Doklady, Soil Science, 13: 1601–1605, Madison, Wisc./USA).

– (1974): Explanatory note to the soil map of Burma. – Repr., 22 p., Ministry of Agric. and Forest, Rangoon.

ROZANOV, B. G. & I. M. ROZANOVA (1961): Soils of the moist monsoon tropical zone of Burma (in Russian). – Pochvovedeniy, SSSR, 12: 75–84 (Abstr. in Engl.), Moskow.

– – (1962): Soils of the arid monsoon tropical zone of Burma (in Russ.). – Pochvovedeniy, SSSR, 3: 73–82 (Abstr. in Engl.), Moskow.

– – (1963): Subtropical mountainous and high soils of Burma (in Russ.). – Vest. moskov. Univ., Bull. Pochvov, 3: 71–78, 1 fig.

RUNDALL, W. H. (1927): Lead in the Southern Shan States. – Bull. Inst. Min. Met.

– (1928): Note on the occurrence of galena in clay, Southern Shan States, Burma. – Trans. Inst. Min. Met. 37: 27–49.

RUTTEN, M. G. (1950): Comparison of Lepidocyclina zeijlmansi TAN from Borneo with Lepidocyclina birmanica RAO from Burmah. – Nederl. Kon. Akad. Wet. Proc., 53, 2: 196–198, 1 fig., Amsterdam.

SAHNI, B. (1931): Revision of Indian fossil plants. Coniferales. – Paleontologia Ind., N.S., 11, 4: 51–124.

– (1933): in L. L. FERMOR: General Report of the Geological Survey of India for the year 1931. – Rec. Geol. Surv. India, 66, 1: 97–98.

SAHNI, M. R. (1936 a): in A. M. HERON: General Report of the Geological Survey of India for the year 1935. – Rec. Geol. Surv. India, 71, 1: 1–104.

– (1936 b): On the supposed Cretaceous Cephalopods from the Red Beds of Kalaw and the age of the Red Beds. – Rec. Geol. Surv. India, 71: 166–169.

– (1936 c): On the geological age of the Namyau, Liu-Wun and Napeng Beds and of certain other formations in Indo China. – Rec. Geol. Surv. India, 71, 1: 217–230.

– (1937 a): Discovery of Orbitolina-bearing rocks in Burma with a description of Orbitolina birmanica, nov. sp. – Rec. Geol. Surv. India, 71, 4: 360–375.

– (1937 b): A Mesozoic coniferous wood Mesembrioxylon shanense, sp. nov. from the Southern Shan States of Burma. – Rec. Geol. Surv. India, 71, 4: 380–388.

– (1940): The Jurassic Brachiopoda of the Namyau Beds of the Northern Shan States, Burma. – Paleontologia Ind., N.S., 30, Mem. 1: 1–39, pl. 4.

– (1962): The Lower Palaeozoic in India and Burma with observations in its faunal anomalies and the age of the Vindhayan System. – Rec. Geol. Surv. India, 91: 357–364.

SAHNI, M. R. & V. V. SASTRI (1937): A monograph of the Orbitolines found in the Indian Continent (Chitral, Gilgit, Kashmir), Tibet and Burma, with observations on the age of the associated volcanic series. – Indian Geol. Surv. Mem., Paleontologia Ind., N.S., 33, 1: 1–50, 6 pls., figs. 4–6.

– – (1958): New Microforaminifera from the *Orbitolina*-bearing Cretaceous rocks of Burma and Tibet. – Rec. Geol. Surv. India, 85, 4: 437–452, pl. 2.

SAITO, R. (1964a): Some considerations on the stratigraphy of Burma (First Report). – On pre-Cambrian rocks of the Mergui Area, Southern Burma. – Kumamoto J. Sci., Ser. B, Sect. 1 Geol., 3, 2: 14–19.

– (1964b): Some considerations on the stratigraphy of Burma (Third Rep.). The structure of the Western border of Southern Shan State. – Kumamoto J. Sci., Ser. B, Sect. 1 Geol., 3, 2: 25–36.

SAITO, R. & S. UOZUMI (1959a): The relation between the red beds and the coal-bearing series in the Southern Shan States. – Kumamoto J. Sci., Ser. B, Sect. 1 Geol., 3, 2: 20–24.

– (1961): Some considerations on the stratigraphy of Burma (Fourth Rep.). The Depositional characters of the coal-bearing formations in Burma. – Kumamoto J. Sci., Ser. B, Sect. 1, 4, 2: 90–105, 8 figs., 1 map.

SALDEN, E. B. (1870): Official narrative of the expedition to explore the trade routes to China via Bhamo. – Sel., Rec. Govt. India, 79: 19–173.

SALE, H. M. & P. EVANS (1940): The geology of the British oil fields; I. The geology of the Assam-Arakan oil region (India & Burma). – Geol. Mag., 77: 337–363.

SANDER, N. J., HUMPHREY, W. E. & J. F. MASON (1975): Tectonic framework of Southeast Asia and Australia: Its significance in the occurrence of petroleum. – Proc. 9. World Petrol. Congr., Prepr. Panel Discussion 7: 1–25, Tokyo.

SANTO, T. (1969): On the characteristic seismicity in South Asia from Hindukush to Burma. – Bull. Int. Inst. Seismol. Earthquake Eng., 6: 81–95.

SARIN, D. D. (1963): Petrography and origin of Peguan sandstones exposed at Chauk, Burma. – J. Sedim. Petrol., 33, 2: 333–342, 5 figs.

– (1964a): Characteristics of channel sediments of the Chindwin River. – J. Burma Res. Soc., 47, 2: 352–365.

– (1964b): Petrography and origin of Taungtha Formation exposed at Taungtha Myingyan District. – J. Burma Res. Soc., 47, 1.

– (1968): Composition of channel sediments of the Chindwin River. – J. Sci. Technol. Burma (1st Burma Res. Congr., 1966), 1, 1: 45–58, 14 figs., Rangoon.

SASTRY, M. V. A. & U. B. MATHUR (1970): Bibliography of Indian Geology. IV. Paleontological Index by T. H. D. LA TOUCHE. Revised and enlarged. Decapod Crustacea. – Geol. Surv. India, Misc. Publ. India, 18: 1–73, 76 figs., 23 pls.

SAWATA, H., APORNSUWAN, S. & V. PISUTHA-BRNOND (1975): Note on geology of Khao Phra, Petchaburi, West Thailand. – J. Geol. Soc. Thailand 1, 1–2: 31–49, Bangkok.

SCHELLMANN, W. (1977): The formation of lateritic silicate bauxites and criteria for the exploration and assessment of deposits. – Nat. Resources & Develop. 5: 119–134, 6 figs., Tübingen.

SCHWEINFURTH, U. & H. SCHWEINFURTH-MARBY (1975): Exploration in the eastern Himalayas and the river Gorge country of southeastern Tibet – Francis (Frank) Kingdom Ward (1885–1958). – Geoecol. Res., 3: 1–114, 1 fig., 2 pls.

SCLATER, J. G. & R. L. FISHER (1974): Evolution of the east central Indian Ocean, with emphasis on the tectonic setting of the Ninetyeast Ridge. – Geol. Soc. Amer., Bull., 85: 683–702.

SCOTT, W. H. (1936): Ruby mines of Burma: Summary of Chemical Consulate Report, by D. H. HOWELL. – Gems and Gemmol., 2: 3–6, 31–34.

SCRIVENOR, J. B. (1934): Note on the correlation of the geology of Burma and Malaya. – In: H. L. CHHIBBER (1934): The Geology of Burma: 519–524.

SEARLE, D. L. (1961): Report on prospecting for radioactive minerals in Burma. – Rep. Internat. Atom. Energy Agency WP/5/20.

SEARLE, D. L. & BA THAN HAQ (1964): The Mogok Belt of Burma and its relationship to the Himalayan orogeny. – 22nd Internat. Geol. Congr., India, 11: 133–161, Abstr.: 184, 3 figs., New Delhi.

SEELY, D. R., VAIL, P. R. & G. G. WALTON (1974): Trench slope model. – in: BURK, C. A. & C. L. DRAKE (eds.): The Geology of Continental Margins. – Berlin: Springer Verlag.

SEIN MYINT (1968): Occurrences of brine deposits in the vicinity of Maymyo. – Sci. Technol. Burma (1st Burma Res. Congr., 1966), 1, 3: 457–465, 2 figs., Rangoon.

SEIN MYINT, HOKE MAW, TIN HTUT & TIN HLAING (1972): Strike-slip faulting in Upper Tenasserim. – Geol. Soc. Malaysia, Newsl. (Malays. Conf. Geol. Southeast Asia, Abstr., Kuala Lumpur, Malaysia, 1972), 34: 52.

SEIN MYINT et al. (1970): Airborne magnetic and photogeological survey of Yinmabin-Monywa-Myinmu Region. – Sci. Technol. Burma (5th Burma Res. Congr., 1970), 3, 2: 319–330, 1 fig.

SEIN SHWE, U. & MYINT KYAW (1970): Estimation of flood over Upper Sittang catchment in connection with meteorological condition. – Union of Burma J. Sci. Technol. (5th Burma Res. Congr., 1968), 3, 1: 219–230, 10 figs., Rangoon.

SENGUPTA, S. (1966): Geological and geophysical studies in western part of Bengal Basin, India. – Amer. Assoc. Petr. Geol. Bull., 50: 1001–1017.

SETHU RAMA RAU, S. (1921): Note on the stratigraphy of Singu-Ye-Nangyat area. – Rec. Geol. Surv. India, 53, 4: 321–330.

– (1930): The geology of the Mergui district. – Mem. Geol. Surv. India, 55, 1.

SHI NAI CHOW & ALI MARACAN (1968): Analysis of soil samples from twelve districts of Burma. – Sci. Technol. Burma (1st Burma Res. Congr., 1966), 1, 1: 15–22, 1 fig., Rangoon.

SHOULS, M. M. (1973): Seismicity and plate tectonics in the Thailand-Burma-Andaman Sea Area. – CCOP Newsletters, 1, 1 (Rep.): 17–19.

SIMPSON, R. R. (1906): The Namma, Man-Sang and Man-se-le coal fields, Northern Shan State, Burma. – Rec. Geol. Surv. India, 33: 125–156.

– (1922): Notes on a visit to the Burma ruby mines. – Trans. Min. Geol. Inst. India, 27, 1: 42–58.

SMITH, A. G. & A. HALLAM (1970): The fit of the southern continents. – Nature, 225: 139–144.

SMITH, M. A. (1940): The amphibians and reptiles obtained by Mr. Ronald KAULBACK in Upper Burma. – Rec. Indian Mus., 42, 3: 465–487, 1 pl.

SMITH, S. (1941): Some Permian corals from the Plateau Limestone of the Southern Shan States, Burma. – Palaeontologia Ind., N.S., 30, Mem. 2: 1–21.

SNELLING, N. J., BIGNELL J. & R. R. HARDING (1968): Ages of Malayan granites. – Geol. en Mijnb., 47: 358–359.

SNOW, A. G. (1905): Tin mining in Lower Burma. – Min. J. 78: 247, London.

SOE WIN (1965): The jade mines of Northern Burma. – The Working People's Daily, 5, 12, 19, 26 Sept.

– (1968a): Copper-nickel-cobalt mineralization at Bawdwin Mine. – J. Sci. Technol. Burma, 2, 1.

– (1968b): The application of geology to the mining of jade. – Sci. Technol. Burma, (1st Burma Res. Congr., 1966), 1, 3: 445–456, 2 figs.

– (1970): Copper, nickel, cobalt mineralization at Bawdwin Mine. – J. Sci. Technol. Burma, 3, 3: 471.

SOE WIN, HTAY LWIN, THAN SEIN & KYAW AUNG (1968): Secondary mineralization at Bawdwin Mine. – J. Sci. Technol. Burma (2nd Burma Res. Congr., 1967), 1, 2: 229–239, 3 figs.

SOMMERLATTE, H. (1942): Die Wolframerzlagerstätten Burmas. – Met. Erz, 39: 2–6, 25–28.

– (1948): Die Erzlagerstätten Burmas. – 187 pp., Diss., Techn. Univ. Berlin-Charlottenburg, Berlin.

– (1959): Die Blei-Zink-Erzlagerstätte von Bawdwin in Nord-Burma. – Z. dtsch. Geol. Ges., 110, 3: 491–504, 8 figs.

SOMPONGSE CHANTARAMEE (1979): Tectonic synthesis of the Langsang Area and discussion of regional tectonic evolution. – Third Reg. Conf. on Geol. and Min. Res. of Southeast Asia, Bangkok, 1978: 117–186.

SONDHI, V. P. (1934): in FERMOR, L. L.: General Report of the Geological Survey of India for the year 1931: – Lower Chindwin and Shwebo Districts. – Rec. Geol. Surv. India, 66, 1: 104.

– (1937): in HERON, A. M.: General Report of the Geological Survey of India for the year 1936: Southern Shan States, Burma. – Rec. Geol. Surv. India, 72, 1: 1–121.

SPATH, L. F. (1934): On a Turonian Ammonite (Mammites daviesi) from Ramri Island, Burma. – Rec. Geol. Surv. India, 68, 1.

STAMP, L. D. (1922a): Some remarks on the tectonics of Burma. – Proc. Congr. Geol. Intern., 13: 1145–1150.

– (1922b): Geology of the part of the Pondaung Ranges, Burma. – Trans. Min. Geol. Inst. India, 17: 161–180.

– (1922c): An outline of the Tertiary geology of Burma. – Geol. Mag., 59: 481–501.

- (1923): Ecology of the riverine tract of Burma. – J. Ecol., **11**: 136–138.
- (1924): Vegetation of Burma. – Univ. Rangoon Res. Monogr., **1**: 7–14.
- (1925): Seasonal rhythm in the Tertiary sediments of Burma. – Geol. Mag., **62**: 515–528.
- (1926a): Some evidence tending to prove a Mesozoic age for the granites of Tenasserim. – Quart. J. Geol. Min. Met. Soc. India, **35**: 155–160.
- (1926b): The igneous complex of Green Island and the Amherst Coast, Lower Burma. – Geol. Mag., **63**: 399–410.
- (1927a): The geology of the oil fields of Burma. – Bull. Amer. Assoc. Petr. Geol., **11**: 557–580.
- (1927b): Note on some fossiliferous localities in the Peguan Rocks of Central Burma. – J. Burma Res. Soc., **17**: 177–189.
- (1927c): The oil fields of Burma; Recent advances of geological knowledge. – Proc. 14th Ind. Sci. Congr.: 227–236.
- (1927d): The conditions governing the occurrence of oil in Burma. – J. Inst. Petr. Techn., **13**: 21–70, Dallas.
- (1928): Structural features connected with the oil deposits of Burma. – J. Inst. Petr. Techn., **14**: 28–63, Dallas.
- (1929): The oil fields of Burma. – J. Inst. Petr. Techn., **15**: 300–349, Dallas.
- (1940): The Irrawaddy River. – Geogr. J., **95**, 5: 329–356, London.

STAMP, L. D. & H. L. CHHIBBER (1927): The igneous and associated rocks of the Kabwet Area (Shwebo and Mandalay Districts) Upper Burma. – Trans. Min. Geol. Inst. India, **21**: 97–128, Calcutta.

STAUFFER, P. H. (1973): Malaysia and Southeast Asia in the pattern of continental drift. – Bull. Geol. Soc. Malaysia, **7**: 89–138.

STEIN, K. (1931): Birma. Grundlagen einer Landeskunde. – Mitt. Geogr. Ges. München.

STEVENSON, J. F. (1863): Account of a visit to the hot springs of Pai in the Tavoy District. – J. Asiat. Soc. Bengal, **32**: 383–386.

STOKES, R. B., TANTISUKRIT, C. T. & K. V. CAMPBELL (1975): Proceedings of the Conference on the Geology of Thailand. – Dep. Geol. Sci. Chiang Mai Univ., Spec. Publ. No. 1.

STOLICZKA, F. (1868): Die Andaman Inseln, Assam, usw. – Verh. K. K. Geol. Reichsanst. Wien: 192–193.

STONELEY, R. (1974): Evolution of the continental margins bounding a former Southern Tethys. – In: C. A. BURK & C. L. DRAKE (eds.): The Geology of Continental Margins. – 1009 pp.: 889–903, Berlin-Heidelberg-New York: Springer.

STONIER, G. A. (1900): Preliminary report on the auriferous tract in the Wuntho District in Burma. – Gen. Rep., Geol. Surv. India (1899–1900: 59–63, Calcutta.

STOPPEL, D. (1974): Bericht über Conodonten-Faunen (Ordoviz) aus Proben der Deutschen Geologischen Mission Burma (Grube Bawsaing). – Rep., Bundesanst. f. Geowiss. u. Rohstoffe, Hannover.

STREETER, G. S. (1889): The ruby mines of Burma. – J. Soc. Arts, **37**: 266–275.

STROVER, G. A. (1873): Memorandum on the metals and minerals of Upper Burma. – Geol. Mag., **10**: 67–73.

STROGANOVA, S. A. (1974): Age of folding in northwestern sector of Southeast Asia. – International Geol. Rev. (USA) (in Russ.: Sovetsk. Geol., SSSR, 1973: 2, 132–136), **16**: 988–992, 1 fig.

STUART, M. (1910a): Geology and prospects of oil in Western Prome and Kama, Lower Burma (including Namayan, Padaung, Taungbogyi, and Ziaing). – Rec. Geol. Surv. India, **38**: 259–270, 1 pl.
- (1910b): Fossil fish teeth from the Pegu System, Burma. – Rec. Geol. Surv. India, **38**, 4: 292–301, 3 pls.
- (1910c): The recorrelation of the Pegu system in Burma with Notes on the horizon of the oil-bearing strata (including the geology of Padaukpin, Bonbyin and Aukmancin). – Rec. Geol. Surv. India, **38**, 4: 271–291, 1 pl.
- (1913): The geology of the Henzada District, Burma. – Rec. Geol. Surv. India, **41**, 4: 240–265.
- (1919): The galena deposits of Northeastern Putao. – Rec. Geol. Surv. India, **50**, 3: 241–254.
- (1923a): Geological traverses from Assam to Myitkyina through the Hukong Valley. – Rec. Geol. Surv. India, **54**, 4: 398–409.
- (1923b): Amber and the dammar of living bees. – Nature, **111**: 83–84.
- (1925a): Suggested origin of the oil-bearing strata of Burma – deduced from the geological history of the Country during Tertiary times. – J. Inst. Petrol. Geol., **11**: 296–304, London.

– (1925 b): The Eocene lignites and amber deposits of Burma and their relationship to certain occurrences of mineral oil. – J. Inst. Petrol. Techn., **11**: 475–486, London.

SUN, Y. C. (1939): On the occurrence of the Fengshanian (late Upper Cambrian) Trilobites fauna in W. Yunnan. – 40th Ann. Pap., Nation., Univ. Peking, **1**: 29–34, 1 pl.

– (1945): The Sino Burmese geosyncline of early Palaeozoic time with special reference to its extend and character. – Bull. Geol. Soc. China. **25**: 1–7.

– (1948): Problems of the Paleozoic stratigraphy of Yunnan. – 50th Ann. Pap., Nation. Peking Univ., Geol. Ser.: 1–27.

SUNTHRALINGAM, T. (1968): Upper Paleozoic stratigraphy of the area west of Kampar, Perak. – Bull. Geol. Soc. Malaysia, **1**: 1–15.

SWINTON, W. E. (1939): New fossil freshwater Tortoise from Burma. – Rec. Geol. Surv. India, 74, 4: 548–551, 2 figs.

SYMES, M. (1800): An account of an Embassy to the Kingdome of Ava, sent by the Governor General of India in the year 1795. – 40, 504 pp., London, repr., Edinburgh 1827.

– (1826): A brief account of the religion and civil institutions of the Burmese; and a description of the Kingdome of Assam to which is added an account of the petroleum wells in the Burma Dominions, extracted from Rangoon up the river Erawaddy to Amaraporah, the present capital of the Burmah Empire. – **80**: 151 pp., Calcutta.

SZALAY, F. S. (1970): Late Eocene *Amphipithecus* and the origin of Catarrhine Primates. – Nature, **227**: 355–357.

Tahal Water Planning (1963 a): Report on water investigation in part of the Dry Zone. – Rep. for Govt. of Burma, Tahal Water Planning Ltd., Tel Aviv.

– (1963 b): Moulmein Water Supply Scheme Feasibility Report. – Rep. for Govt. of Burma, Tahal Water Planning Ltd., Tel Aviv.

TAINSH, H. R. (1950): Tertiary geology and principal oil fields of Burma. – Bull. Amer. Assoc. Petr. Geol., **34**: 823–855, 7 tables, 10 figs.

TALBOT, F. S. (1920): Mining the ruby in Burma. – The World's Work, May: 594–607.

TANATAR, J. J. (1907): Beitrag zur Kenntnis der Rubinlagerstätte von Nanyazeik. – Z. prakt. Geol., **15**: 316–320.

TAVERNE, L. (1977): On the Actinopterygian fishes from the Cenomanian of the Kyi River (Burma, Pakokku District). – Geol. Jb., **B 23**: 47–59, 5 figs., 2 pls.

TATSUMI, T. & T. WATANABE, (1971): Geological environment of formation of the Kuroko-type deposits. – Soc. Min. Geol. Japan, Spec. Issue: 216–229, Tokyo.

TEILHARD de CHARDIN, P. (1965): Comparisons des Formations Plio-Pleistocènes de Birmanie et des Contrées avoisinantes. – Anthropologie, France, **69**, 3–4: 209–218, 2 pls.

THA HLA & BA THAN HAQ (1960): The petrography of the rocks of the Kyaukse Hill, Upper Burma. – J. Burma Res. Soc., 15th Annivers. Conf., **1**: 441–472.

THACPAW, S. C. (1968): Mineralization and paragenesis of ore minerals at Shweminbon. – J. Sci. Technol. Burma, **1**, 1: 33–43, 6 figs.

THAKUR, V. C. & A. K. JAIN (1974): Tectonics of the region of Eastern Himalayan Syntaxis. – Curr. Sci., **43**: 483–785.

THAW TINT (1972): A critical review of the Paleozoic stratigraphy of the Northern Shan States and the new finds in the Paleozoic Paleontology, Burma. – Geol. Soc. Malaysia, Newsl. (Reg. Conf. Geol. Southeast Asia, Abstr.; Kuala Lumpur, Malaysia, 1972), **34**: 57–58.

THAW TINT & HLA WAI (1970): The Lower Devonian Trilobite fauna from the East Medaw Area, Maymyo District. – Union of Burma J. Sci. Technol. (15th Burma Res. Congr., 1970), **3**, 2: 283–306, 9 figs.

THEERAPONGS THANASUTHIPITAK (1979): Geology of Uttaradit and its implications on tectonic history of Thailand. – Third Reg. Conf. on Geol. and Min. Res. of Southeast Asia, Bangkok, Thailand 1978: 187–197.

THEIN, M. L. (1967): Nautiloid Cephalopods from the area east of Kyaukse, Burma. – Union of Burma J. Sci. Technol. **1**: 76–76, Rangoon.

– (1973): On the occurrence of *Dadonella* facies from the Upper Chindwin area, Western Burma. – Union of Burma J. Sci. Technol., **3** (1970): 277–282, 7 figs., Rangoon.

THENIUS, E. (1959): in LOTZE, F.: Handbuch der stratigraphischen Geologie, 3, 2, Tertiär, Wirbeltierfaunen: 328 pp., 32 tables, 12 figs. 10 pls., Stuttgart: Enke.

THEOBALD, W. (1862): Notes on stone implements from Bundlekhund and the Andamans. – J. Asiat. Soc. Bengal, **31**: 323–327.

– (1865): Note on the discovery of stone implements in Burma. – Proc. Asiat. Soc. Bengal: 125–127.

– (1869): On the beds containing silicified wood in Eastern Prome, British Burmah. – Rec. Geol. Surv. India, **2**, 4: 79–86.

– (1870a): On the alluvial deposits of the Irrawadi, more particularly as contrasted with those of the Ganges. – Rec. Geol. Surv. India, **3**, 1: 17–27.

– (1870b): Note on petroleum in Burmah. – Rec. Geol. Surv. India, **3**, 3: 72–73.

– (1870c): Exhibition of a stone implement from Prome. – Proc. Asiat. Soc. Bengal: 220–222.

– (1871): The axial group in Western Prome, Bristish Burma. – Rec. Geol. Surv. India, **2**: 33–34.

– (1872a): A few additional remarks on the axial group of Western Prome. – Rec. Geol. Surv. India, **5**, 3: 79–82.

– (1872b): A brief notice of some recently discovered petroleum localities in Pegu. – Rec. Geol. Surv. India, **5**, 4: 120–122.

– (1873a): On the Salt Springs of Pegu. – Rec. Geol. Surv. India, **6**, 3: 67–73.

– (1873b): Stray notes on the metalliferous resources of British Burmah. – Rec. Geol. Surv. India, **6**, 4: 90–95.

– (1873c): Sitsayan Shales, in the geology of Pegu. – Mem. Geol. Surv. India, **10**, 3.

– (1973d): The geology of Pegu. – Mem. Geol. Surv. India, **10**, 2: 189–359.

– (1880): Geology and economic mineralogy of Lower Burma. – Brit. Burma Gazette, **1**: 32–67.

– (1895): Note on Dr. Fritz NOETLING's Paper on the Tertiary system in Burma in the Records of the Geological Survey of India for 1895, Pt. 2, Rec. Geol. Surv. India, **28**: 150–151.

TIN AYE & KYAW NYEIN (1970): Review of tin and tungsten deposits of Burma. – Sci. Technol. Burma (1st Burma Res. Congr. 1966), **3**, 1: 39–74, 5 figs.

TIN MYINT (1970): Microstratigraphy and facies study of the type section of Shwezetaw Formation. – Myanma Oil Corp. Burma Paleont. Note **113**, Rangoon.

TIN NYUNT (1969): The case history of Yenangyaung oil field. – Sci. Technol. Burma (3rd Burma Res. Congr. 1968), **2**, 3: 487–524, 26 figs.

TIPPER, G. H. (1906): Preliminary notes on the Trias of Lower Burma. – Rec. Geol. Surv. India, **34**, 2: 134 pp.

– (1907): Further note on the Trias of Lower Burma and on the occurrence of *Cardita beaumonti* d'ARCH. in Lower Burma. – Rec. Geol. Surv. India, **35**, 2: 119 pp.

– (1911): The geology of the Andaman Islands. – Mem. Geol. Surv. India, **35**, 4: 195–216.

TOBIEN, H. (1975): The structure of the Mastodon molar (Proboscidae, Mammalia), Part 2; The Zygodont and Zygobunodont patterns. – Mainz. Geowiss. Mitt., **4**: 195–233, 32 figs.

TOIT, A. DU (1937): Our wandering continents. – Edinburgh: Oliver & Boyd.

TOOMEY, D. F. (1968): Middle Devonian (Eifelian) Foraminifera from Padaukpin, Northern Shan States, Burma. (Abstr.). – Geol. Soc. Amer., Progr. Ann. Meeting: 299.

TOZER, E. T. (1973): The earliest marine Triassic rocks: their definition, Ammonoid fauna, distribution and relationship to underlying formations. – In: LOGAN, A. & L. V. HILLS: The Permian and the Triassic Systems and their mutual Boundary. – Canad. Soc. Petrol. Geol. Mem. 2: 549–556, Calgary, Alberta, Canada.

TRAPP, W. (1975): Stratigraphie und Fauna des Ordoviziums von Thailand unter besonderer Berücksichtigung der Ostracoden. – Unpubl. Rep.

TRAUTH, F. (1930): Upper Triassic fossils from the Burma-Siamese frontier on some fossils from the Kamawkala Limestone. – Rec. Geol. Surv. India, **63**, 1: 174–176.

TREMENHEERE, G. B. (1841a): Report on the tin of the Province of Mergui. – J. Asiat. Soc. Bengal, **10**: 845–851, **11**: 24, 289–290.

– (1841b): Report on the manganese of Mergui Province. – J. Asiat. Soc. Bengal, **10**: 852–853.

– (1843): Report of a visit to the Pakchun River, and of some tin localities in the Southern portion of the Tenasserim Provinces. – J. Asiat. Soc. Bengal, **12**: 523–534.

– (1845): Report, etc. with information concerning the Prince of the tin ore of Mergui. – J. Asiat. Soc. Bengal, **14**: 329–332.

TREMENHEERE, G. B. & C. LEMON (1846): Report on the tin of the Province of Mergui in Tenasserim, in the northern part of the Malayan Peninsula. – Trans. Roy. Geol. Soc. Cornwall, **6**: 68–75.

TREMENHEERE, G. B., O'RILEY, E. & T. OLDHAM (1852): Report on the tin and other mineral productions of the Tenasserim Provinces. – Sel. Rec. Bengal Govt., 6: 21–44.

TUN HLAING (1970): Washaung Irrigation Project. – Union of Burma J. Sci. Technol. (3rd Burma Res. Congr., 1968), 3, 1: 115–132, 6 pls., Rangoon.

TUN U MAUNG & AUNG TIN U (1969): The importance of integrating geological data in geophysical interpretation for oil exploration. – Union of Burma J. Sci. Technol. (2nd Burma Res. Congr., 1967), 2, 1: 27–52, 12 figs., Rangoon.

TURNER, W. B. (1842): Note on fireclay at Moulmein. – Calcutta J. Nat. Hist., 2: 596–597.

TWEEN, A. (1863): Memorandum on the composition of water from the Hot Springs of Pai, Tavoy District. – J. Asiat. Soc. Bengal, 32: 386.

UNESCO (1971): Geological Map of Asia and the Far East, 1:5 000 000. – Unesco, Paris.

UN-ECAFE (1952): Coal and iron ore resources of Asia and the Far East. – United Nations Publ. 1952, II. F. 1., Bangkok, 155 pp.

UNICEF (1980): Assignment report of hydrogeological-geophysical consultancy with special reference to hydrogeology of the Dry Zone. – Rangoon, April 1980.

United Nations (1966): Survey of lead and zinc mining and smelting in Burma. – Gen. Rep. prepared for the Government of Burma by the United Nations acting as Executing Agency for the United Nations Development Progr., 69 pp., 1 map.

UN ESCAP CCOP (1974): The offshore hydrocarbon potential of East Asia. – 67 pp., Bangkok.

UN ESCAP (1975): Oil and Natural Gas Map of Asia (2nd Ed.). – Bangkok.

Union of Burma (1973): Report on Geological Survey of the Monywa area (Unpubl.). – Metal Min. Agency, Overseas Techn. Coop. Govt. of Japan, Vol. I.

UNO, Y. & S. FUKUSHI (1967): Properties of some soils in Burma and Laos. – Soil. Sci. Plant Nutrit., 13, 1: 17–24.

URE, A. (1843): Analysis of iron ore from Tavoy and Mergui, and of limestone from Mergui. – J. Asiat. Soc. Bengal, 12: 236–239.

VALDIYA, K. S. (1976): Himalayan transverse faults and folds and their parallelism with subsurface structures of North Indian plains. – Tectonophys. 32: 353–386, Amsterdam.

VEEVERS, J. J. (1977): Models of the evolution of the Eastern Indian Ocean. – In: Indian Ocean Geology and Biostratigraphy (Studies following Deep-Sea Drilling legs 22–29): 151–163; Amer. Geophys. Union, Washington.

VEEVERS, J. J., JONES, J. G. & J. A. TALANT (1971): Indo-Australian stratigraphy and the configuration and dispersal of Gondwanaland. – Nature, 229: 383–388.

VERMA, R. K., AHLUWALIA, M. S. & M. MULEHOPADHYAY (1976): Seismicity, gravity and tectonics of N.E. India & Burma. – Abstr. 25th Internat. Geol. Congr. Sydney 1976, 3: 724.

VERMA, R. K., MANYOMUKHOPADHYAY & M. S. AHLUWALIA (1976): Earthquake mechanisms and tectonic features of Northern Burma. – Tectonophys., 32, 3: 387–399, Amsterdam.

VINAYAK, R. M. (1922): Note on the oil-shales of Mergui. – Rec. Geol. Surv. India: 342–343.

VOLKER, A. (1966): The deltaic area of the Irrawaddy River in Burma. – UNESCO, Rech. Zone Tropic hum., 6: 373–379, 1 fig., Paris.

VREDENBURG, E. W. (1904): Gem sands from Burma. – Rec. Geol. Surv. India, 31, 1: 45.

– (1921a): Results of a revision of some portions of Dr. NOETLING's Second Monograph of the Tertiary Fauna of Burma. – Rec. Geol. Surv. India, 51: 224–302, 1 fig.

– (1921b): Comparative diagnoses of Conidae and Cancellariidae from the Tertiary Formations of Burma. – Rec. Geol. Surv. India, 53, 2: 130–141, 1 pl.

– (1921c): A Zone Fossil from Burma: Ampullina (Megatylotus) birmanica. – Rec. Geol. Surv. India, 53, 4: 359–369.

– (1921d): Illustrated comparative diagnoses of fossil Terebridae from Burma. – Rec. Geol. Surv. India, 51: 339–361.

– (1921e): Comparative diagnoses of Pleurotomidae from the Tertiary Formations of Burma. – Rec. Geol. Surv. India, 53: 83–129, 3 pls.

– (1921f): Analysis of the Singu Fauna founded on RAO BAHADUR SETHU RAMA RAU's Collections. – Rec. Geol. Surv. India, 53, 4: 331–342, 1 pl.

– (1923): Oligocene Echinoidea collected by RAO BAHADUR SETHU RAMA RAU, in Burma. – Rec. Geol. Surv. India, 54: 412–415, 1 pl.

VREDENBURG, E. W. & B. PRASHAD (1921): Unionidae from the Miocene of Burma. – Rec. Geol. Surv. India, 51: 371–374.

VREDENBURG, E. W. & M. STUART (1909): On the occurrence of *Ostrea latimarginata,* a characteristic species in the "Yenangyaung stage" of Burma. – Rec. Geol. Surv. India, **38**: 127–132.

WADIA, D. N. (1967): The Himalayan Geosyncline. – National Inst. Sci. India, Proc., Part A, **32**, 5–6: 523–531.

WALDIE, D. (1970): Analysis of a new mineral from Burma (O'Rileyite). – Asiat. Soc. Bengal: 279–283; Chem. News, **23**: 4–5.

WALKER, F. W. (1922): Amherst District. – Rec. Geol. Surv. India, **54**, 1.

WALKER, T. L. (1902): The geology of Kalahandi State, Central Provinces. – Mem. Geol. Surv. India, **33**: 22 pp.

WALTER, H. (1955): Die Klimadiagramme als Mittel zur Beurteilung für oekologische, vegetationskundliche und landwirtschaftliche Zwecke. – Ber. dt. Bot. Ges., **68**: 331–344, Stuttgart.

WALTERS, H. (1833): On geological specimens from Arrakan. – J. Asiat. Soc. Bengal, **2**: 263–264.

WANG, K. P. (1965): The mineral industry of Burma. – Miner. Yearb. USA, **1963**, 4, 372: 1261–1266.

WARTH, F. J. (1916): Note on the soil of the experimental farms. – Bull. Dept. of Agric., Burma, 13.

WARTH, F. J. & PO SAW (1919): Absorption of lime by soils. – Pusa Mem., Chem. Ser., **5**: 157–172.

WARTH, F. J. & PO SHIN (1918): The phosphate requirements of Lower Burma paddy soils. – Pusa Mem., Chem. Ser. **5**: 132–156.

WASHINGTON, H. S. (1924): The lavas of Barren Island and Narcondam. – Amer. J. Sci., 5th ser., **7**: 441–456.

WATERHOUSE, J. B. (1973): Permian Brachiopod-correlations for South East Asia. – Geol. Soc. Malaysia, Bull. **6**: 187–210.

WATKINS, N. W. (1932): The ruby mines of Upper Burma. – Gemmologist, **1**: 263–272, 325–342.

– (1936): Rubies and sapphires in Burma. – Gemmologist, **5**: 154–157.

WEGENER, A. (1929): Die Entstehung der Kontinente und Ozeane. – Die Wissenschaft, 66, 4. Aufl., 231 pp., Braunschweig.

WEIR, J. (1930): Upper Triassic fossils from the Burma-Siamese frontier; Brachiopoda Lamellibranchia from the Thaungyin River. – Rec. Geol. Surv. India, **63**, 1: 168–173.

WELLS, A. J. (1969): The crush zone of the Iranian Lagros Mountains and its implications. – Geol. Mag., **106**: 385–394.

WHITE, J. T. O. (1886): Analysis of a specimen of jade from Mogaung, found in the Palace at Mandalay. – Chem. New, **54**: 20.

WHITTAKER, J. E., ZANINETTI, L. & D. ALTINER (1979): Further remarks on the micropalaeontology of the Late Permian of eastern Burma. – Notes Lab. Palaeont., Univ. Genève, **5**, 1: 11–23, 1 fig., 3 pls.

WHITTINGTON, H. B. (1953a): New evidence for the Permian age of the Moulmein System. – In: CLEGG, E. L. G.: The Mergui, Moulmein and Mawchi Series. – Rec. Geol. Surv. India, 78, 2: 193, 194.

– (1953b): Geological reconnaissance between Hailem and Ke-shi Mansam, Southern Shan States. – Rec. Geol. Surv. India, **78**, 2: 203–216, 1 map.

– (1973): Ordovician Trilobites. – Atlas of Palaeobiogeography, ed. by A. HALLAM: 13–18, New York.

WHITTINGTON, H. B. & C. P. HUGHES (1972): Ordovician geography and faunal provinces deduced from trilobite distribution. – Phil. Trans. Roy. Soc. London, B, Biol. Sci., **263**, 850: 235–278.

WILLIAMS, D. (1843): Report on the appearance of a volcanic island off the Coast of Arracan. – J. Asiat. Soc. Bengal, **12**: 832–833.

– (1845): Note on a supposed submarine eruption off the Coast of Arracan. – J. Asiat. Soc. Bengal, **14**, Proc.: 24–25.

– (1846): Note on the eruption of a mud-volcano in the Island of Ramree. – J. Asiat. Soc. Bengal, **15**, Proc.: 92–93.

WIN SWE (1970): Strike slip at the Sagaing-Tagaung Ridge. – 1970 Burma Res. Congr., Abstr., **3**: 101.

– (1972): Strike-slip faulting in Central Belt of Burma. – Geol. Soc. Malaysia, Newsl. (Reg. Conf. Geol. Southeast Asia), Abstr., No. 34: 59, Kuala Lumpur.

WINGATE, J. B. & R. M. THURLEY (1908): Report on the Conditions of the Local Salt Industry in Burma and allied questions. – Govt. Print. Burma.

WININGER, D. C. (1974): The mineral industry of Burma. – In: Minerals Year book 1972, 3, Area Rep., Intern. (ed. by A. E. SCHRECK): 179–186, US Bur. Mines, Washington, D.C.

– (1976): The mineral industry of Burma. – US Bur. Mines, Miner. Yearb., **3**: 189–193.

WINKLER, H. G. F. (1974): Petrogenesis of metamorphic rocks. – Springer study ed., Geol., 320 pp.

WINSTON, D. & H. NICOLIS (1967): Late Cambrian and Early Ordovician faunas from the Wilberns Formation of Central Texas. – J. Paleont., 41, 6: 66–96.

WNENDT, W. (1979): Geologie und lagerstättengeologische Bewertung der Zinn-Wolframlagerstätte Hermyingyi/Birma. – Diss. Univ. Bochum.

WOLFART, R. (1970): Fauna, Stratigraphie und Paläogeographie des Ordoviziums in Afghanistan. – Beih. geol. Jb., 89: 125 pp.

– (1974a): Die Fauna (Brachiopoda, Mollusca, Trilobita) des älteren Ober-Kambriums (Ober-Kushanian) von Dorah Shah Dad, Südost Iran, und Surkh Bum. Zentral Afghanistan. – Geol. Jb., B 8: 71–184.

– (1974b): Die Fauna (Brachiopoda, Mollusca, Trilobita) aus dem Unter-Kambrium von Kerman, Südost Iran. – Geol. Jb., B 8: 5–70.

WOLFART, R. & H. WITTEKINDT (1980): Geologie von Afghanistan. – Beiträge zur regionalen Geologie der Erde. – Vol. 12, Berlin, Stuttgart: Gebr. Borntraeger.

WOODTHORPE, R. G. (1888): Explorations on the Chindwin River, Upper Burma. – Proc. Roy. Geogr. Soc., N.S., 11: 197–216.

World Atlas of Agriculture. – Under the aegis of the International Association of Agricultural Economists, Vol. 1–4; Novara 1969–1976: 1. Europe, USSR, Asia Minor, 1969, XIII, 527 pp., figs., tables; 2. Asia and Oceania, 1973, XI, 671 pp., figs., tables, 3. Americas, 1970, VIII, 497 pp., figs., tables; 4. Africa, 1976, XI, 761 pp., figs., tables; Land utilisation maps and relief maps, 1969, 62 pls.

World Bank: Development in Burma (1976): Issues and Prospects. – World Bank, Rep. No. 1024–B4.

WUNDERLICH, H. G. (1974): Bau der Erde. Geologie der Kontinente und Meere II, Asien, Australien. – Mannheim: Bibliographisches Institut.

WYNNE, T. T. (1897): The ruby mines of Burma. – Trans. Inst. Min. Met., 5: 161–175.

YANSHIN, A. L. (1966): Tectonic Map of Eurasia, 1:5 000 000 (in Russ.). – Geol. Inst. Acad. Nauk. SSSR and Ministry Geol. SSR, Moscow.

YEW, C. C. (1971): The geology and mineralization of the eastern Kuala Lumpur area, West Malaysia. – Unpubl. B.Sc. Thesis, Dep. of Geol., Univ. Malaysia.

YIN, TSAN-HSUN (1966): China in the Silurian Period. – J. Geol. Soc. Austr. 13: 277–297.

YOUNG, B. & W. JANTARANIPA (1970): The discovery of Upper Palaeozoic fossils in the "Phuket Series" of Peninsular Thailand. – Proc. Geol. Soc. London, No. 1662: 5–7.

ZALOKAR, B. & I. JURKOVIC (1961): Copper ore occurrence at Kyaukse, Burma. – Geol. Vjesn. Jugoslav, 14 (1960): 265–270, 2 figs., (Abstr. in Serbo Croat.).

– – (1962): Copper deposits of the Southern Shan States, Burma. – Geol. Vjesn. Jugoslav, 15 (1961), 1: 229–248, 7 figs., 1 map (Abstr. in Serbo Croat.).

ZANINETTI, L., WHITTAKER, J. E. & D. ALTINER (1979): The occurrence of *Shanita amosi* BRÖNNIMANN, WHITTAKER & ZANINETTI (Foraminiferida) in the Late Permian of the Tethyan region. – Notes Lab. Paleont. Univ. Genève, 5, 1: 1–9, 1 fig., 2 pls., Genève.

Locality Index

Identification of localities is not always clear. Many village names are used for more than one place. In such cases, all of the villages are listed that could be in the vicinity of the location of interest. Some localities could not be identified and the name of the nearest village is given.

Different spellings are included in the index in brackets. Coordinates in brackets usually are for the location of the village if a nearby location is referred to in the text.

Name	Coordinates N	E	Pages
Akyab	20° 08′	92° 54′	1, 7, 20, 30, 97, 159, 213, 220
Alechaung	19° 22′	94° 52′	206
Allenmyo	19° 22′	95° 14′	131, 132
Amherst	16° 05′	97° 34′	81, 173, 201, 206, 207
Anatholin range	near Mergui		173
Arakan coast	16° 06′	94° 17′	1, 104–106, 128, 153, 156, 158, 159, 169, 170
Arakan yoma	20° 44′	93° 50′	9, 17, 19, 24–26, 28–30, 32, 34, 38, 77, 81, 82, 90, 91, 97, 101, 106, 110, 149, 160, 173–175, 194, 195, 204, 205, 213, 215, 218, 220, 221, 223, 225
Ataran	16° 06′ (16° 12′	98° 02′ 97° 56′)	215
Aungban	20° 39′	96° 38′	79, 203
Ayadaw	21° 07′	94° 48′	162
Balu chaung	12° 15′	98° 54′	64
Banmauk	24° 23′	95° 51′	177, 205
Bassein	16° 47′	94° 45′	1, 5, 23–25, 34, 82, 87, 130, 158, 196, 205, 207, 214, 220, 221
Bawdwin	23° 05′	97° 15′	11, 51, 53–56, 116, 124–126, 144, 174, 176–180, 182, 196, 198, 201, 202
Bawetaung	20° 57′	96° 46′	144, 181
Bawgyo	22° 35′	97° 16′	116, 118, 205
Bawhningon	20° 31′	96° 37′	203
Bawlaba	19° 11′	93° 42′	
Bawlake	19° 10′	97° 20′	50
Bawsaing	20° 58′	96° 50′	54, 58, 59, 118, 144, 174, 176, 177, 179, 181, 182, 194, 203
Bernardomyo	23° 00′	96° 27′	47, 208
Bhamo	24° 15′	97° 14′	5, 22, 33, 39, 41, 49, 50, 88, 110, 139, 141, 148, 170, 176, 212, 213
Bhopi vum	23° 16′	93° 56′	130, 198
Bin byai (Binbye)	18° 24′	96° 44′	215

Subject Index